PROTEIN BIOCHEMISTRY, SYNTHESIS, STRUCTURE AND CELLULAR FUNCTIONS

CALCIUM SIGNALING

PROTEIN BIOCHEMISTRY, SYNTHESIS, STRUCTURE AND CELLULAR FUNCTIONS

Additional books in this series can be found on Nova's website
under the Series tab.

Additional E-books in this series can be found on Nova's website
under the E-books tab.

PROTEIN BIOCHEMISTRY, SYNTHESIS, STRUCTURE AND CELLULAR FUNCTIONS

CALCIUM SIGNALING

MASAYOSHI YAMAGUCHI
EDITOR

Nova Science Publishers, Inc.
New York

Library of Congress Cataloging-in-Publication Data

Calcium signaling / editor, Masayoshi Yamaguchi.
     p. ; cm.
 Includes bibliographical references and index.
 ISBN 978-1-61324-313-8 (hardcover)
 1. Cellular signal transduction. 2. Intracellular calcium. I. Yamaguchi, Masayoshi.
 [DNLM: 1. Calcium Signaling. QU 55.2]
 QP517.C45C35 2011
 571.7'4--dc22
                              2011010116

Published by Nova Science Publishers, Inc. ✛ New York

CONTENTS

PREFACE

Signal transduction plays a pivotal role in cellular regulation. The second messenger is generated in cells once the first messenger binds to the receptors of plasma membranes. The second messenger, which was found firstly, is cyclic adenosine monophosphate. After that, intracellular calcium ion (Ca2+) has been demonstrated to play a role as a second messenger for hormonal stimulation in cells. The pivotal role of Ca2+ in cellular regulation was established with the finding of calmodulin and protein kinase C that modulate the effect of Ca2+ in the regulation of cellular functions. Calcium signaling is probably the most ubiquitous cellular signal that mediates the action of many hormones, cytokines, and neurotransmitters. This book provides an in depth examination of calcium signaling.

Chapter 1 - In unicellular organisms, like in cells of the higher eukaryotes, the mechanisms of Ca^{2+} signaling consist of structural and functional elements capable for reception, modulation, and translation of the triggering calcium signal. As the cytosol Ca^{2+} acceptors, not only receptor proteins of the calmodulin (CaM) type serve, but also signal proteins that are a part of the signal protein kinase cascades. Comparative analysis of the Ca^{2+} messenger systems in cells of the lower and higher eukaryotes revealed many similar components (Ca^{2+} channels, EF-hand proteins, Ca^{2+} ATPases, protein kinases, Ca^{2+} transporters and others). Their domain architecture and highly conserved structure is a molecular basis for Ca^{2+} signaling for all eukaryote cells. It is obvious that mechanisms of Ca^{2+} signaling in the lower and higher eukaryotes have also the numerous differences caused by peculiarities of their molecular evolution. Multiple Ca^{2+}-dependent pathways have been demonstrated in lower eukaryotes. In this chapter last data about Ca^{2+} signaling mechanisms in unicellular eukaryotes is generalised. Mechanisms of Ca^{2+} influx and of transduction of Ca^{2+} signals are analyzed and also the role of Ca^{2+} in regulation of physiological processes is considered. The special attention is given consideration to evolutionary aspects of participation of Ca^{2+} and Ca^{2+} sensor proteins in regulation of cell events.

Chapter 2 - Store-operated calcium entry is a major mechanism for calcium influx in non-excitable, but also in excitable, cells. This event is regulated by the filling state of the intracellular calcium stores, which, upon depletion, evoke the opening of calcium permeable channels in the plasma membrane. Recent studies have revealed that the protein STIM1 (Stromal Interaction Molecule-1) acts as a calcium sensor that communicates information about the amount of calcium stored to the plasma membrane. STIM1 has also been located in the plasma membrane where it can regulate calcium channel gating. A number of studies have revealed that the calcium channels in the plasma membrane involve the protein Orai1, which shows a high selectivity for calcium. Since certain store-operated currents are not calcium selective, other channels, such as those of the TRP family, have been presented as candidates to conduct calcium entry. Store-operated calcium entry plays an important role in the regulation of a number of cellular functions and dysregulation of this mechanism leads to a number of dysfunctions.

Chapter 3 - As the research is growing in the field of signal transduction, it is widely accepted that most of the signaling pathways exist as non-linear complex network. Calcium, a ubiquitous second messenger, plays regulatory role by positioning itself at the core of various signaling pathway in eukaryotes and act as "Hub" in multiple signaling pathways. In plants, this versatile cation is involved in environmental stresses, including both biotic and abiotic, and developmental processes. Recent studies suggest that in plants, perception of multiple stimuli, leads to generation of a 'calcium signature', which enables it to define the nature and level of response. However, each cell has to be well versed with the mechanism to decode this signature. The decoding process starts with sensing of fluctuation in intracellular calcium levels by Ca^{2+} sensors and responders, which result in alteration in their structural and functional properties. This leads to modulation of the expression and function of downstream targets. These targets ultimately modulate various cellular, biochemical and physiological processes. In this review, author are making an attempt to discuss the regulation of calcium homeostasis and various signaling pathways in plants and their role in encoding and generating a specific and overlapping response. Ultimately, this concerted interplay of calcium signaling enables the plant to adapt under different environmental stresses and developmental processes.

Chapter 4 - The cysteine-rich extracellular calmodulin (CaM) binding protein (CaMBP) cyrA from *Dictyostelium* possesses 4 tandem EGF-like (EGFL) domains in its C-terminus. cyrA is secreted during development where it comprises one of many extracellular matrix proteins in the sheath of the multicellular pseudoplasmodium or slug. Proteolysis causes the release of cleavage products of cyrA of various sizes enriched in the EGFL domains and at least one of these domains (EGFL1) can feed back to inhibit cyrA proteolysis. cyrA shares sequence similarity to mammalian tenascin C (tenC) with EGFL1 showing high sequence identity to Ten14, an EGFL domain from tenC. Like Ten14, EGFL1 of cyrA enhances cell movement. In addition, EGFL1 has been shown to enhance both random cell motility and cAMP-mediated chemotaxis by activation of downstream calcium-dependent signalling via binding to the surface of *Dictyostelium* cells. Thus cyrA represents a true matricellular protein from lower eukaryotes and the only matricellular CaMBP identified to date for any organism. In keeping with the extracellular locale of cyrA, extracellular CaM has also been demonstrated in *Dictyostelium*. Extracellular CaM has been discovered in a small number of other organisms but its functions have not been well defined. Addition of CaM antagonists to *Dictyostelium* inhibits extracellular cyrA breakdown suggesting a role for extracellular CaM in protecting cyrA proteolysis. Using GFP-constructs as well as polyclonal antibodies against both the C- and N-termini of cyrA, the intracellular locales of this CaMBP and the ways it is processed to release fragments containing EGFL domains is coming to light. Here author review past work and present some novel data on this interesting novel matricellular CaMBP.

Chapter 5 - Nucleomorphin (NumA1) was isolated and verified as a calmodulin-binding protein from multicellular development of *Dictyostelium*. NumA1 shares many attributes with mammalian nucleolar proteins including multiple nuclear and nucleolar localization signals (NLSs and NoLSs) and an extensive acidic glu/asp repeat. Initial studies showed that deletion of the glu/asp repeat resulted in the formation of multiple nuclei suggesting a role for NumA1 in regulating nuclear number and continued work suggests a cell cycle role for this protein. Yeast two hybrid (Y2H) and co-immunoprecipitation studies revealed that calcium-binding protein 4a (CBP4a /cbpD1) binds to the acidic glu/asp domain of NumA1 in a calcium-dependent manner. NumA1 was subsequently verified as a nucleolar protein when treatments with actinomycin-D (AM-D) caused the loss of NumA1 patches adjacent to the nuclear envelope. In keeping with its nucleolar localization, immunolocalization studies revealed that NumA1 disappears from the nucleolar patches during prophase when the nucleolus dissociates and reappears during telophase with nucleolar reformation. As a NumA1-binding partner, CBP4a also localizes to the nucleolus and, similarly, AM-D treatment causes its loss from the nucleolus. CBP4a also disappears and reappears during prophase and telophase respectively, as befits a nucleolar protein. During mitosis, CBP4a exists as multiple islands throughout the nucleoplasm. Treatment of cells with BAPTA, a chelator of calcium, causes CBP4a to be dislodged from the nucleolus which fits with its role as a calcium-dependent NumA1-binding protein. A second NumA1-binding protein was also revealed by Y2H which does not require calcium: *Dictyostelium* puromycin-sensitive aminopeptidase (DdPsa). DdPsa localizes predominantly to the nucleus. The relationship of NumA1 to mammalian nucleolar proteins and further insight into the role of calcium signalling is discussed in the light of additional data.

Chapter 6 - It is well established that Ca^{2+} signalling mediates the effects of mechanotransduction in chondrocytes of mature articular cartilage. However, little is known about the precise regulation of Ca^{2+} homeostasis in differentiating cells of developing hyaline cartilage. Therefore, the author's research group is committed to characterise the Ca^{2+} homeostasis and to map the 'Ca^{2+}toolkit' of differentiating chondrogenic mesenchymal cells. High density cell culture system (HDC) established from chondrogenic mesenchymal cells isolated from limb buds of 4-day-old chicken embryos is a well-known model of *in vitro* cartilage differentiation, in which a spontaneous cartilage formation occurs in 6 days. Author measured cytosolic free Ca^{2+} concentration ($[Ca^{2+}]_i$) in cells of HDC on different days of culturing. After an initial value of 80 nM, a significant transient elevation was detected in Fura-2-loaded cells on day 3 of culturing, when the majority of cells differentiate into chondroblasts and chondrocytes. This 140 nM peak of cytosolic Ca^{2+} concentration is a result of increased Ca^{2+} influx and is found to be indispensable to proper chondrogenesis, because elimination of extracellular Ca^{2+} abolished the Ca^{2+} peak of day 3 and inhibited cartilage formation. Uncontrolled Ca^{2+} influx evoked by a Ca^{2+} ionophore (A23187) exerted dual effects on chondrogenesis in a concentration-dependent manner; low concentration of the ionophore increased $[Ca^{2+}]_i$ up to 150 nM and facilitated cartilage formation, whereas high concentration of this compound elevated it over 250 nM and almost totally blocked cartilage formation. Proliferation of chondrogenic cells was more sensitive to modulation of $[Ca^{2+}]_i$ then the viability of cells. Although chondrogenic cells express both IP$_3$ and ryanodine receptors and can release Ca^{2+} from intracellular stores, these stores proved to play a minor role in the Ca^{2+} homeostasis of these cells. As the inhibition of the Ca^{2+}-calmodulin sensitive protein phosphatase calcineurin impeded the $[Ca^{2+}]_i$-peak in chondrogenic cells and reduced cartilage formation, author propose its contribution in the regulation of $[Ca^{2+}]_i$ in these cells. Author also found that chondrogenic cells secreted ATP and administration of ATP to the culture medium evoked Ca^{2+} transients exclusively in the presence of extracellular Ca^{2+} and on day 3 of culturing. Moreover, ATP caused elevated protein expression of the chondrogenic transcription factor

Sox9 and also stimulated cartilage matrix production. ATP may exert these functions via acting through purinergic receptors; and indeed, expression of both ionotropic (P2X) and metabotropic (P2Y) purinergic receptors were detected. Metabotropic purinergic receptor agonist UTP caused a low level (60 nM) transient elevation of $[Ca^{2+}]_i$ in 3-day-old HDC, without having an influence on cartilage matrix production. Application of suramin, which blocks all P2X receptors but not P2X$_4$, did not impede the effects of ATP; furthermore, P2X$_4$ appeared in the plasma membrane of differentiating cells only from day 3. In summary, chondrogenic cells possess a set of different molecules which enable them to modulate their Ca^{2+} homeostasis and $[Ca^{2+}]_i$ was found to be kept in a narrow range during chondrogenesis. Author present evidence on a significant new regulatory mechanism of chondrogenesis with revealing the role of Ca^{2+} influx of chondrogenic cells via P2X$_4$ purinergic receptors.

Chapter 7 - Voltage-gated Ca^{2+} (Ca$_v$) channels allow the entry of Ca^{2+} into a cell when the membrane is depolarized. In the nervous system, Ca$_v$ channels control a broad array of functions including neurotransmitter release, neurite outgrowth, synaptogenesis, neuronal excitability, activity-dependent gene expression, and neuron survival, differentiation, and plasticity. Ca$_v$2.1 (P/Q-type) channels have a dominant, specific role in synaptic transmission at central excitatory synapses. In *CACNA1A* gene mutations, changes to the pore-forming α1 subunit of the Ca$_v$2.1 channel cause several autosomal-dominant neurological disorders in humans including familial hemiplegic migraine type 1 (FHM1), episodic ataxia type 2 (EA2), spinocerebellar ataxia type 6 (SCA6), and epilepsy. Mice with mutations in the *Cacna1a* gene are a useful tool for obtaining insights into disease processes and defining channel functions. Mouse Ca$_v$2.1 mutants include FHM1 model strains, including R192Q and S218L knockin mice, a SCA6 model strain carrying additional CAG repeats in the *cacna1a* locus of knockin mice, a knockout strain lacking Ca$_v$2.1 currents, and spontaneous strains that include *rocker*, *tottering, rolling Nagoya, leaner, tottering-4j, tottering-5j*, and *wobbly* mice. This chapter summarizes the human disease phenotypes and functional consequences of disease-causing mutations expressed in cell culture models, and overviews the results that Ca$_v$2.1 mutant mice have provided regarding the disease mechanisms of Ca$_v$2.1 channelopathy and the functions of Ca$_v$2.1 channels.

Chapter 8 - Regucalcin was discovered by Yamaguchi in the year of 1978 as a calcium-binding protein that does not contain EF-hand motif of calcium-binding domain. The name regucalcin was proposed for this calcium-binding protein, which can regulate various Ca^{2+}-dependent enzyme activations in liver cells. The regucalcin gene is localized on the chromosome X, and the organization of the regucalcin gene consists of seven exons and six introns. AP-1, NF1-A1, and RGPR-p117 bind to the promoter region of the rat regucalcin gene and enhance transcription activity of regucalcin gene expression that is mediated through calcium signaling. Regucalcin plays a pivotal role in the keep of intracellular calcium ion (Ca^{2+}) homeostasis due to activating Ca^{2+} pump enzymes in the plasma membrane (basolateral membrane), microsomes (endoplasmic reticulum), mitochondria, and nuclei of many cell types. Regucalcin has a suppressive effect on calcium signaling from the cytoplasm to the nucleus in the proliferative cells. Regucalcin has also been demonstrated to transport to the nucleus, and it can inhibit Ca^{2+}-dependent protein kinase and protein phosphatase activities, Ca^{2+}-activated deoxyribonucleic acid (DNA) fragmentation, and DNA and ribonucleic acid synthesis in the nucleus. Overexpression of regucalcin suppresses cell death and apoptosis in the cloned rat hepatoma cells induced by various signaling factors. Regucalcin can inhibit the enhancement of cell proliferation due to hormonal stimulation.

Regucalcin plays an important role as a suppressor protein in cell signaling system, and it is proposed to play a pivotal role in keep of cell homeostasis and function.

Chapter 9 - Cancer represents a major field of investigation for the modern medicine. Anti-cancer drugs induce tumor cell death. While some anti-cancer drugs are genotoxic, directly targeting nucleic acids, others interfere with the signaling pathways to trigger programmed cell death. For the induction and the execution of apoptosis, intracellular calcium signals are important. Much interest is now oriented in discovering new combinatorial strategies that are now tested in clinical trials in order to improve standard procedures. The author's results indicate that cisplatin (cis-di-amino-dichloride-platin) and arsenic trioxide modify intracellular Ca^{2+} by different mechanisms, supporting the hypothesis that calcium signaling plays an important role in chemical-induced neuroblastoma cell toxicity. In addition, understanding how specific drug combinations work to result in enhanced effects on neuroblastoma cell death might represent a new therapeutic strategy for neuroblastoma as well as ways to overcome drug resistance. There may be a strong link between drug resistance, cell death and changes in intracellular Ca^{2+}. Future experiments have to identify the specific mechanisms to illuminate possible targets and to override drug resistance.

In: Calcium Signaling
Editor: Masayoshi Yamaguchi

ISBN: 978-1-61324-313-8
©2012 Nova Science Publishers, Inc.

Chapter 1

CA^{2+} SIGNALING IN LOWER EUKARYOTES

I. V. Shemarova[*]

Sechenov Institute of Evolutionary Physiology and Biochemistry,
Russian Academy of Sciences, St. Petersburg, Russia,

ABSTRACT

In unicellular organisms, like in cells of the higher eukaryotes, the mechanisms of Ca^{2+} signaling consist of structural and functional elements capable for reception, modulation, and translation of the triggering calcium signal. As the cytosol Ca^{2+} acceptors, not only receptor proteins of the calmodulin (CaM) type serve, but also signal proteins that are a part of the signal protein kinase cascades. Comparative analysis of the Ca^{2+} messenger systems in cells of the lower and higher eukaryotes revealed many similar components (Ca^{2+} channels, EF-hand proteins, Ca^{2+} ATPases, protein kinases, Ca^{2+} transporters and others). Their domain architecture and highly conserved structure is a molecular basis for Ca^{2+} signaling for all eukaryote cells. It is obvious that mechanisms of Ca^{2+} signaling in the lower and higher eukaryotes have also the numerous differences caused by peculiarities of their molecular evolution. Multiple Ca^{2+}-dependent pathways have been demonstrated in lower eukaryotes. In this chapter last data about Ca^{2+} signaling mechanisms in unicellular eukaryotes is generalised. Mechanisms of Ca^{2+} influx and of transduction of Ca^{2+} signals are analyzed and also the role of Ca^{2+} in regulation of physiological processes is considered. The special attention is given consideration to evolutionary aspects of participation of Ca^{2+} and Ca^{2+} sensor proteins in regulation of cell events.

Keywords: Calcium signaling, Ca^{2+}-dependent protein kinases, Ca^{2+}-binding proteins, calcium channels, calmodulin, Ca^{2+} pumps, EF-hand proteins, lower eukaryotes, PMCA-type Ca^{2+} ATPases, TRP channels, VGCC-like channels

---

[*] e-mail: shem@iephb.ru

INTRODUCTION

Mechanisms of intracellular signaling involving Ca^{2+} ions as a second messenger have an ancient evolutionary origin. Even prokaryotes have been shown to contain such important functional components of the information intracellular systems as Ca^{2+} channels, primary and secondary transmembrane Ca^{2+} carriers, CaM-like proteins mediating participation of Ca^{2+} in regulation of cell division, growth, and chemotaxis.

The tendency for use of Ca^{2+} in processes of intracellular signaling was further developed in cells of the lower eukaryotes. In protists and lower fungi, a multicomponent system for accumulation, regulation, and utilization of intracellular Ca^{2+} is formed for generation of the calcium signal. This system is similar to that of higher eukaryotes, it includes mechanisms of voltage- and receptor-activated Ca^{2+} influx through cell membrane, mechanisms of calcium release and store-operated Ca^{2+} entry (SOC) with use of second messengers such as inositol 1,4,5-trisphosphate (IP_3), arachidonic acid and cyclic ADP-ribose (cADPR) as well as the mechanism of the Ca^{2+}-induced Ca^{2+} release" (CICR).

In unicellular eukaryotes, like in cells of the higher eukaryotes, the Ca^{2+} messenger system consist of structural and functional elements capable for reception, modulation, and translation of the triggering calcium signal. As the cytosol Ca^{2+} acceptors, not only receptor proteins of the CaM type serve, but also signal proteins that are a part of the signal protein kinase cascades, including Ca^{2+}-dependent protein kinases (CDPK), phosphatases, proteinases and many other target proteins.

Comparative analysis of the Ca^{2+} signaling mechanisms in cells of the lower and higher eukaryotes revealed many similar features indicating the universal and single principles underlying the structural-functional organization of all eukaryote Ca^{2+}-dependent signaling systems, although there also are differences, whose nature appears to be due to peculiarities of molecular evolution of the signal mechanisms. By the present time, some molecular Ca^{2+} signaling mechanisms in the lower eukaryotes have already been deciphered. Pathways of activation and propagation of the Ca^{2+} signal are described, proteins modulating activity of calcium signal are determined, and the role of Ca^{2+} in regulation of physiological processes in lower eukaryote cells is established. However, in the evolutionary aspect, the problem of Ca^{2+} signaling has not been earlier considered.

Ca^{2+} INFFLUX AND Ca^{2+} RELEASE

The triggering Ca^{2+} signal is formed on the surface of the cell membrane during a sharp change of environmental conditions as well as under the effect of mechanic, stress, and alimentary factors and is propagated into a cell through specialized structures - cation channels or extracellular Ca^{2+}-sensing receptors (Fig. 1). A considerable amount of information coming into cells of the lower eukaryotes is realized via Ca^{2+} channels that are included in a specialized Ca^{2+} messenger system providing reception, transduction, and transmission of the Ca^{2+} signals. Some of them will be considered below.

Ca^{2+} Channels

In cells of all organisms, Ca^{2+} channels are a part of the Ca^{2+}-dependent signal pathways using Ca^{2+} sources for formation of the calcium signal that represents a local rise in intracellular Ca^{2+} concentration ($[Ca^{2+}]_i$). An increase of the $[Ca^{2+}]_i$ is a trigger signal for the beginning of numerous physiological processes both in the higher eukaryotic cells and in microorganisms including prokaryotes [1].

For realization of fast cellular reactions, extracellular Ca^{2+} is preferentially used. This mechanism of the voltage-gated Ca^{2+} influx operates, when cells respond to change of membrane potential by an increase in the Ca^{2+} influx through the voltage-gated Ca^{2+} channels (VGCCs). This mechanism is characteristic of electrically excitable cells of multicellular organisms [2]. In ciliated protozoa, microalgae, yeast cation channels sensitive to change of membrane potential have also been revealed. This group of channels include nonselective hyperpolarization-activated cation channels, warm and cold sensor channels, transient receptor potential channels (TRP) and VGCC-like channels [3-7]. Presence of such channels assumes that they participate in the behavioral control and in regulation of the fast signaling processes, activated in response to chemical stimuli or mating pheromones.

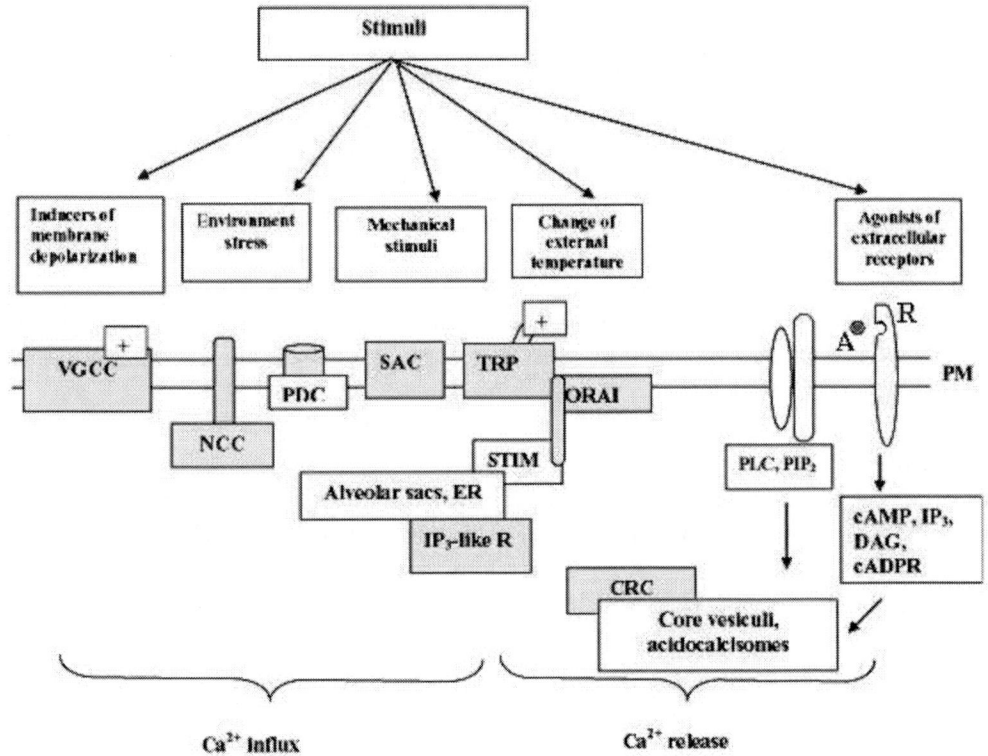

Figure 1. Mechaninisms of calcium influx and release. VGCC - voltage-gated Ca^{2+} channel, PDC – passive diffusion of cations, NCC - non-selective cation channel, SAC - stretch-activated channel, CRC - Ca^{2+} release channel, TRP - transient receptor potential channel. R - receptor, A – agonist. Other signatures in text. The Ca^{2+}-permeable channels are illuminated in grey.

To provide initiation of slow or short-lasting metabolic and signaling processes, intracellular Ca^{2+} sources as a rule are used. In this case, the external triggering signals realize their action through activation of the receptor-opereted Ca^{2+} channels (ROCs) localized in plasma membranes or membranes of subcellular organelles, such as Golgi complex, endoplasmic reticulum (ER), and vacuoles. This mechanism of the receptor-dependent Ca^{2+} mobilization is typical of unexcitable metazoan cells and most of lower eukaryotes.

Ca^{2+}-permeable stretch-activated channels (SACs), which are activated in reply to mechanical stimulus and take part in membrane repolarization at end of potential of action, it is possible to include in special group. They are found in membrane of the *Chlamidomonas, Dictyostelium, Saccharomyces* and *Paramecium*.

Non-Selective Cation Channel (NCC)

NCC has been revealed in membrane of ciliated protozoa and yeast. Calcium-activated conductance in a non-selective cation channel activated by bath applications of elevated Ca^{2+} concentrations was found in *Paramecium* cells [8]. The channel became very active when the Ca^{2+} concentration was above 3.2 µM. The channel was also activated by depolarization. The voltage dependency was steep upon depolarization, whereas upon hyperpolarization the channel activity barely changed. This channel had poor selectivity for monovalent alkali cations. Using the Goldman-Hodgkin-Katz equation for the reversal potential, the permeability ratios with respect to K^+ for Na^+, Rb^+, Cs^+ and Li^+ were nearly 1. Authors notice that, although the permeability ratios were similar for each cation, the single channel conductances differed. The single channel conductances were 467 pS with K^+ as the charge carrier, 406 pS with Na^+, 397 pS with Rb^+, 253 pS with Cs^+ and 198 pS with Li^+ upon depolarization in 100 mM cation solutions. Functional significance of these channels, apparently, consists in maintenance of ionic Ca^{2+} homeostasis in the conditions of the chemical, osmotic, temperature stresses leading to membrane depolarization [8].

Non-selective cation channels of type NSC1 (hyperpolarization-activate cation channel 1) in *Saccharomyces cerevisiae* serve as regulators of the activity of the potassium transporters, Trk1p and Trk2p and are important for growth of cells. An increase of their Ca^{2+} permeability leads to supression Trk1p and Trk2p and to inhibition cell grown [9 -11].

Warm and Cold Sensor Channels

These cation channels, activated by depolarization and hyperpolarization have been discovered in ciliary membrane of *Paramecium tetraurelia*. They differed by pharmacological and physiological properties, ionic specificity and mechanisms of activation and inactivation. Sensor channels include heat-activated Ca^{2+} channels and cold-activated Ca^{2+} channels [12]. These channels have electrophysiological characteristics resembling those of non-selective cation channels revealed in vertebrate excitatory cells. They serve for fast transmission of Ca^{2+} signals on effector proteins. On the other hand, they can be involved in transmission of GTP-activated oscillating inward currents with a periodicity similar to that of the membrane-potential and behavioral responses that can specify their participation in transfer of slower periodic signals [13]. Probably, these channels are variants of TRP channels (see more low).

VGCC-Like Channels

VGCC-like channels have also been revealed in some species of the lower eukaryotes. There are evidences of their presence in the ciliary membrane *Paramecium* [2, 5, 14], in apicomplexan parasite *Toxoplasma gondii* [15], in tonoplasts of the budding yeast *Saccharomyces cerevisiae* [4, 16-17] and in choanoflagellate *Monosiga brevicollis* [18]. The electrophysiologic properties of voltage-gated channels have been studied in the greatest detail in *Paramecium sp.* [2, 19, 20].

Jennings [21] observed behavior of the ciliate *Paramecium caudatum* responding to contact, light, elevation of temperature, and chemical stimulation. These observations stimulated subsequent neurobiological investigations that have enabled concluding that *Paramecium* are unique cell forms combining simultaneously functions both of biosensor and of effector; therefore, they were figuratively called "swimming sensor cells" or "swimming neurons" [20]. The *Paramecium* body is covered with thousands of cilia containing sensor proteins and cation channels including VGCC-like channels. Like mammalian VGCCs, the voltage-gated channels in *Paramecium* are activated by factors initiating membrane depolarization. There are some cations, organic repellents, mechanical stimuli, elevated environmental temperature [22]. The membrane depolarization, in turn, evokes the action potential opening VGCCs for Ca^{2+} influx. The [Ca^{2+}]$_i$ increases and the swimming cell behavior changes practically instantaneously, which is manifested as the physiological "reaction of escape" caused by reversion of ciliary beating. The events providing this Ca^{2+}-dependent type of cell behavior were studied in detail for the 1960s–1970s [23, 24].

VGCCs are characterized by that they are activated during membrane depolarization, whereas at the rest potential (approximately from -70 to -80 mV) they are in the inactive state. In mammalian excitable cells, L-, T-, N- and P/Q-types VGCCs have been described in the greatest detail; they differ in molecular structure, localization, voltage dependence, conductivity, sensitivity to pharmacological agents, and some other characteristics [2, 26, 27]. In neurons, skeletal and cardiac myocytes, the L-type VGCCs are most numerous [27, 28]. They are blocked by the L-type calcium channel inhibitors such as nimodipine, nifedipine, verapamil, diltiazem and Cd^{2+} [2, 26].

Research of action of L-type calcium channel inhibitors on *Paramecium* behaviour has shown that they or are inefficient or insufficiently effective [5, 29]. However, such inhibitors, as a Ca^{2+} chelator (EGTA), general calcium channel blockers (such as lanthanides), inhibitors of intracellular Ca^{2+} release (ruthenium red and neomycin), and inhibitors of T-type calcium channels (NNC 55-0396, 1-octanol and Ni^{2+}) completely inhibited galvanotactic movement of *Paramecium bursaria* cells, that can testify to an involvement in paramecia Ca^{2+} signaling a calcium release channels and T-type Ca^{2+} channels, activated by lower voltage [29].

The different channel conductivity in the media containing divalent and monovalent cations can specify in presence of the intraspecies heterogeneity in population of the voltage-gated Ca^{2+} channels in *Paramecium*. Thus, the conductivity of two different voltage-gated Ca^{2+} channels in the media containing Ca^{2+}, Ba^{2+}, Mg^{2+}, and monovalent cations was estimated in *Paramecium* cilia incorporated into planar lipid bilayers [30]. Both channels were much more permeable to divalent than univalent cations, and one of them discriminated significantly among the divalent cations. The highest conductivity of the latter channel was found in the Ca^{2+}-containing medium. The selectivity and voltage dependence of this channel are comparable to those of conventional VGCCs.

An important criterion for properties of ion channels is their inactivation mechanism. Already the first measurements of calcium currents in *Paramecium* under conditions of reliable turn off of outward currents have shown that the time course of their inactivation is essentially similar to that of calcium currents in metazoans. Under depolarizing voltage clamp of *Paramecium* an inward Ca^{2+} current developed and subsequently relaxed within 10 milliseconds. It was shown that the relaxation was substantially slowed when the most of extracellular Ca^{2+} replaced with either Sr^{2+} or Ba^{2+}. Evidence is presented that the relaxation is not accounted by a drop in electromotive force acting on Ca^{2+}, or by activation of a delayed K^+ current. Relaxation of the current must, therefore, result from an inactivation of the Ca^{2+} channel. This inactivation persisted after a pulse, as manifested by a reduced Ca^{2+} current during subsequent depolarization. Inactivation was retarded by procedures that reduce net entry of Ca^{2+}, and was independent of membrane potential [31]. The obtained data unequivocally indicate that the Ca^{2+} channel undergoes inactivation as a consequence of Ca^{2+} entry during depolarization (the "current-dependent inactivation" unlike the "voltage-dependent inactivation" for canonical VGCCs), but not by changes themselves of the membrane potential. The mechanism of block of Ca^{2+} channels by intracellular Ca^{2+} in the lower eukaryotes, like in higher eukaryote cells, appears to be due to participation of the Ca^{2+}-CaM complex in binding with the channel [5].

Methods of molecular cloning and mutagenesis are used more and more often for studying structure, regulation, and properties of Ca^{2+} channels. A large array of mutants *Paramecium* displaying altered behaviour have been isolated and mapped to over 20 loci. Detailed electrophysiological analyses have been carried out on several classes of mutants revealing defects in specific ion channels in some cases. Mutants have proven very useful to analyze channel properties, to unravel interactions between channels and to discover the function of these channels in a variety of cellular processes. Interesting studies have been carried out on *pawn* and *CNRs* strains of mutant *P. tetraaurelia* negative for several functions of VGCCs. Using these mutants, it was proven that defects of the transmembrane Ca^{2+} signal propagation (the absence of inward Ca^{2+} current under depolarizing voltage clamp of *Paramecium*) led to disturbance of ciliary motility and cell locomotion [32].

The *pawn* and *CNRs* mutants have become widely used in experiments on separation of membrane currents [33], studying of the ionic basis of the mechanoreceptor potential [34], monitoring of the Ca^{2+} and Ba^{2+} influx via Ca^{2+} channels *in vivo* [35], and in studying of properties of membrane vesicles known as cortical alveoli [36, 37].

Good candidates for the role of model organisms to study properties of voltage-gated Ca^{2+} channels are the so called "dancer mutants" of *P. tetraaurelia*. Mutation of *Dn* gene has been established to cause a pronounced change of characteristics of Ca^{2+} current in *Paramecium*. Unlike wild-type cells that generate leveling action potentials in response to depolarizing stimuli, the "dancer mutants" are more excitable and evoke action potential on the "all or nothing" principle [38].

Little is known at present about the structure of VGCC-like channels in lower eukaryotes cells. There are data that Cch1 protein of the yeast *Saccharomyces cerevisiae* is a homologue of the pore-forming α1 subunit of mammalian voltage-gated Ca^{2+} channels. Putative pore loop located between the S5 and S6 transmembrane segments is presented positively charged amino acid residue arginine or lysine that serves for voltage sensing. Cch1 protein constitutes a high-affinity Ca^{2+} influx system with stretch-activated Ca^{2+} permeable channel Mid1. Teng and co-workers (2008) characterized the kinetic property of a putative Cch1-Mid1 Ca^{2+}

channel overexpressed in *S. cerevisiae* cells, and showed that the L-type VGCC blockers nifedipine and verapamil partially inhibited Cch1-Mid1 activity, but typical P/Q-, N-, R- and T-type VGCC blockers did not inhibit activity. In contrast, a third L-type VGCC blocker, diltiazem, increased Cch1-Mid1 activity. Diltiazem did not increase Ca^{2+} uptake in the *cch1* Δ and *mid1*Δ single mutants and the *cch1*Δ *mid1*Δ double mutant cells (lacking both Cch1 and Mid1), indicating that the diltiazem-induced increase in Ca^{2+} uptake is completely dependent on Cch1-Mid1.

These results suggest that Cch1 is pharmacologically similar to L-type VGCCs, but the interactions between Cch1 and the L-type VGCC blockers are more complicated than those in metazoan cells [7].

Data about the VGCCs in other taxa of the lower eukaryotes are fragmentary and so far cannot yet be systematized.

Stretch-Activated Calcium Channels (SACs)

Mechanosensitive channels appear ubiquitous but they have not been well characterized in lower eukaryotes. Tension-sensitive channel currents were identified in *Chlamydomonas reinhardtii*, a protist that shows a marked behavioral response to mechanical stimulation [39]. When a negative pressure was applied to the cell body with a patch clamp electrode, single-ion-channel currents of 2.4 pA in amplitude were observed. The currents were inhibited by 10 μm gadolinium, a general blocker of mechanosensitive channels. It was supposed that the currents were most likely due to Ca^{2+} influxes because the current was absent in Ca^{2+}-free solutions and the reversal potential was 98 mV positive to the resting potential. The distribution of channel-open times conformed to a single exponential component and that of closed times to two exponential components. At comparison of channels from a body and a flagella it has been established that they have similarity on following parameters: both channels were inhibited by Gd^{3+} at 10 μm but not at 1 μm both passed Ca^{2+} and Ba^{2+}; their kinetic parameters for channel opening were similar. These observations raise the possibility that *Chlamydomonas* mechanosensitive channels are calcium channels and they may function both in the behavioral control through the mechanoreception by the flagella and in the regulation of cellular physiology in response to mechanical perturbation on the cell body [39]. Recent genetic studies in *Chlamydomonas reinhardtii* show that mechanosensitive channel named CrPKD2 belong to transient receptor potential (TRP) superfamily and functions in a pathway linking cell-cell adhesion and Ca^{2+} signaling [40].

The involvement of stretch-activated calcium channels in *Dictyostelium* movement has been revealed Lombardi and co-workers [41]. They observed small, brief, Ca^{2+} transients in randomly moving wild-type cells that were dependent on both intracellular and extracellular sources of calcium. Treatment of cells with Gd^{3+} inhibited transients and decreased cell speed, consistent with the involvement of stretch-activated calcium channels in regulating *Dictyostelium* motility. Additional support for SAC activity was given by the increase in frequency of Ca^{2+} transients when *Dictyostelium* cells were moving on a more adhesive substratum or when they were mechanically stretched. These results indicate that mechano-chemical signaling via SACs plays a major role in maintaining the rapid movement of *Dictyostelium* cells [41].

The yeast Mid1 protein functions as a stretch-activated Ca^{2+} permeable channel when expressed in mammalian cells [42]. This protein with molecular mass of 100 kDa is required for Ca^{2+} influx stimulated by the mating pheromone and by a capacitative calcium entry, like mechanism acting in response to Ca^{2+} depletion from the endoplasmic reticulum. In protease protection experiments and by indirect fluorescence microscopy were found that Mid1 is present in the plasma membrane as well as ER membrane [42]. Mid1 is required for successful mating; the mutant phenotype is rescued by elevated extracellular calcium. Homologs of the *MID1* gene are found in fungi *Basidiomycetes* and *Ascomycetes,* which morphologically are comparable with yeast [43].

In yeast *Saccharomyces cerevisiae* there is at least one more channel protein, which is activated by a stretch force. This protein is coded through a gene containing a TRP-like sequence. The protein forms an ion channel in the vacuolar membrane and is therefore called Yvc1 (yeast vacuolar conductance 1) [44]. Yvc1 can clearly be activated by hypertonic shock *in vivo* and by stretch force under patch clamp. Like its animal counterparts, Yvc1 is polymodal, being gated by membrane stretch force and by cytoplasmic Ca^{2+} [45].

TRP family channels are found out in two protozoan species: *Paramecium tetraurelia* and choanoflagellate *Monosiga brevicollis* [18, 46]. Their primary structures are diverse with sequence similarities only in some short amino acid sequence motifs mainly within sequences covering TM5, TM6, and adjacent domains. The biology of these channels in protists is for the present studied very poorly. I will remind, that its mammalian homologues are components of phospholipase C-driven signaling pathways and can be modified by traditional inhibitors of receptor-dependent calcium entry, including phospholipase A2 inhibitor, and W-7, a calmodulin antagonist. In protozoans a regulator of these channels, apparently, can be CaM.

Ligand-Gated Ca²⁺ Channels

This class of Ca^{2+} channels can include both true Ca^{2+} channels and low selective cation channels that are permeable first of all to monovalent cations, but also are able to transport Ca^{2+} and thereby to control its intracellular activity.

The group of the true Ca^{2+} channels includes the channels whose receptor agonist either performs the function of a channel or directly interacts with the channel structure. The prototype of these channels was the mammalian nicotinic acetylcholine receptor itself is directly a nonselective cation channel. The nicotinic acetylcholine receptor channel is permeable to Ca^{2+}, but under physiological conditions it transports predominantly Na^{+} and K^{+}. Its selectivity to monovalent cations is 3–5 times as great as to Ca^{2+}. Results of study of the mechanisms involved to Ca^{2+} signaling of *Trypanosoma cruzi* epimastigotes suggests that the receptor-like membrane structures of these protozoan parasites can serve evolutionary predecessors of the mammalian nicotinic acetylcholine receptor. These molecular species of parasite membrane structures were able to bind nicotinic ligands; therefore, nicotine interaction could lead to the activation of the mechanisms involved in intracellular calcium concentration increase in the parasite [47].

The mechanism of receptor-depended intracellular Ca^{2+} release implies the presence in eukaryote cells of special structures depending on Ca^{2+}. Indeed, in mammalian cells, many proteins were revealed, their function and activity being regulated by $[Ca^{2+}]_i$. They include, in particular, some protein kinases taking part in transduction of proliferative and chemotaxis signals. These proteins are components of signaling pathways, in which Ca^{2+} plays the role of second messenger.

The Ca^{2+} messenger system in metazoan cells is activated by several mechanisms including stimulation of the G protein-coupled receptors (GPCRs) and phospholipase C

(PLC) [48, 49]. As a result of such activation the intracellular Ca^{2+} mobilization occurs and local calcium signal is formed. In unicellular eukaryotes, pathways of Ca^{2+} mobilization by activation of the G protein-coupled receptors, PLC and phosphoinositide (PI) metabolites were also revealed [50]. This activation appears to lead to stimulation of two independent mechanisms of Ca^{2+} mobilization, one of them related to the external Ca^{2+} influx, the other coupled to release of intracellular Ca^{2+} via channels activated by second messengers. In 1984, Bumann and co-workers were the first to establish that in the slime mold *Dictyostelium discoideum* the agonist-dependent Ca^{2+} influx associated with activation of chemoattractant receptors cARs [51]. The authors have shown that *Dictyostelium* exhibit a transient uptake of extracellular Ca^{2+} approximately 5 seconds (s) after activation of surface folate or cAMP receptors. The cells maintained an external level of 3-8 µM Ca^{2+} until the beginning of aggregation and then started to take up Ca^{2+}. The attractants, folic acid, cyclic AMP, and cyclic GMP, induced a transient uptake of Ca^{2+} by the cells. The response was detectable within 6 s and peaked at 30 s. Half-maximal uptake occurred at 5 nM cyclic AMP or 0.2 µM folic acid, respectively. The apparent rate of uptake amounted to 2 x 10^7 Ca^{2+} per cell per min. Following uptake, Ca^{2+} was released by the cells with a rate of 5 x 10^6 ions per cell per min. Specificity studies indicated that the induced uptake of Ca^{2+} was mediated by cell surface receptors. The amount of accumulated Ca^{2+} remained constant as long as a constant stimulus was provided. The cyclic AMP-induced uptake of Ca^{2+} increased during differentiation and was dependent on the external Ca^{2+} concentration. Saturation was found above 10 µM external Ca^{2+}. The time course and magnitude of the attractant-induced uptake of external Ca^{2+} agree with a role of Ca^{2+} during contraction. During development the extracellular Ca^{2+} level oscillated with a period of 6-11 min. The change of the extracellular Ca^{2+} concentration during one cycle would correspond to a 30-fold change of the cellular free Ca^{2+} concentration.

For the further studying these Ca^{2+} entry systems, ^{45}Ca^{2+} uptake by resting and activated amoebae *Dictyostelium discoideum* has been analysed [52]. Like the surface chemoreceptors, folate- and cAMP-induced Ca^{2+} uptake responses were developmentally regulated; the former response was evident in vegetative but not aggregation-competent cells, whereas the latter response displayed the opposite pattern of expression. In contrast, other characteristics of these Ca^{2+}-uptake pathways were remarkably similar. Both systems exhibited comparable kinetic properties, displayed a high specificity for Ca^{2+}, and were inhibited effectively by Ruthenium Red, sodium azide, and carbonylcyanide m-chlorophenyl-hydrazone. These results, together with the finding that vegetative cells transformed with a plasmid expressing the surface cAMP receptor exhibit a cAMP-induced Ca^{2+} uptake, suggest that different chemoreceptors activate a single Ca^{2+} entry pathway via ligand-gated Ca^{2+}-channels [52].

Ca^{2+} influx in *Dictyostelium discoideum* is stimulated also by platelet-activating factor (PAF, 1-*O*-alkyl-2-acetyl-*sn*-glycero-3-phosphocholine), which is produced during development and in response to cAMP [53]. At the cellular level, PAF modulates intracellular second messengers, such as cAMP, cGMP, Ca^{2+} and IP$_3$, via a G protein-coupled membrane receptor [54, 55]. Modulatory activities linked to intracellular receptors have also been suggested, based on differential effects of PAF antagonists, but they are less well characterized [53, 56].

The mechanism of PAF action has been analysed by Schaloske and co-workers [57]. The authors found that PAF activity was confined to the period of spike-shaped oscillations and suggest that the role of PAF is to augment cAMP relay. PAF showed a reduced response in the Gβ(-) strain LW14 and was unable to induce Ca^{2+} influx in the Gα2(-) strains HC85 and JM1. The latter expresses the cAMP receptors cAR1 constitutively, and exhibits cAMP-induced Ca^{2+} influx, albeit at a reduced level. In order to decide whether the inability of PAF to elicit a Ca^{2+} response in JM1 cells was due to the lack of differentiation and/or the lack of Gα2, the IP_3-dependent pathway have been blocked by compound U73122 and found that Ca^{2+} entry was blocked, whereas a closely related inactive compound, U73343, did not alter the response. In agreement with this, NBD-Cl, an inhibitor of Ca^{2+} uptake into the IP_3-sensitive store in *Dictyostelium*, also abolished PAF activity. The latter was not inhibited in the presence of the plasma membrane antagonists BN-52021 or WEB 2170. On the basis of the above-stated proofs the conclusion that PAF operate intracellularly via the IP_3-signaling pathway at or upstream of the IP_3-sensitive store has been made.

Further, it has been established that IP_3-sensitive store in *Dictyostelium* is blocked by means of xestospongin C, which, as it is known, binds to the IP_3-sensitive store in mammalian cells and inhibits IP_3- and thapsigargin-induced Ca^{2+} release. In addition, xestospongin C inhibited Ca^{2+} uptake into purified vesicle fractions and induced Ca^{2+} release. This suggests that, in the case of *Dictyostelium*, xestospongin C opens rather than plugs the IP_3 receptor channel as was proposed for mammalian cells [58].

The existence of IP_3-like receptor in *Dictyostelium* was confirmed by subsequent molecular studies. Now it is precisely established that IP_3-like receptor *Dictyostelium* named iplA Ca^{2+} channel is a ligand-gated channel governing Ca^{2+} efflux from ER stores to the cytosol [59].

In opportunistic yeast *Candida albicans*, two pathways of Ca^{2+} mobilization were revealed, one of them induced by IP_3, the other by internal positive potential. These pathways differ by the amount of released Ca^{2+}, by the nature of the calcium response, and by pharmacological characteristics [60].

A similar IP_3-dependent pathway of Ca^{2+} mobilization from vacuoles was revealed in *Neurospora crassa* and *Saccharomyces cerevisae* [61, 62]. Existence of this pathway in yeasts implies that the calcium response is mediated through IP_3-like or ryanodine receptors. However, molecular studies have shown that in yeasts no homologues of ryanodine or IP_3 receptors [63].

It has been established that a transient increase in cytosolic Ca^{2+} is mediated through Yvc1p, a vacuolar membrane protein with homology to TRP channels (see above). After this release, low cytosolic Ca^{2+} is restored and vacuolar Ca^{2+} is replenished through the activity of Vcx1p, a Ca^{2+}/H^+ exchanger. These studies reveal a novel mechanism of internal Ca^{2+} release and establish a new function for TRP channels [63].

Functional evidence for Ca^{2+} release in response to ryanodine or IP_3 receptor agonists has been described in several protozoans. Treatment of permeabilized *Plasmodium chabaudi* parasites with IP_3 results in Ca^{2+} release, which is inhibited by the IP_3 receptor antagonist heparin [64]. Another apicomplexan parasite, *Toxoplasma gondii*, responds to agonists and antagonists of both, ryanodine and IP_3 receptors, by mediating increases in $[Ca^{2+}]_i$. The carbachol-induced receptor-dependent Ca^{2+} release from intracellular stores was revealed in hemoflagellate *Trypanosoma cruzi* [47, 65]. A biphasic increase of the IP_3 level preceded the intracellular Ca^{2+} mobilization. It was fast and short-term at the early phase of stimulation of the trypanosomes with carbachol, with a maximum after 1 min and return to the basal level by 6 min. At the second phase, the IP_3 level rose at the 10–12th min, while it decreased at the 20th min after carbachol stimulation. The pretreatment of trypanosomes with a PLC inhibitor

(U73122, 10 μM) led to complete inhibition of triggered by carbachol IP$_3$ synthesis, whereas there was not substantial change on the maximum elevation in [Ca^{2+}]$_i$. The first peak of IP$_3$ synthesis was completely abolished when the cells were incubated with phorbol 12-myristate 13-acetate ester before carbachol stimulation. There was also a detectable increment of DAG at 1 min with a maximum at 3 min, this level remaining elevated until at least 10 min [65]. Thus, the data obtained in this study can testify to presence at trypanosomes of complex mechanism of intracellular Ca^{2+} release with participation of PKC, IP$_3$ and DAG.

The receptor-dependent Ca^{2+} mobilization coupled with activation of PLC and an increase of the IP$_3$ level was found in parasitic amoeba *Entamoeba histolytica*. It is to be noted that stimulation of membrane receptors by fibronectin in these microorganisms leads not only to intracellular mobilization, but also to extracellular Ca^{2+} influx [66]. Interesting data have been obtained about the role of arachidonic acid, an important phospholipid messenger in mammalian cells, in regulation of extracellular Ca^{2+} influx and intracellular Ca^{2+} release through channels of plasma membrane and acidocalcisomes, respectively, in hemoflagellate *T. cruzi* [67]. In mammalian cells, the principal sources of arachidonic acid are phosphoinositides and diacylglycerol formed after their hydrolysis. In protists, most of free arachidonic acid seems to be formed during hydrolysis of intracellular phospholipid sources including diacylglyceropyrophosphate [68].

Activation of PLC in cells initiates a signal cascade leading to simultaneous formation of several second messengers that are activators of Ca^{2+} release from intracellular stores [49]. Metabolite of nicotinamide dinucleotide, cyclic adenosine diphosphate ribose (cADPR), is considered as a trigger of the intracellular Ca^{2+} release independent of products of phosphoinositide metabolism [69]. In mammalian cells, the cADPR-messenger system is suggested to act through ryanodine receptors (RyR). It has turned out to be unexpected that cADPR functions already in the lower eukaryotes [70]. However, its physiological role in protists is unclear.

The complex mechanism of intracellular Ca^{2+} release with participation of IP$_3$ and cyclic ADP-ribose as second messengers has also been reveled in *Euglena gracilis*. Masuda and co-workers [70] have shown that IP$_3$ and cADPR released Ca^{2+} from *Euglena* microsome fraction in dose-dependent manners. Caffeine, which also induced Ca^{2+} release from the microsomes, caused desensitization of the Ca^{2+} response to cADPR, although the Ca^{2+} response to IP$_3$ was not affected by caffeine. Further, Ruthenium Red inhibited the Ca^{2+} release induced by cADPR, but not by IP$_3$. These results suggest that in *E. gracilis* cADPR functions as an endogenous messenger to activate a caffeine-sensitive, Ca^{2+}-release mechanism, whereas IP$_3$ induces Ca^{2+} release by a distinct mechanism [70].

Most full by this time the mechanism of the receptor-dependent Ca^{2+} release in *Paramecium* is studied. This mechanism is mainly carried out through IP$_3$R. IP$_3$R cloned from *Paramecium* (GenBank accession number CAI39148.1) displays strongest sequence similarities to the rat type 3 IP$_3$R, about an overall identity of 19% and similarity of 34% [71]. To date, a database search of the *Paramecium* genome reveals 34 genes related to Ca^{2+} release channels (CRCs) of the IP$_3$ or ryanodine receptor type (IP$_3$R, RyR). Phylogenetic analyses show that these Ca^{2+} release channels can be subdivided into six groups (*Paramecium tetraurelia* CRC-I to CRC-VI), each one with features in part reminiscent of IP$_3$Rs and RyRs.

Recently *P. tetraurelia* CRC-IV-1 gene family was characterized, whose relationship to IP$_3$Rs and RyRs is restricted to their C-terminal channel domain [71]. CRC-IV-1 channels localize to cortical Ca^{2+} stores (alveolar sacs) and also to the endoplasmic reticulum. This is in contrast to a recently described true IP_3 channel, a group II member (*P. tetraurelia* IP_3R_N-1), found associated with the contractile vacuole system [72]. Silencing of either one of these CRCs results in reduced exocytosis of dense core vesicles (trichocysts), although for different reasons. Knockdown of *P. tetraurelia* IP_3R_N affects trichocyst biogenesis, while CRC-IV-1 channels are involved in signal transduction since silenced cells show an impaired release of Ca^{2+} from cortical stores in response to exocytotic stimuli [71].

Similar Ca^{2+} release channels were identified in choanoflagellate *Monosiga brevicollis* [20]. It is interesting that *M. brevicollis* contains 4 homologues of IP_3Rs, whereas *C. elegans* and humans possess 1 copy and 3 copies of IP_3Rs, respectively. Compared with the *Paramecium* IP_3R, *M. brevicollis* IP_3Rs show higher sequence similarity to metazoan IP_3Rs [20]. Moreover, Cui [20] in these organisms for the first time identified the genes coding store-operated channel Orai and the ER sensor protein STIM. The presence of homologues of the Orai channel subunit and the STIM Ca^{2+} sensor in *M. brevicollis* provides the first evidence that store-operated Ca^{2+} entry could operate in unicellular organisms. This observation suggests that store-operated Ca^{2+} entry represents a primordial Ca^{2+} entry pathway [20]. Choanoflagellate Orai possesses the intragenic repeat patterns and critical residues identified in multicellular animals [73]. Similar to non-Chordata STIM molecules, choanoflagellate STIM contains a single N-terminal EF-hand domain and a sterile α-motif domain in the ER lumen, a single transmembrane segment, and a cytoplasmic coiled-coil domain but lacks a C-terminal lys-rich tail [74].

Thus, the lower eukaryotes contain not only structural components necessary for formation of the calcium signal, but also the mechanisms of agonist-induced influx and intracellular Ca^{2+} mobilization, which served as evolutionary predecessors of the Ca^{2+} signal mechanisms in the higher eukaryote cells.

CA^{2+}-BINDING PROTEINS

Ca^{2+}-binding proteins (CaBPs) are calcium sensors, or primary Ca^{2+} receptors, as a result of interaction with Ca^{2+} modify the conformation and acquire the capability for activation of numerous enzymes and regulatory proteins to affect thereby dynamics of various cell processes. In this section, we consider structure and functions of CaBPs, as they play the key role in modulation and realization of calcium signals in lower eukaryote cells.

In the lower eukaryotes, CaBPs participate in regulation of the cell cycle and change of the activities of various enzymes and signal proteins, transmembrane Ca^{2+} ion transport, and the state of cytoskeleton or play a role of intracellular Ca^{2+} buffer (calsequestrin). It is important to notice that calcium buffers can modulate intracellular calcium transients, by changing their time course and spatial spreading, and can thereby modify calcium-dependent intracellular signaling. Most of CaBPs revealed in lower eukaryotes belong to proteins of the EF-hand superfamily (Table 1).

The EF-hand domain classified in more than 120 families, and includes calcium sensors such as troponin C (TnC), myosin, S100s, and the ubiquitous calmodulin; proteases like calpains; calcium buffers such as parvalbumin and calbindin D_{9k}; and many other calcium-binding proteins. EF-hand (EFh) proteins, characterized by a helix- loop-helix motif paired in functional domains. EF-hand proteins may be viewed as molecular switches activated by calcium concentration transients.

Table 1. Ca^{2+}-binding proteins in lower eukaryotes

Species of the microorganism	Name	Function	Reference
Amoeba proteus	CaM	Activation of regulatory enzymes	[81]
Phusarum polycephalum	EF-hand calcium-binding protein (CBP40)	Proteinase	[82, 155]
Euglena gracilis	CaM	Gravitaxis	[83]
	Calreticulin	Calcium reservoir Signaling	[129]
Acanthamoeba healyi	EF-hand calcium-binding protein (AhABP)	Actin bundling	[156]
Entamoeba histolytica	EF-hand calcium-binding protein (EhCaBP1/2)	Parasite growth Signaling	[87, 90, 15 7, 160]
	Granin-1	Granule discharge	[88]
	Granin-2	Endo- and phagocytosis	[88]
Zoothamnium geniculatum	Spasmin-like protein	Contraction of the spasmoneme	[78]
Carchesium colypinum	Spasmin-like protein	Contraction	[80]
Trypanosoma cruzi	Calflagin (F29)	Calcium sensor Parasite motility Calcium reservoir	[84, 85, 158]
	Calreticulin (TcCRT)	Promotes infectivity Activation of regulatory enzymes	[167]
	CaM	Modulating calcium-dependent cellular processes Regulator of flagellar function and assembly	[169]
	Flagellar calcium-binding protein (FCaBP)	Calcium sensor	[170]
Trypanosoma brucei	Calflagin	Calcium sensor	[86, 159]
	CaM	Modulating calcium-dependent cellular processes	[159]
Trypanosoma brucei rhodesiense	CaM	Modulating calcium-dependent cellular processes	[171]
Trypanosoma carassii	Calreticulin (TcaCRT)	Calcium reservoir Binds host complement component C1q	[168]
Chlamydomonas reinhardtii	EF-hand calcium-binding protein (p72)	Regulation of flagellar motility	[87]
Dictyostelium discoideum	Centrin-like protein (DdCrp)	Centrosome duplication Cell cycle progression	[93]
	Centrin-like protein (DdCenB)	Aggregation Sporulation	[98]
	CaM-like protein (calB)	Modulator of activities of the actin system	[107]
	Calnexin	Calcium reservoir	[130]
	Calreticulin	Phagocytosis	[130]
Naegleria gruberi	Centrin	Component of centrosome-associated flagellar roots Contraction	[96]
Leishmania donovani	Centrin	Centrosome duplication and segregation Activation of regulatory enzymes	[97]
	CaM	Modulating calcium-dependent cellular processes	[172]

Table 1. (continued)

Species of the microorganism	Name	Function	Reference
Leishmania infantum	EF-hand calcium-binding protein (p11)	Calcium sensor	[173]
Paramecium multimicronucleatum	Centrin CaM	Contraction Activation of regulatory enzymes Modulating calcium-dependent cellular processes Phagocytosis	[161] [165]
Paramecium caudatum	Centrin	Contraction	[161]
Paramecium tetraurelia	Calpactin-like annexin Copine CaM	Trichocyst docking Membrane trafficking Ion channel subunit Activation of regulatory enzymes Modulating calcium-dependent cellular processes	[123] [164] [46, 166, 180]
Giardia lamblia	Centrin Annexin (α8-giardin) Annexin E1(α14 giardin) CaM	Formation of flagella Trophozoite motility and growth Cytoskeletal rearrangement Calcium carrier Activation of regulatory enzymes	[162] [125] [124] [174]
Saccharomyces cerevisiae	Centrin CaM	Calcium carrier NAD kinase activator Growth Signaling	[162] [105]
Agaricus campestris	CaM	NAD kinase activator	[105]
Coprinus lagopus	CaM	NAD kinase activator	[105]
Schizosaccharomyces pombe	CaM (cam1) Ncs1p	Vegetative growth Negatively regulates sporulation	[106] [185]
Tetrahymena thermophila	EF-hand calcium-binding protein (TCBP-25) CaM	Maintenance of cell shape Ciliary motility Cytokinesis Phagocytosis	[110] [163]
Tritrichomonas suis	EF-hand calcium-binding protein (Ts-p41) Calnexin (Ts-p66)	Flagellar mobility Folding of newly synthesized proteins and glycoproteins Signaling	[115] [126]
Trichomonas vaginalis	Centrin CaM	Component of anchoring arms Calcium sensor	[175] [176]
Plasmodium falciparum	CaM (Pfcam)	Calcium sensor	[177]
Toxoplasma gondii	CaM	Gliding mobility Invasion Activator of actin-myosin motor	[178]
Cryptosporidium parvum	CaM	Cell signaling	[134]
Eimeria tenella	Calmodulin-domain protein kinase (EtCDPK)	Protein secretion Invasion Differentiation	[179]

These proteins are based on the EFh motif with a 12 or 14 amino acid long interhelical loop able to bind a calcium ion. The minimal structural and functional unit of EF-hand proteins is a domain comprised of a pair of EFh motifs tethered together by a linker of variable length. The most represented family group in the EF-hand superfamily collects the

CaM-like proteins, endowed with a tandem of EF-hand domains able to bind up to four calcium ions. They appear to have originated from a single motif by two cycles of gene duplication and fusion, which occurred very early in evolution, leading to cooperativity, i.e., a free-energy coupling, between the two calcium binding sites belonging to the same domain. The affinity for calcium of each EF-hand domain is, indeed, higher than that of a single EFh motif [75].

It was long believed that CaBPs existed only in metazoan cells; however, proteins with a high affinity for Ca^{2+} ions, including CaM-like proteins, were also detected in the lower eukaryotes and even bacteria. Moreover, it was shown that some of them revealed in prokaryotes included two and more motifs similar structurally with the EFh [76].

Nevertheless, these proteins differ markedly from metazoan CaBPs [1]. Quite a few CaBPs belonging to EF-hand domain-containing proteins have been revealed in the lower eukaryotes [77]. Structurally, they are closer to the calcium-binding proteins of multicellular organisms than to CaM-like proteins of bacteria. Information about properties and functions of CaBPs in the lower eukaryotes is fragmentary. Described in the greatest detail are biochemical properties of the CaM-like proteins revealed in unicellular organisms of various taxonomic groups.

The presence of CaBPs in the lower eukaryotes was first reported in the early 1970s. Acidic, low-molecular high-affinity Ca^{2+}-binding proteins were extracted from ciliate *Zoothamnium geniculatum* [78]. It has been established that these proteins (spasmin-like proteins) to bind Ca^{2+} at concentrations ranging from 10^{-8} to 10^{-6} M and may be involved directly in the calcium-induced contraction of the spasmoneme.

Later, proteins (mol. weights of 16, 18, and 22 kDa) with similar properties were isolated from ciliate *Carchesium colypinum* [79]. In the immunoprecipitation reaction, antibodies to these CaBPs cross reacted with spasmins of some ciliates, *Stentor*, *Spirostomum*, and *Blefarisma*, but not with mammalian calmodulin and troponin C [80]. Subsequently, CaBPs were detected in microalgae *Euglena gracilis*, parasitic amoeba *Amoeba proteus*, and acellular slime mold *Phusarum polycephalum* [81-83].

African trypanosomes express a family of dually acylated, EF-hand calcium-binding proteins called the calflagins [84]. Originally, the 24-kDa calcium-binding protein of haemoflagellate *Trypanosoma cruzi*, was recognized by antisera from both humans and experimental animals infected with this organism. It has appeared that near its C-terminus are two regions that have sequence similarity with several Ca^{2+}-binding proteins and that conform to the EF-hand structure [85]. The protein's low Ca^{2+}-binding capacity (less than 2 mol of Ca^{2+}/mol of protein) and high Ca^{2+}-binding affinity (apparent Kd less than 50 μM Ca^{2+}) are consistent with binding of Ca^{2+} via the EF-hand domains. Immunofluorescence assays using a mouse antiserum directed against the fusion protein localized the native protein to the trypanosome's flagellum. The protein's abundance, Ca^{2+}-binding property, and flagellar localization have allowed to assume that it participates in molecular processes associated with the high motility of the parasite [85].

Some years later in a closely related species *Trypanosoma brucei*, the 22, 24, and 38 kDa CaBPs were found, with their biochemical and immunological properties differing from those of known proteins belonging to annexins and of EF-hand domain-containing proteins. It is of interest that the 22 kDa CaBPs in the N-terminal region have a significant homology (58%) with that of the mammalian p21 tumor protein [86].

By this time structure and properties of CaBPs from trypanosomes are well studied. It is established that these proteins associate with lipid raft microdomains in the flagellar membrane, where they putatively function as calcium signaling proteins. It is shown that their expression is regulated the life cycle stage of the parasite, with protein levels approximately 10-fold higher in the mammalian bloodstream form than in the insect vector procyclic stage that can specify in the important role of these proteins in mammalian infection [84].

The flagellar p72 protein of the microalgae *Chlamydomonas reinhardtii* has three domains, two of which are EF-hand domains. The target of this protein is nucleoside diphosphate kinase requiring GTP for its activation, which indicates involvement of p72 protein in the GTP-mediated signaling [87].

In unicellular eukaryotes, examples of proteins containing three or four cation-binding centers are TCBP23 and TCBP25 proteins, centrin, and caltractin, as well as calmodulins and CaM-like proteins. Group of proteins with properties of EF-hand includes granin-1 and granin-2 isolated from cytoplasmic granules of the protozoan parasite *Entamoeba histolytica*. These proteins are involved *in vivo* in processes of endo- and phagocytosis [88].

Among the proteins containing four domains of EF-hand, the simplest is the low molecular EhCaBP protein also isolated from *Entamoeba histolytica* [89]. The 3D structure of Ca^{2+}-bound EhCaBP has been derived using multidimensional nuclear magnetic resonance (NMR) spectroscopic techniques. This protein characterized by a low structural homology (30%) relative to other known CaBPs, contains two globular domains, each consisting of two pairs of the spiral-loop-spiral sites similar topologically with those in the mammalian CaM. EhCaBP is a 14.7 kDa (134 residues) monomeric protein thought to play a role in the pathogenesis of amebiasis. By means of a NMR it has been established that EhCaBP exists in two forms: apo-EhCaBP and holo-EhCaBP. Ca^{2+}-free form (apo-EhCaBP) exists in a partially collapsed form compared to the Ca^{2+}-bound (holo) form, which has an ordered structure (PDB ID) [90]. Deuterium exchange studies on the partially structured apo-EhCaBP reveal that the C-terminal domain is better structured than the N-terminal domain. The protein can be reversibly folded and unfolded upon addition of Ca^{2+} and EGTA, respectively. Titration shows a slow initial folding of the apo form with increasing Ca^{2+} concentration, followed by a highly cooperative folding to its final state at a certain threshold of Ca^{2+}. Ca^{2+} and the EGTA titration taken together show that site II in the N-terminal domain has the highest affinity for Ca^{2+} contrary to earlier studies. The study of Mukherjee and co-workers [90] has thrown light on the relative Ca^{2+}-binding affinity and specificity of each site in the intact protein. Large conformational changes seen in transforming from the apo to holo form of EhCaBP suggest that this protein should be functioning as a molecular switch and might have a significant role in host-parasite recognition.

CaBPs from lower eukaryotes also include highly specialized centrin and spasmin-like proteins participating in formation of the filament system during contraction, mitosis and cytokinesis [79, 80, 91-95].

It is established that centrin, a approximately or equal to 20 kDa CaBP also known as caltractin, is a component of centrosome-associated flagellar roots capable of calcium-mediated contraction, and is also found in the centrosomes of vertebrate cells [96].

The analysis of a centrin gene from amoeboflagellate *Naegleria gruberi*, reveals conserved features that distinguish centrins from calmodulin. Antibodies to bacterially expressed *Naegleria* centrin, which also recognize yeast Cdc31p, were employed to localize centrin immunoreactivity in organisms possessing specialized microtubule-organizing centers

(MTOCs) or accessory structures. In the simplest associations, as found in *Naegleria* flagellates and vertebrates tracheal epithelium, centrin is intimately associated with the cylinder of the basal bodies. In cells with unfocused mitotic spindles no localization of centrin was detected [96].

In *Saccharomyces cerevisiae* cells, which lack centrioles, centrin immunoreactivity was observed as punctate cytoplasmic bodies but not associated with spindle pole MTOCs. In *Paramecium multimicronucleatum*, centrin immunoreactivity is localized to the infraciliary lattice, previously shown to exhibit calcium-mediated contraction. In *Vorticella microstoma*, known for the calcium-induced rapid contraction of its stalk, centrin immunoreactivity is localized to the contractile spasmoneme and myonemes. Similar antigens from *Paramecium* and *Vorticella* are detected by anti-centrin and anti-spasmin. The pattern of localization of centrin immunoreactivity supports the conjecture that a contractile system involving centrin, initially associated with centriolar structures, was recruited during evolution to build specialized organelles in different organisms and cell types [96].

It was shown on the example of the protozoan parasite *Leishmania donovani* that centrin took part in processes of centrosome duplication and segregation, which indicates a functional role of this protein in control of cell cycle progression [97]. *Dictyostelium discoideum* centrin-related protein (DdCrp) is the most divergent member of the centrin family. Most strikingly it lacks the first two EF-hand consensus motifs, whereas a number of other centrin-specific sequence features are conserved. Southern and Northern blot analysis and the data presently available from the *Dictyostelium* genome and cDNA projects suggest that DdCrp is the only centrin isoform present in *Dictyostelium*. Immunofluorescence analysis with anti-DdCrp antibodies revealed that the protein is localized to the centrosome, to a second, centrosome-associated structure close to the nucleus and to the nucleus itself. Confocal microscopy resolved that the centrosomal label is confined to the corona surrounding the centrosome core. Unlike for other centrins the localization of DdCrp is cell cycle-dependent. Both the centrosomal and the centrosome-associated label disappear during prometaphase, most likely in concert with the dissociation of the corona at this stage. The striking differences of DdCrp to all other centrins may be related to the distinct structure and duplication mode of the *Dictyostelium* centrosome [93].

Recently the novel centrin has been characterised from *Dictyostelium discoideum* (DdCenB) [98]. DdCenB is localized at three points in the cell cycle: interphase, mitosis and cytokinesis. In interphase DdCenB primarily localizes to the nuclear envelope (NE). Although the NE remains intact during mitosis and cytokinesis in *Dictyostelium*, DdCenB disappears from the NE at these two stages of the cell cycle. In addition to localization at the NE, weak localization of DdCenB took place in the nucleoplasm and cytoplasm (weakest). Although the nucleoplasmic concentration appears constant throughout the cell cycle, the very faint localization in the cytoplasm does appear to increase to the level of the nucleoplasm during mitosis and cytokinesis. Unlike most centrins, characterized by this time, it has not been presented proofs of DdCenB localization of at the centrosome at any point in the cell cycle [98].

In the lower eukaryotes, the best studied among calcium-binding proteins, containing four motifs of EF-hand, is calmodulin (CaM). In unicellular organisms, like in metazoan cells, CaM is a regulator of diverse cell functions and acts preferentially through modulation of different enzyme activities, including adenylyl and guanylyl cyclases, phosphodiesterases, calcineurin, Ca^{2+}-ATPase, phospholipase A2, and some others [99-104]. The widespread of

CaM in unicellular organisms seems to be due to that this protein, as a mediator of calcium signaling, is needed for various Ca^{2+}-dependent functions. It has been established that in eukaryotic microorganisms the CaM-Ca^{2+} activates cell mobility, regulates cell cycle, and initiates cytokinesis, as well as plays an important role in viability and in cell invasion.

In pioneer studies a comparison of calmodulins isolated from vegetative mycelia of *Basidiomycetes* fungi, *Agaricus campestris* and *Coprinus lagopus* has shown that they have similar mobilities to those of animal calmodulins on nondenaturing polyacrylamide gel electrophoresis in the presence or absence of Ca^{2+}. The molecular weights of both calmodulins were determined to be 16 kDa. Both calmodulins activated muscle myosin light chain kinase and NAD^+ kinase in a Ca^{2+}-dependent manner, and the activities were inhibited by trifluoperazine or chlorpromazine [105].

Further, the gene encoding calmodulin from the fission yeast, *Schizosaccharomyces pombe* has been cloned [106]. It is established that a 1.6kb DNA fragment contains a gene whose deduced product possesses 74% amino acid homology with bovine calmodulin. This gene, which is unique in the *S. pombe* genome and is named cam1, encodes 149 amino acids excluding the first methionine and is transcribed into mRNA of 1.2-kb length. It has an intron that apparently starts immediately after the initiation codon and is 126 bp long. *S. pombe* calmodulin exhibits more homology to vertebrate calmodulin than to that of the budding yeast, *Saccharomyces cerevisiae*. Gene disruption experiments revealed that *cam1* gene function is essential for vegetative growth of *S. pombe*. Spores bearing disrupted cam1 halt growth soon after germination and rarely carry out the first cell division, indicating that CaM does not exist in excess in those cells [106].

A gene, encoding CaM-like protein, named *calB*, was cloned and characterized in slime mold *Dictyostelium discoideum* [107]. A relationship to CaM is suggested by sequence identity (50%), similar exon-intron structure and cross-reactivity with anti-CaM sera. The level of calB mRNA is developmentally regulated with maxima during aggregation and in spore [107]. *D. discoideum* contains at least two more CaM-like proteins which have specific functions during distinct steps of development [108, 109].

The *Tetrahymena* Ca^{2+}-binding protein of 25 kDa (TCBP-25) is a calmodulin family protein containing four EF-hand type calcium-binding motifs. TCBP-25 is localized in the whole cell cortex and around both the migratory and stationary pronuclei at the pronuclear exchange stage during conjugation. TCBP-25 plays an important role in the pronuclear exchange and in the maintenance of cell shape [110].

The presented examples of the regulatory CaM action on various intracellular processes in the lower eukaryotes indicate that this protein can modulate activity of numerous Ca^{2+}-dependent enzymes and proteins-effectors. The question arises of how does CaM recognize and regulate activity of tens of different protein targets? It has turned out that the mechanisms of its effect on intracellular substrates of eukaryote microorganisms are basically the same as in the higher eukaryote cells. The basis for ligand-mediated modulation of activity by calmodulin consists of three processes: recognition of substrate on the principle of complementarity (the presence of amphiphilic α-spirals in proteins), activation of protein-target by phosphorylation-dephosphorylation, and, finely, inhibition of activity of the CaM-dependent enzymes as a result of Ca^{2+} removal from the calcium-binding sites and subsequent dissociation of CaM from a enzyme complex [77, 111-114].

Calcium-binding proteins with the most complex conformation have the EF-hand motifs that are usually tandemly repeated, and the number of repeats ranges from five to eight. In mammals, this group of CaBPs includes, in particular, calbindin isolated from intestinal cells and calretinin found in neurons. The domain organization of these proteins involves six EF-hand motifs condensed into a single domain. Their functions have remained so far poorly understood. The CaBP also containing six canonic EF-hand motifs was found in protozoan parasite *Tritrichomonas suis* (Table 1). It is supposed that functions of this protein are connected with regulation of the flagellar mobility [115].

The so-called CaM-like domain protein kinases (CDPKs), originally identified in plants, also have been revealed in apicomplexan and ciliated protozoa [116-118]. The phylogenetic analysis of CDPK genomic sequences have shown that the introns shared between protist and plant CDPKs presumably originated before the divergence of plants from Alveolates. Additionally, the calmodulin-like domains of protozoan CDPKs have intron positions in common with animal and fungal calmodulin genes. These results, together with the presence of a highly conserved phase zero intron located precisely at the beginning of the calmodulin-like domain, suggest that the ancestral CDPK gene could have originated from the fusion of protein kinase and calmodulin genes facilitated by recombination of ancient introns [117].

Unlike true CaM-kinases, activation of the CDPKs requires Ca^{2+} ions, but not phospholipids and calmodulin [119]. In parasitic protozoans, these protein kinases play an important role in motility, including entry of and egress from host cells [120, 121].

Among other Ca^{2+}-binding proteins, annexins were detected in unicellular eukaryotes [122-125]. Alpha-14 giardin (annexin E1), a member of the α-giardin family of annexins, has been shown to localize to the flagella of the intestinal protozoan parasite *Giardia lamblia* [124]. Alpha giardins show a common ancestry with the annexins, a family of proteins most of which bind to phospholipids and cellular membranes in a Ca^{2+}-dependent manner and are implicated in numerous membrane-related processes including cytoskeletal rearrangements and membrane organization. It has been proposed that alpha-14 giardin may play a significant role during the cytoskeletal rearrangement during differentiation of *Giardia*. Unlike most annexin structures, which typically possess N-terminal domain, alpha-14 giardin (α14-giardin) is composed of only a core domain, followed by a C-terminal extension that may serve as a ligand for binding to cytoskeletal protein partners in *Giardia*. In the Ca^{2+}-bound structure were detected five bound calcium ions, one of which is a novel, highly coordinated calcium-binding site not previously observed in annexin structures. This novel high-affinity calcium-binding site is composed of seven protein donor groups, a feature rarely observed in crystal structures. In addition, phospholipid-binding assays suggest that α14-giardin exhibits calcium-dependent binding to phospholipids that co-ordinate cytoskeletal disassembly/assembly during differentiation of the parasite [124]. Alpha8-giardin (α8-giardin)-other representative of the multi-gene α-giardin family in the *Giardia lamblia* [125]. This protein is located on the plasma membrane and flagella. Reduction of α8-giardin transcript levels by ribozyme-mediated cleavage decreased trophozoite motility and growth rate, indicating the functional importance of α8-giardin to *Giardia* trophozoite biology [125]. In *Paramecium*, these Ca^{2+}-dependent phospholipid-binding proteins are selectively recognized by antibody against the common annexin sequences and are located in sites of anchoring trichocysts and cytoprocts [123].

Apart from annexins, proteins belonging to a new group of Ca^{2+}-dependent phospholipid-binding proteins copines were revealed in *Paramecium* [123]. These proteins contain Ca^{2+}-binding domains composed approximately of 120 amino acids. The functional role and the exact location of copines in *Paramecium* are so far unknown.

Calnexins, Ca^{2+}-binding proteins with chaperone activity in the endoplasmic reticulum, were found in parasitic protozoans *Tritrichomonas suis, Leishmania major* [126-128], in microalga *Euglena gracilis* [129], as well as in cell slime mold *Dictyostelium discoideum* [130]. Their function in microorganisms yet is not quite clear. It is supposed that at *Dictiostelium* this protein together with calreticulin – another endoplasmic reticulum lumenal CaBP, is important for phagocytosis (Table 1) [130]. A dramatic decline in the rate of phagocytosis was observed in double mutants lacking calreticulin and calnexin, whereas only mild changes occurred in single mutants. *Dictyostelium* cells are professional phagocytes, capable of internalizing particles by a sequence of activities: adhesion of the particle to the cell surface, actin-dependent outgrowth of a phagocytic cup, and separation of the phagosome from the plasma membrane. It is noticed that in the double-null mutants, particles still adhered to the cell surface, but the outgrowth of phagocytic cups was compromised. Green fluorescent protein-tagged calreticulin and calnexin, expressed in wild-type cells, revealed a direct link of the ER to the phagocytic cup enclosing a particle, such that the Ca^{2+} storage capacity of calreticulin and calnexin might directly modulate activities of the actin system during particle uptake [130].

The above examples testify to the essential contribution of CaBPs in the control of cellular functions and allow to consider the calcium ions activating these proteins, as the important components of lower eukaryotes Ca^{2+} signaling.

Thus, the mechanism of calcium signal modulation by CaBPs, which operates in prokaryotes, has got its further development in eukaryotic microorganisms and, by acquiring new regulatory elements of input signal in the course of evolution, has become the structural-functional basis for formation of such mechanism in the higher eukaryote cells.

CA^{2+} EFFLUX AND CA^{2+} ACTIVE EXPORT FROM CELL

A peculiarity of the calcium signaling in eukaryote cells is a short-term local calcium signal, as the long-term $[Ca^{2+}]_i$ increase that is not associated with formation of directed calcium wave leads to an cellular pathology and cell death [131]. Therefore, the mechanism of extinguishment of the calcium signal is the central component of the calcium signal system. In inactivation of a calcium signal take part calcium transporters that are included into the regulatory-transport cell system providing the $[Ca^{2+}]_i$ recovery to the basal level. Like metazoan cells, unicellular eukaryotes have effective mechanisms of extinguishment of Ca^{2+} signal with the aid of protein-carriers. The cell membrane of eukaryotic microorganisms contains several carriers responsible for Ca^{2+} export from a cell (Table 2). One of them has the highest affinity for Ca^{2+} and has been revealed in some studied unicells. The carrier represents a plasma membrane Ca^{2+} ATPase (PMCA), an enzyme that removes Ca^{2+} from cell against a pronounced concentration gradient existing on the membrane.

The CaM-dependent PMCA-type Ca^{2+} ATPase found in *Paramecium tetraurelia* is the closest to the mammalian PMCA Ca^{2+} ATPase. In early works it has been noticed that, like in cells of the higher eukaryotes, *Paramecium* PMCA-type Ca^{2+} ATPase reacts to even a slight increase of the intracellular Ca^{2+} concentration [132, 133]. By this time there are biochemical proofs of existence of high-affinity CaM-dependent Ca^{2+} ATPases in different protozoan parasites [134].

Biochemical evidence for calmodulin stimulation has been reported for Ca^{2+} ATPases from *Trypanosoma cruzi* [135], *Leishmania braziliensis*, and *Leishmania donovani* [136]. The exclusive dependence of the PMCA-type Ca^{2+} ATPases on only from Ca^{2+} for their activity and their positive allosteric modulation by Mg^{2+} distinguish these protozoan enzymes from animal plasma membrane Ca^{2+} ATPases [137].

The protozoan PMCA-type Ca^{2+} ATPases described to date at the molecular level [134] appear to lack a typical regulatory calmodulin-binding domain conserved in animal PMCAs; this might suggest the presence of a different domain able to bind calmodulin, or a non-calmodulin-dependent regulation. Absence of a calmodulin-binding domain also has been established concerning the PMCA-type Ca^{2+} ATPase (PAT1) from a slime mold *Dictyostelium discoideum* [138]. Immunofluorescence analysis has shown that PAT1 colocalizes with bound calmodulin to intracellular membranes, components of the contractile vacuole complex. The presence of PAT1 on the contractile vacuole suggests that in *Dictyostelium* this organelle might function in Ca^{2+} homeostasis as well as in water regulation [138, 139].

The genes coding PMCA-type Ca^{2+} ATPases have been found in apicomplexan parasites [134, 140] and choanoflagellate *Monosiga brevicollis* [18]. In last case, motif scan searches indicate that the PMCA homologue in *M. brevicollis* exhibits a calmodulin-binding domain at the equivalent position in the C-terminal tail of animal PMCAs. From this it is possible to assume that the choanoflagellate PMCA-type Ca^{2+} ATPases are phylogenetically closer to that of multicellular animals, than PMCA from Protozoa.

In genomes of lower fungi there are no the genes coding PMCA-type Ca^{2+} ATPases. In budding yeast structural genes for two Ca^{2+} pumps, PMC1 and PMR1, which are required for Ca^{2+} sequestration into the vacuole and secretory organelles have been identified [141]. The function of either Ca^{2+} pump is sufficient for yeast viability, but deletion of both genes is lethal because of elevation of [Ca^{2+}]$_i$ and activation of calcineurin, a Ca^{2+}- and CaM-dependent protein phosphatase. Calcineurin activation decreases Ca^{2+} sequestration in the vacuole by a putative Ca^{2+} antiporter and may also increase Ca^{2+} pump activity [141].

As a Ca^{2+} ATPase has originally been described the cation-transporting P-type ATPase (CTA3 ATPase) of *Schyzosaccharomyces pombe* [142, 143], but a further study showed that by sequence and function CTA3 was a Na$^+$/K$^+$-ATPase (ENA ATPase) [144]. Structurally, the ENA (from *exitus natru*: exit of sodium) ATPase is very similar to the sarco/endoplasmic reticulum Ca^{2+} (SERCA) ATPase, and it probably exchanges Na$^+$ (or K$^+$) for H$^+$. Benito and co-workers [144] assumed that other fungi are furnished with Na$^+$-efflux ATPases, which also pump Na$^+$. Really, ENA genes of these enzymes are by this time found out in 65 fungal species [145]. The protozoan parasites *Leishmania* and *Trypanosoma* have ATPases phylogenetically related to fungal K$^+$-ATPases, which are probably functional homologues of the fungal enzymes [144-146]. The ENA ATPase is necessary at high external pH values, where the antiporters cannot mediate uphill Na$^+$ efflux. This occurs in some fungal environments and at some points of protozoa parasitic cycles, which makes the ENA ATPase a possible target for controlling fungal and protozoan parasites.

Table 2. Ca^{2+} ATPases and cation/Ca^{2+} exchangers in lower eukaryotes

Species of the microorganism	Name	Type	Function	Reference
Paramecium tetraurelia	CaM-dependent Ca^{2+} ATPase	PMCA	Ca^{2+} homeostasis	[132, 133]
Sterkiella histriomuscorum	ShPMCA	PMCA	Excystment	[184]
Euplotes crassus	NCX	Na^+/Ca^{2+} exchanger	Ca^{2+} homeostasis	[147]
Trypanosoma brucei	TbPMC2	PMCA	Ca^{2+} homeostasis and growth	[134, 181]
Trypanosoma cruzi	PMCA-type Ca^{2+} ATPase	PMCA	Ca^{2+} pump, active export Ca^{2+} from cell	[135]
Leishmania braziliensis	PMCA-type Ca^{2+} ATPase	PMCA	Ca^{2+} pump	[136]
Leishmania donovani	CaM-dependent Ca^{2+} ATPase	PMCA	Ca^{2+} pump	[136, 137]
Toxoplasma gondii	TgA1	PMCA	Polyphosphate storage, intracellular Ca^{2+} homeostasis and virulence	[182,183]
Plasmodium falciparum	PfCHA	Ca^{2+}/H^+ exchanger	Ca^{2+} homeostasis	[154]
Entamoeba histolytica	PMCA-type Ca^{2+} ATPase	PMCA	Ca^{2+} pump	[184]
Dictyostelium discoideum	PAT1	PMCA	Calcium and water regulation	[138, 139]
Monosiga brevicollis	PMCA-type Ca^{2+} ATPase	PMCA	Ca^{2+} pump	[18]
	NCX	Na^+/Ca^{2+} exchanger	Ca^{2+} homeostasis	[18]
Saccharomyces cerevisiae	Vcx1p/Hum1p	Ca^{2+}/H^+ exchanger	Ca^{2+} sequestration into the vacuole	[152]
Chlamydomonas reinhardtii	CrCAX1	Ca^{2+}/H^+ exchanger	Ca^{2+} homeostasis	[153]

The true Na^+/Ca^{2+} exchange in the unicellular organisms is studied insufficiently. Biochemical methods had been proved presence of the Na^+/Ca^{2+} exchanger in ciliate *Euplotes crassus* [147]. Cells were loaded with fura-2/AM or SBF1/AM for fluorescence measurements of cytosolic Ca^{2+} and Na^+ respectively. Ouabain pre-treatment and external sodium ions $[Na^+]_o$ substitution in fura-2/AM-loaded cells elicited a bepridil-sensitive $[Ca^{2+}]_i$ rise followed by partial recovery, indicating the occurrence of Na^+/Ca^{2+} exchanger working in reverse mode. In experiments on prolonged effects, ouabain, $[Na^+]_o$ substitution, and bepridil all caused $[Ca^{2+}]_o$-dependent $[Ca^{2+}]_i$ increase, showing a role for Na^+/Ca^{2+} exchange in Ca^{2+}

homeostasis. In addition, by comparing the effect of orthovanadate (affecting not only Ca^{2+} ATPase, but also Na^+/K^+ ATPase and, hence, Na^+/Ca^{2+} exchange) to that of bepridil on $[Ca^{2+}]_i$, it was shown that Na^+/Ca^{2+} exchange contributes to Ca^{2+} homeostasis. In electrophysiological experiments, no membrane potential variation was observed after bepridil treatment suggesting compensatory mechanisms for ion effects on cell membrane voltage, which also agrees with membrane potential stability after ouabain treatment. These data has convincingly shown the existence of a Na^+/Ca^{2+} exchanger in the plasma membrane of *E. crassus* [147]. The molecular-genetic confirmation of presence of the Na^+/Ca^{2+} exchanger in lower eukaryotes has been received on an example of choanoflagellate *Monosiga brevicollis* [18].

The mechanism of the Ca^{2+} signal extinguishment includes elements of a system for maintenance of calcium homeostasis, which are already present in prokaryotes. The Ca^{2+}/H^+ antiporter that stimulates Ca^{2+} efflux from cells represents these elements. The Ca^{2+}/H^+ exchanger was found in eukaryotic microorganisms of various taxonomic groups, including protists and fungi [148-151]. Unlike Ca^{2+}/H^+ exchanger in prokaryotes, in lower eukaryotes this transporter rather play a critical role in sequestering Ca^{2+} into the Ca^{2+}-storing vacuoles and may function in coordination with Ca^{2+} release channels, to shape stimulus-induced cytosolic Ca^{2+} elevations.

In particular, it has been established that the Ca^{2+}/H^+ exchanger Vcx1p/Hum1p facilitates Ca^{2+} sequestration into the vacuole of the *Saccharomyces cerevisiae* and by that attenuates the propagation of intracellular Ca^{2+} signals [152].

It is established by this time that many unicellular organisms Ca^{2+}/H^+ antiporter function is carried out by the cation exchanger (CAX). Recently CrCAX1 gene encoding a Ca^{2+}/H^+ and Na^+/H^+ exchanger from the unicellular green alga *Chlamydomonas reinhardtii* has been cloned and characterized [153]. CrCAX1 was more closely related to fungal cation exchanger (CAX) genes than those from higher plants but has structural characteristics similar to plant Ca^{2+}/H^+ exchangers including a long N-terminal tail. When CrCAX1-GFP was expressed in *Saccharomyces cerevisiae*, it was localized at the vacuole. CrCAX1 could suppress the Ca^{2+}-hypersensitive phenotype of a yeast mutant and mediated proton gradient-dependent Ca^{2+}/H^+ exchange activity in vacuolar membrane vesicles. Ca^{2+} transport activity was increased following N-terminal truncation of CrCAX1, suggesting the existence of an N-terminal auto-regulatory mechanism. CrCAX1 could also provide tolerance to Na^+ stress when expressed in yeast or *Arabidopsis thaliana* because of Na^+/H^+ exchange activity. This Na^+/H^+ exchange activity was not regulated by the N-terminus of the CrCAX1 protein. CrCAX1 was transcriptionally regulated in *Chlamydomonas* cells grown in elevated Ca^{2+} or Na^+. Thus, this proton-coupled transporter, CrCAX1, transports both monovalent and divalent cations and plays a role in cellular cation homeostasis by the transport of Ca^{2+} and Na^+ into the vacuole [153].

Recently similar divalent cation exchanger (PfCHA) has been revealed in malaria parasite *Plasmodium falciparum* [154]. Presence of this enzyme at various unicellular systems can serve as evidence of a common origin of molecular mechanisms of regulation of Ca^{2+} homeostasis in the lower eukaryotes.

Other mechanisms of Ca^{2+} removal from lower eukaryote cells also exist, but their contribution to the Ca^{2+} signal extinguishment is less significant compared with the above mechanisms.

CONCLUSION

Calcium is one of universal regulators of intracellular processes occurring not only in cells of higher organisms, but also in unicellular eukaryotes. Analysis of the data accumulated by the present time indicates that the main direction of both molecular and functional evolution of mechanisms of intracellular signaling in eukaryotes is associated with development of the calcium messenger system that is revealed practically in all current representatives of ancient species of unicellular eukaryotes.

Apparently, as molecular basis for formation of the Ca^{2+} signaling in unicellular eukaryotes the membrane transport mechanisms and elements of the system of maintenance of calcium homeostasis already arisen in bacteria served. They include Ca^{2+} channels, transmembrane Ca^{2+} carriers, and primitive forms of Ca^{2+}-binding proteins.

The structural-molecular organization of the system of transmission of triggering information from the cell surface to intracellular targets was modified in the course of progressive evolution. Apart from the Ca^{2+} sensitive sensor molecules existing already in bacteria, the unicellular eukaryotes acquired highly specialized mechanisms of voltage- and receptor-mediated transduction of the trigger signal to a local Ca^{2+} signal and its subsequent translation to postreceptor structures. In the lower eukaryotes, for the first time in evolution, the system of modulation of calcium homeostasis appears with the aid of highly specialized Ca^{2+}-binding proteins. To provide this function, primitive mechanisms of Ca^{2+} mobilization from the intracellular stores are formed in eukaryotic microorganisms. Some of these mechanisms obtained their evolutionary development in cells of the higher eukaryotes.

Like in metazoan cells, the mechanisms of Ca^{2+} signaling in the lower eukaryotes seem to be based on processes of formation and directed propagation of the wave-like increase of the intracellular Ca^{2+} concentration (the calcium wave) and possibly local $[Ca^{2+}]_i$ oscillations affecting development of programmed cell processes. However, information on this issue is not yet sufficient. The fact that calcium oscillations accompany the process of conjugation in ciliates and of differentiation in slime mold cells can be an argument in favor of this suggestion.

At present, the total structural and functional organization of Ca^{2+} signaling mechanisms has been described sufficiently well in the lower eukaryotes, in particular, mechanisms have been studied of Ca^{2+} influx into cells, functional characteristics and ways of regulation of Ca^{2+} channels have been described, various mechanisms of Ca^{2+} release from intracellular stores have been studied, and systems of emergency and long-term Ca^{2+} removal from cells have been established. However, protein kinase pathways of signal transduction with participation of the Ca^{2+} messenger system in the lower eukaryotes are still far from their complete understanding.

ACKNOWLEDGMENT

I thank Dr. V. P. Nesterov for critically reading the manuscript.

REFERENCES

[1] Shemarova, I. V., & Nesterov, V. P. (2005). Evolution of Ca^{2+} signaling mechanisms. Part 1. Role of Ca^{2+} ions in signal transduction in prokaryotes. *Zh. Evol. Biokhim. Fiziol.*, 41, 12–17.

[2] Tsien, R. W. (1983). Calcium channels in excitable cell membranes. *Annu. Rev. Physiol.*, 45, 341–358.

[3] Saimi, Y., Martinac, B., Gustin, M. C., Culbertson, M. R., Adler, J., & Kung, C. (1988). Ion channels in *Paramecium*, yeast and *Escherichia coli*. *Trends Biochem. Sci.*, 13, 304–309.

[4] Bertl, A., Gradmann, D., & Slayman, C. L. (1992). Calcium and voltage-dependent ion channels in *Saccharomyces cerevisiae*. *Philos. Trans. R. Soc. Lond. B Biol. Sci.,* 338, 63–72.

[5] Plattner, H., & Klauke, N. (2001). Calcium in ciliated Protozoa: sources, regulation, and calcium regulated cell functions. *Int. Rev. Cytol.*, 201, 115–208.

[6] Iida K., Teng J., Tada T., Saka A., Tamai M., Izumi-Nakaseko H., Adachi-Akahane S., & Iida H. (2007). Essential, completely conserved glycine residue in the domain III S2-S3 linker of voltage-gated calcium channel alpha1 subunits in yeast and mammals. *J. Biol. Chem.*, 282, 25659–25667.

[7] Teng, J., Goto, R., Iida, K., Kojima, I., & Iida, H. (2008). Ion-channel blocker sensitivity of voltage-gated calcium-channel homologue Cch1 in *Saccharomyces cerevisiae*. *Microbiology.*, 154, 3775–3781.

[8] Saitow, F., Nakaoka, Y., & Oosawa, Y. (1997). A calcium-activated, large conductance and non-selective cation channel in *Paramecium* cell. *Biochim. Biophys. Acta.*, 1327, 52–60.

[9] Bihler, H., Slayman, C. L., & Bertl, A. (1998). NSC1: a novel high-current inward rectifier for ations in the plasma membrane of *Saccharomyces cerevisiae*. *FEBS Lett.*, 432, 59–64.

[10] Roberts, S. K., Fischer, M., Dixon, G. K., & Sanders, D. (1999). Divalent cation block of inward currents and low-affinity K$^+$ uptake in *Saccharomyces cerevisiae*. *J. Bacteriol.*, 181, 291–297.

[11] Bihler, H., Slayman, C. L., & Bertl, A. (2002). Low-affinity potassium uptake by *Saccharomyces cerevisiae* is mediated by NSC1, a calcium-blocked non-specific cation channel. *Biochim. Biophys. Acta.,* 1558, 109–118.

[12] Imada, C., & Oosawa, Y. (1999). Thermoreception of *Paramecium*: Different Ca^{2+} channels were activated by heating and cooling. *J. Membr. Biol.*, 168, 283–287.

[13] Clark, K. D., Hennessey, T. M., Nelson, D. L., & Preston, R. R. (1997). Extracellular GTP causes membrane-potential oscillations through the parallel activation of Mg^{2+} and Na$^+$ currents in *Paramecium tetraurelia*. *J. Membr. Biol.,* 157, 159–167.

[14] Gonda, K., Oami, K., & Takahashi, M. (2007). Centrin controls the activity of the ciliary reversal-coupled voltage-gated Ca^{2+} channels Ca^{2+}-dependently. *Biochem. Biophys. Res. Commun.*, 362, 170–176.

[15] Nagamune, K., & Sibley, L. D. (2006). Comparative genomic and phylogenetic analyses of calcium ATPases and calcium-regulated proteins in the apicomplexa. *Mol. Biol. Evol.*, 23, 1613–1627.

[16] Paidhungat, M., & Garrett, S. A. (1997). Homolog of mammalian, voltage-gated calcium channels mediates yeast pheromone-stimulated Ca^{2+} uptake and exacerbates the cdc1(Ts) growth defect. *Mol. Cell. Biol.*, 17, 6339–6347.

[17] Bonilla, M., & Cunningham, K. W. (2002). Calcium release and influx in yeast: TRPC and VGCC rule another kingdom. *Sci STKE.*, 127, pe17.

[18] Cai, X. (2008). Unicellular Ca^{2+} signaling 'toolkit' at the origin of metazoa. *Mol. Biol. Evol.*, 25, 1357–1361.

[19] Machemer, H., & de Peyer, J. (1977). Swimming sensory cells. Electrical membrane parameters, receptor properties and motor control in ciliated protozoa. *Verh. Disch. Zool. Ges.*, 86–110.

[20] Machemer, H. (1998). Mechanisms of graviperception and response in unicellular systems. *Adv. Space Res.,* 21, 1243–1251.

[21] Jennings, H. S. (1976). *Behavior of the lower organisms*. London: Indiana Univ., Bloomington.

[22] Seravin, L. N. (1967). *Motornye sistemy prosteishikh* (Motor systems of Protozoa). Leningrad: Nauka., 175–183.

[23] Naitoh, Y., & Eckert, R. (1969). Ionic mechanisms controlling behavioral responses of *Paramecium* to mechanical stimulation. *Science.*, 164, 963–965.

[24] Naitoh, Y., & Eckert, R. (1972). Electrophysiology of ciliated Protozoa. *Exp. Physiol. Biochem.*, 5, 17–31.

[25] Lee, A., Westenbroek, R. E., Haeseleer, F., Palczewski, K., Scheuer, T., & Catterall, W. A. (2002). Differential modulation of Ca(v)2.1 channels by calmodulin and Ca^{2+}-binding protein 1. *Nat. Neurosci.*, 5, 210–217.

[26] Reuter, H. (1985). A variety of calcium channels. *Nature.*, 316, 391.

[27] Won Y. J., Ono F., & Ikeda S. R. (2011). Identification and modulation of voltage-gated Ca^{2+} currents in zebrafish Rohon-Beard neurons. J. Neurophysiol., 105, 442–453.

[28] Ertel, E. A., Campbell, K. P., Harpold, M. M., Hofmann, F., Mori, Y., Perez-Reyes, E., Schwartz, A., Snutch, T. P., Tanabe, T., Birnbaumer, L., Tsien, R. W., & Catterall, W. A. (2000). Nomenclature of voltage-gated calcium channels. *Neuron.*, 25, 533–535.

[29] Aonuma, M., Kadono, T., & Kawano, T. (2007). Inhibition of anodic galvanotaxis of green paramecia by T-type calcium channel inhibitors. *Z. Naturforsch.*, 62, 93–102.

[30] Ehrlich, B. E., Finkelstein, A., Forte, M., & Kung, C. (1984). Voltage-dependent calcium channels from *Paramecium* cilia incorporated into planar lipid bilayers. *Science.*, 225, 427–428.

[31] Brehm, P. & Eckert, R. (1978). Calcium entry leads to inactivation of calciumchannel in *Paramecium*. *Science.*, 202, 1203–1206.

[32] Hinrichsen, R. D., Saimi, Y., & Kung, C. (1984). Mutants with altered Ca^{2+}-channel properties in *Paramecium tetraurelia*: isolation, characterization and genetic analysis. *Genetics.*, 108, 545–558.

[33] Oertel, D., Schein, S. J., & Kung, C. (1977). Separation of membrane currents using a *Paramecium* mutant. *Nature.*, 268, 120–124.

[34] Satow, Y., Murphy, A. D., & Kung, C. (1983). The ionic basis of the mechanoreceptor potential of *Paramecium tetraurelia*. *J. Exp. Biol.*, 103, 253–264.

[35] Ling, K. J. & Kung, C. (1980). Ba^{2+} influx measures the duration of membrane excitation in *Paramecium*. *J. Exp. Biol.* 84, 73–87.

[36] Thiele, J. & Schultz, J. E. (1981). Ciliary membrane vesicles of *Paramecium* contain the voltage-sensitive calcium channel. *Proc. Natl Acad. Sci. USA.*, 78, 3688–3691.

[37] Otter, T. & Salmon, E. D. (1979). Hydrostatic pressure reversibly blocks membrane control of ciliary motility in *Paramecium*. *Science.*, 206, 358–361.

[38] Hinrichsen, R. D. & Saimi, Y. (1984). A mutation that alters properties of the calcium channel in *Paramecium tetraurelia*. *J. Physiol.*, 351, 397–410.

[39] Yoshimura, K. (1998). Mechanosensitive channels in the cell body of *Chlamydomonas*. *J Membr. Biol.*, 166, 149–155.

[40] Tsiokas, L. (2009). Function and regulation of TRPP2 at the plasma membrane. *Am. J. Physiol. Renal. Physiol.*, 297, F1–F9.

[41] Lombardi, M. L., Knecht, D. A., & Lee J. (2008). Mechano-chemical signaling maintains the rapid movement of *Dictyostelium* cells. *Exp. Cell Res.*, 314, 1850–1859.

[42] Yoshimura, H., Tada, T., & Iida, H. (2004). Subcellular localization and oligomeric structure of the yeast putative stretch-activated Ca^{2+} channel component Mid1. *Exp. Cell Res.*, 293, 185–195.

[43] Lew, R. R., Abbas, Z., Anderca, M. I., & Free, S. J. (2008). Phenotype of a mechanosensitive channel mutant, mid-1, in a filamentous fungus, *Neurospora crassa*. *Eukaryot. Cell.*, 7, 647–655.

[44] Chang, Y., Schlenstedt, G., Flockerzi, V., & Beck, A. (2010). Properties of the intracellular transient receptor potential (TRP) channel in yeast, Yvc1. *FEBS Lett.*, 584, 2028–2032.

[45] Su, Z., Zhou, X., Loukin, S. H., Saimi, Y., & Kung, C. (2009). Mechanical force and cytoplasmic Ca^{2+} activate yeast TRPY1 in parallel. *J. Membr. Biol.*, 227, 141–150.

[46] Saimi, Y., & Kung, C. (2002). Calmodulin as an ion channel subunit. *Annu Rev. Physiol.*, 64, 289–311.

[47] Bollo, M., Venera, G., de Jimenez Bonino, M. B., & Machado-Domenech, E. (2001). Binding of nicotinic ligands to and nicotine-induced calcium signaling in *Trypanosoma cruzi*. *Biochem. Biophys. Res. Commun.*, 281, 300–304.

[48] Rathouz, M. M. & Berg, D. K. (1994). Synaptic type acetylcholine receptors raise intracellular calcium levels in neurons by two mechanisms. *J. Neurosci.*, 14, 6935–6945.

[49] Berridge, M. J. (1993). Inositol trisphosphate and calcium signalling. *Nature.*, 361, 315–325.

[50] Mikoshiba, K. & Hattori, M. (2000). IP$_3$ receptor-operated calcium entry. *Sci. STKE.*, 26, PE1.

[51] Bumann, J., Wurster, B., & Malchow, D. (1984). Attractant-induced changes and oscillations of the extracellular Ca^{2+} concentration in suspensions of differentiating *Dictyostelium* cells. *J. Cell. Biol.*, 98, 173–178.

[52] Milne, J. L., & Coukell, M. B. (1991). A Ca^{2+} transport system associated with the plasma membrane of *Dictyostelium discoideum* is activated by different chemoattractant receptors. *J. Cell Biol.*, 112, 103–110.

[53] Sordano, C., Cristino, E., Bussolino, F., Wurster, B., & Bozzaro, S. (1993). Platelet-activating factor modulates signal transduction in *Dictyostelium*. *J. Cell. Sci.*, 104, 197–202.

[54] Braquet, P., Touqui, L., Shen, T. Y. & Vargaftig, B. B. (1987). Perspectives in platelet-activating factor research. *Pharmacol. Rev.*, 39, 97–145.

[55] Snyder, F. (1989). Biochemistry of platelet-activating factor: a unique class of biologically active phospholipids. *Proc. Soc. Exp. Biol. Med.,* 190, 125–135.

[56] Hwang, S. B. (1990). Specific receptors of platelet-activating factor, receptor heterogeneity and signal transduction mechanisms. *J. Lipid Mediators.,* 2, 123–158.

[57] Schaloske, R., Sordano, C., Bozzaro, S., & Malchow, D. (1995). Stimulation of calcium influx by platelet activating factor in *Dictyostelium. J. Cell. Sci.,* 108, 1597–1603.

[58] Schaloske, R., Schlatterer, C., & Malchow, D. (2000). A xestospongin C-sensitive Ca^{2+} Store is required for cAMP-induced Ca^{2+} influx and cAMP oscillations in *Dictyostelium. J. Biol. Chem.,* 275, 8404–8408.

[59] Lam, D., Kosta, A., Luciani, M. F., & Golstein, P. (2008). The inositol 1,4,5-trisphosphate receptor is required to signal autophagic cell death. *Mol. Biol. Cell.,* 19, 691–700.

[60] Calvert, C. M. & Sanders, D. (1995). Inositol trisphosphate-dependent and Independent Ca^{2+} mobilization pathways at the vacuolar membrane of *Candida albicans. J. Biol. Chem.,* 270, 7272–7280.

[61] Cornelius, G., Gebauer, G., & Techel, D. (1989). Inositol trisphosphate induces calcium release from *Neurospora crassa* vacuoles. *Biochem. Biophys. Res. Commun.,* 162, 852–856.

[62] Belde, P. J., Vossen, J. H., Borst-Pauwels, G.W., & Theuvenet, A. P. (1993). Inositol 1,4,5-trisphosphate releases Ca^{2+} from vacuolar membrane vesicles of *Saccharomyces cerevisiae, FEBS Lett.,* 323, 113–118.

[63] Denis, V., & Cyert, M. S. (2002). Internal Ca^{2+} release in yeast is triggered by hypertonic shock and mediated by a TRP channel homologue. *J. Cell Biol.,* 156, 29–34.

[64] Passos, A. P., Garcia, C. R. (1998). Inositol 1,4,5-trisphosphate induced Ca^{2+} release from chloroquine-sensitive and -insensitive intracellular stores in the intraerythrocytic stage of the malaria parasite *P. chabaudi. Biochem. Biophys. Res. Commun.,* 245, 155–160.

[65] Marchesini, N., Bollo, M., Hernandez, G., Garrido, M. N., & Machado-Domenech, E. E. (2002). Cellular signalling in *Trypanosoma cruzi*: biphasic behaviour of inositol phosphate cycle components evoked by carbachol. *Mol. Biochem. Parasitol.,* 120, 83–91.

[66] Carbajal, M. E., Manning-Cela, R., Pina, A., Franco, E., & Meza, I. (1996). Fibronectin-induced intracellular calcium rise in *Entamoeba histolytica* trophozoites: effect on adhesion and the actin cytoskeleton. *Exp. Parasitol.,* 82, 11–20.

[67] Catisti, R., Uyemura, S. A., Docampo, R., & Vercesi, A. E. (2000). Calcium mobilization by arachidonic acid in *Trypanosomatids. Mol. Biochem. Parasitol.,* 105, 261–271.

[68] Shemarova, I. V. (2007). Phosphoinositide signaling in unicellular eukaryotes. *Crit. Rev. Microbiol.,* 33, 141–156.

[69] Lee, H. C., Walseth, T. F., Bratt, G. T., Hayes, R. N., & Clapper, D. L. (1989). Structural determination of a cyclic metabolite of NAD^+ with intracellular Ca^{2+}-mobilizing activity. *J. Biol. Chem.,* 264, 1608–1615.

[70] Masuda, W., Takenaka, S., Tsuyama, S., Tokunaga, M., Yamaji, R., Inui, H., Miyatake, K., & Nakano, Y. (1997). Inositol 1,4,5-trisphosphate and cyclic ADP-ribose mobilize Ca^{2+} in a Protist, *Euglena gracilis. Comp. Biochem. Physiol. C Pharmacol. Toxicol. Endocrinol.,* 118, 279–283.

[71] Ladenburger, E. M., Sehring, I. M., Korn, I., & Plattner, H. (2009). Novel types of Ca^{2+} release channels participate in the secretory cycle of *Paramecium* cells. *Mol. Cell Biol.,* 29, 3605–3622.

[72] Ladenburger, E. M., Korn, I., Kasielke, N., Wassmer, T., & Plattner, H. (2006). An Ins(1,4,5)P3 receptor in *Paramecium* is associated with the osmoregulatory system. *J. Cell Sci.*, 119, 3705–3717.

[73] Cai, X. 2007a. Molecular evolution and structural analysis of the Ca^{2+} release-activated Ca^{2+} channel subunit, *Orai. J. Mol. Biol.*, 368, 1284–1291.

[74] Cai, X. 2007b. Molecular evolution and functional divergence of the Ca^{2+} sensor protein in store-operated Ca^{2+} entry: stromal interaction molecule. *PLoS ONE.*, 2, e609.

[75] Capozzi, F., Casadei, F., & Luchinat, C. (2006). EF-hand protein dynamics and evolution of calcium signal transduction: an NMR view. *J. Biol. Inorg. Chem.*, 11, 949–962.

[76] Michiels, J., Xi, C., Verhaert, J., & Vanderleyden, J. (2002). The Functions of Ca^{2+} in bacteria: a role for EF-hand proteins? *Trends Microbiol.*, 10, 87–93.

[77] Moorthy, A. K., Singh, S. K., Gopal, B., Surolia, A., & Murthy, M. R. (2001). Variability of calcium binding to EF-hand motifs probed by electrospray ionization mass spectrometry. *J. Am. Soc. Mass Spectrom.*, 12, 1296–1301.

[78] Amos, W. B., Routledge, L. M., & Yew, F. F. (1975). Calcium-binding proteins in a vorticellid contractile organelle. *J. Cell. Sci.,* 19, 203–213.

[79] Yamada, K. & Asai, H. (1982). Extraction and some properties of the protein, spastin B, from the spasmoneme of *Carchesium polypinum. J. Biochem.* (Tokyo), 91, 1187–1195.

[80] Ochiai, T., Kato, M., Ogawa, T., & Asai, H. (1988). Spasmin-like proteins in various ciliates revealed by antibody to purified spasmins of *Carchesium polypinum. Experientia.*, 44, 768–771.

[81] Scheibel, L. W. (1992). Role of calcium/calmodulin-mediated processes in Protozoa. *Int. Rev. Cytol.*, 134, 165–242.

[82] Kuźnicki, J., Kuźnicki, L., & Drabikowski, W. (1979). Ca^{2+}-binding modulator protein in protozoa and myxomycete. *Cell Biol. Int. Rep.*, 3, 17–23.

[83] Daiker, V., Lebert, M., Richter, P., & Häder, D. P. (2010). Molecular characterization of a calmodulin involved in the signal transduction chain of gravitaxis in *Euglena gracilis. Planta*, 231, 1229–1236.

[84] Emmer, B. T., Daniels, M. D., Taylor, J. M., Epting, C. L., & Engman, D. M. (2010). Calflagin inhibition prolongs host survival and suppresses parasitemia in *Trypanosoma brucei* infection. *Eukaryot. Cell*, 9, 934–942.

[85] Engman, D. M., Krause, K. H., Blumin, J. H., Kim, K. S., Kirchhoff, L. V., & Donelson, J. E. (1989). A novel flagellar Ca^{2+}-binding protein in *Trypanosomes. J. Biol. Chem.*, 264, 18627–18631.

[86] Haghighat, N. G. & Ruben, L. (1992). Purification of novel calcium-binding proteins from *Trypanosoma brucei*: properties of 22-, 24- and 38- kilodalton proteins. *Mol. Biochem. Parasitol.*, 51, 99–110.

[87] Patel-King, R. S., Benashski, S. E., & King, S. M. (2002). A Bipartite Ca^{2+}-regulated nucleoside diphosphate kinase system within the *Chlamydomonas* flagellum. The regulatory subunit p72. *J. Biol. Chem.*, 277, 34271–34279.

[88] Nickel, R., Jacobs, T., Urban, B., Scholze, H., Bruhn, H., & Leippe, M. (2000). Two novel calcium-binding proteins from cytoplasmic granules of the protozoan parasite *Entamoeba histolytica*. *FEBS Lett.*, 486, 112–116.

[89] Sahu, S. C., Bhattacharya, A., Chary, K. V., & Govil, G. (1999). Secondary structure of a calcium-binding protein (CaBP) from *Entamoeba histolytica*. *FEBS Lett.*, 459, 51–56.

[90] Mukherjee, S., Kuchroo, K., & Chary, K. V. (2005). Structural characterization of the apo form of a calcium binding protein from *Entamoeba histolytica* by hydrogen exchange and its folding to the holo state. *Biochemistry*, 44, 11636–11645.

[91] Vigues, B., Blanchard, M. P., & Bouchard, P. (1999). Centrin-like filaments in the cytopharyngeal apparatus of the ciliates *Nassula* and *Furgasonia*: evidence for a relationship with microtubular structures. *Cell. Motil. Cytoskeleton*, 43, 72–81.

[92] Brugerolle, G., Bricheux, G., & Coffe, G. (2000). Centrin protein and genes in *Trichomonas vaginalis* and close relatives. *J. Eukaryot. Microbiol.*, 47, 129–138.

[93] Daunderer, C., Schliwa, M., & Graf, R. (2001). *Dictyostelium* centrin-related protein (DdCrp), the most divergent member of the centrin family, possesses only two EF-hands and dissociates from the centrosome during mitosis. *Eur. J. Cell. Biol.*, 80, 621–630.

[94] Itabashi, T., Terasaki, T., & Asai, H. (2004). Novel nuclear and cytoplasmic proteins detected by anti-*Zoothamnium arbuscula* (Protozoa) spasmin 1 antibody in mammalian cells are dependent on the cell cycle. *J. Biochem.*, 136, 651–657.

[95] Zhao, Y., Song, L., Liang, A., & Yang, B. (2009). Characterization of self-assembly of *Euplotes octocarinatus* centrin. *J. Photochem. Photobiol. B.*, 95, 26–32.

[96] Levy, Y. Y., & Lai, E. Y., Remillard, S. P., Heintzelman, M. B., & Fulton, C. (1996). Centrin is a conserved protein that forms diverse associations with centrioles and MTOCs in *Naegleria* and other organisms. *Cell Motil. Cytoskeleton*, 33, 298–323.

[97] Selvapandiyan, A., Duncan, R., Debrabant, A., Bertholet, S., Sreenivas, G., Negi, N. S., Salotra, P., & Nakhasi, H. L. (2001). Expression of a mutant form of *Leishmania donovani* centrin reduces the growth of the parasite. *J. Biol. Chem.*, 276, 43253–43261.

[98] Mana-Capelli, S., Gräf, R., & Larochelle, D. A. (2010). *Dictyostelium* centrin B localization during cell cycle progression. *Commun. Integr. Biol.*, 3, 39–41.

[99] Klumpp, S., Gierlich, D., & Schultz, J. E. (1984). Adenylate cyclase and guanylate cyclase in the excitable ciliary membrane from *Paramecium*: separation and regulation. *FEBS Lett.*, 171, 95–99.

[100] Nanthakumar, N. N., Dayton, J. S., & Means, A. R. (1996). Role of Ca^{2+}/Calmodulin-bnding proteins in *Aspergillus nidulans* cell cycle regulation. *Prog. Cell. Cycle Res.*, 2, 217–228.

[101] Hellstern, S., Dammann, H., Husain, Q., & Mutzel, R. (1997). Overexpression, purification and characterization of *Dictyostelium* calcineurin A. *Res. Microbiol.*, 148, 335–343.

[102] Elwess, N. L. & van Houten, J. L. (1997). Cloning and molecular analysis of the plasma membrane Ca^{2+}-ATPase gene in *Paramecium* cells. *J. Cell. Biol.*, 127, 935–945.

[103] Kuo, W. N., McNabb, M., Kanadia, R. N., Dopson, N., & Morgan, R. (1997). The presence/absence of Bcl2, Ca^{2+}/calmodulin-dependent protein kinase IV, calretinin and p53 in baker's yeast and wheat germ. *Cytobios.*, 91. 7–13.

[104] Kato, K., Sudo, A., Kobayashi, K., Tohya, Y., & Akashi, H. (2008). Characterization of *Plasmodium falciparum* protein kinase 2. *Mol. Biochem. Parasitol.*, 162, 87–95.

[105] Nakamura, T., Fujita, K., Eguchi, Y., & Yazawa, M. (1984). Properties of calcium-dependent regulatory proteins from fungi and yeast. *J. Biochem.* (Tokyo), 95, 1551–1557.

[106] Takeda, T. & Yamamoto, M. (1987). Analysis and in vivo disruption of the gene coding for calmodulin in *Schizosaccharomyces pombe*. *Proc. Natl. Acad. Sci. USA*, 84, 3580–3584.

[107] Rösel, D., Pûta, F., Blahůsková, A., Smýkal, P., & Folk, P. (2000). Molecular characterization of a calmodulin-like dictyostelium protein CalB. *FEBS Lett.*, 473, 323-327.

[108] Coukell, B., Moniakis, J., & Grinberg, A. (1995). Cloning and expression in *Escherichia coli* of a cDNA encoding a developmentally regulated Ca^{2+}-binding protein from *Dictyostelium discoideum*. *FEBS Lett.*, 362, 342–346.

[109] André, B., Noegel, A. A., & Schleicher, M. (1996). *Dictyostelium discoideum* contains a family of calmodulin-related EF-hand proteins that are developmentally regulated. *FEBS Lett.*, 382, 198–202.

[110] Nakagawa, T., Fujiu, K., Cole, E. S., & Numata, O. (2008). Involvement of a 25 kDa *Tetrahymena* Ca^{2+}-binding protein in pronuclear exchange. *Cell Struct. Funct.*, 33, 151–162.

[111] Gauthier, M. L. & O'Day, D. H. (2001). Detection of calmodulin-binding proteins and calmodulin-dependent phosphorylation linked to calmodulin-dependent chemotaxis to folic and cAMP in *Dictyostelium*. *Cell. Signal.*, 13, 575–584.

[112] Tan, J. L. & Spudich, J. A. (1991). Characterization and bacterial expression of the *Dictyostelium* myosin light chain kinase cDNA. Identification of an autoinhibitory domain. *J. Biol. Chem.*, 266, 16044–16049.

[113] Shimizu, K., Murata, T., Tagawa, T., Takahashi, K., Ishikawa, R., Abe, Y., Hosaka, K., & Kubohara, Y. (2004). Calmodulin-dependent cyclic nucleotide phosphodiesterase (PDE1) is a pharmacological target of differentiation-inducing factor-1, an antitumor agent isolated from *Dictyostelium*. *Cancer Res.*, 64, 2568–2571.

[114] Catalano, A., & O'Day, D. H. (2008). Calmodulin-binding proteins in the model organism *Dictyostelium*: a complete & critical review. *Cell Signal*, 20, 277–291.

[115] Felleisen, R. S., Hemphill, A., & Gottstein, B. (2001). A novel EF-hand calcium-binding protein in the flagellum of the protozoan *Tritrichomonas suis*. Parasitology, 122, 125–132.

[116] Kim, K., Messinger, L. A., & Nelson, D. L. (1998). Ca^{2+}-dependent protein kinases of Paramecium-cloning provides evidence of a multigene family. *Eur. J. Biochem.*, 251, 605–612.

[117] Zhang, X. S. & Choi, J. H. (2001). Molecular evolution of calmodulin-like domain protein kinases (CDPKs) in plants and protists. *J. Mol. Evol.*, 53, 214–224.

[118] Zhao, Y., Pokutta, S., Maurer, P., Lindt, M., Franklin, R. M., & Kappes, B. (1994). Calcium-binding properties of a calcium-dependent protein kinase from *Plasmodium falciparum* and the significance of individual calcium-binding sites for kinase activation. *Biochemistry*, 33, 3714-3721.

[119] Kieschnick, H., Wakefield, T., Narducci, C. A., & Beckers, C. (2001). *Toxoplasma gondii* attachment to host cells is regulated by a calmodulin-like domain protein kinase. *J. Biol. Chem.*, 276, 12369–12377.

[120] Ishino, T., Orito, Y., Chinzei, Y., & Yuda, M. (2006). A calcium-dependent protein kinase regulates *Plasmodium* ookinete access to the midgut epithelial cell. *Mol. Microbiol.*, 59, 1175–1184.

[121] Sugi, T., Kato, K., Kobayashi, K., Pandey, K., Takemae, H., Kurokawa, H., Tohya, Y., & Akashi, H. (2009). Molecular analyses of *Toxoplasma gondii* calmodulin-like domain protein kinase isoform 3. *Parasitol. Int.*, 58, 416-423.

[122] Raynal, P. & Pollard, H. B. (1994). Annexins: the problem of assessing the biological role for a gene family of multifunctional calcium and phospholipid-binding proteins. *Biochim. Biophys. Acta*, 1197, 63–93.

[123] Knochel, M., Kissmehl, R., Wissmann, J. D., Momayezi, M., Hentschel, J., Plattner, H., & Burgoyne, R. D. (1996). Annexins in *Paramecium* Cells. Involvement in Site-specific positioning of secretory organelles. *Histochem. Cell. Biol.*, 105, 269–281.

[124] Pathuri, P., Nguyen, E. T., Ozorowski, G., Svärd, S. G., & Luecke, H. (2009). Apo and calcium-bound crystal structures of cytoskeletal protein alpha-14 giardin (annexin E1) from the intestinal protozoan parasite *Giardia lamblia*. *J. Mol. Biol.,* 385, 1098–1112.

[125] Wei, C. J., Tian, X. F., Adam, R. D., & Lu. S. Q. (2010). *Giardia lamblia*: intracellular localization of alpha8-giardin. *Exp. Parasitol.*, 126, 489–496.

[126] Felleisen, R. S., Hemphill, A., Ingold, K., & Gottstein, B. (2000). Conservation of calnexin in the early branching protozoan *Tritrichomonas suis*. *Mol. Biochem. Parasitol.*, 108, 109–117.

[127] Croan, D. J. & Ellis, J. (2000). The *Leishmania major* RNA polymerase II largest subunit lacks a carboxy-terminus heptad repeat structure and its encoding gene is linked with calreticulin gene. *Protist*, 15, 57–68.

[128] Kima, P. E., & Dunn, W. (2005). Exploiting calnexin expression on phagosomes to isolate *Leishmania* parasitophorous vacuoles. *Microb. Pathog.*, 38, 139–145.

[129] Navazio, L., Nardi, M. C., Pancaldi, S., Dainese, P., Baldan, B., Fitchette-Laine, A. C., Faye, L., Meggio, F., Martin, W., & Mariani, P. (1998). Functional conservation of calreticulin in *Euglena gracilis*. *J. Eukaryot. Microbiol.*, 45, 307–313.

[130] Muller-Taubenberger, A., Lupas, A. N., Li, H., Ecke, M., Simmeth, E., & Gerisch, G. (2001). Calreticulin and calnexin in the endoplasmic reticulum are important for phagocytosis. *EMBO J.*, 20, 6772–6782.

[131] Berridge, M. J., Bootman, M. D., & Lipp, P. (1998). Calcium - a life and death signal. *Nature*, 395, 645–648.

[132] Wright, M. V. & van Houten, J. L. (1990). Characterization of a putative Ca^{2+}-transporting Ca^{2+}-ATPase in the pellicles of *Paramecium tetraurelia*. *Biochim. Biophys. Acta,* 1029, 241–251.

[133] Wright, M. V., Elwess, N., & van Houten, J. (1993). Ca^{2+} transport and chemoreception in *Paramecium*. *J. Comp. Physiol.* [B], 163, 288–296.

[134] Moreno, S. N., & Docampo, R. (2003). Calcium regulation in protozoan parasites. *Curr. Opin. Microbiol.*, 6, 359–364.

[135] Benaim, G., Moreno, S. N., Hutchinson, G., Cervino, V., Hermoso, T., Romero, P. J., Ruiz, F., de Souza, W. & Docampo R. (1995). Characterization of the plasma-membrane calcium pump from *Trypanosoma cruzi*. *Biochem. J.*, 306, 299–303.

[136] Benaim, G., Bermudez, R., & Urbina, J. A. (1990). Ca^{2+} transport in isolated mitochondrial vesicles from *Leishmania braziliensis* promastigotes. Mol. Biochem. Parasitol., 39, 61–68.

[137] Mazumder, S., Mukherjee, T., Ghosh, J., Ray, M., & Bhaduri, A. (1992). Allosteric modulation of *Leishmania donovani* plasma membrane Ca^{2+}-ATPase by endogenous calmodulin. *J. Biol. Chem.*, 267, 18440–18446.

[138] Moniakis, J., Coukell, M. B., & Forer, A. (1995). Molecular cloning of an intracellular P-type ATPase from *Dictyostelium* that is up-regulated in calcium-adapted cells. *J. Biol. Chem.*, 270, 28276–28281.

[139] Moniakis, J., Coukell, M. B., & Janiec, A. (1999). Involvement of the Ca^{2+} ATPase PAT1 and the contractile vacuole in calcium regulation in *Dictyostelium discoideum*. *J. Cell. Sci.*, 112, 405–414.

[140] Nagamune, K., Moreno, S. N., Chini, E. N., & Sibley, L. D. (2008). Calcium regulation and signaling in apicomplexan parasites. *Subcell. Biochem.*, 47, 70–81.

[141] Cunningham, K. W., & Fink, G. R. (1994). Ca^{2+} transport in *Saccharomyces cerevisiae*. *J. Exp. Biol.*, 196, 157–166.

[142] Ghislain, M. Goffeau, A., Halachmi, D., & Eilan, Y. (1990). Calcium homeostasis and transport are affected by disruption of cta3, a novel gene encoding Ca^{2+}-ATPase in *Schizosaccharomyces pombe*. *J. Biol. Chem.*, 265, 18400–18407.

[143] Halachmi, D., Ghislain, M., & Eilam, Y. (1992). An intracellular ATP-dependent calcium pump within the yeast *Schizosaccharomyces pombe*, encoded by the gene cta3. *Eur. J. Biochem.*, 207, 1003–1008.

[144] Benito, B., Garciadeblás, B., Rodríguez-Navarro, A. (2002). Potassium- or sodium-efflux ATPase, a key enzyme in the evolution of fungi. *Microbiology*, 148, 933–941.

[145] Rodríguez-Navarro, A., & Benito, B. (2010). Sodium or potassium efflux ATPase a fungal, bryophyte, and protozoan ATPase. *Biochim. Biophys. Acta*, 1798, 1841–1853.

[146] Iizumi, K., Mikami, Y., Hashimoto, M., Nara, T., Hara, Y., & Aoki, T. (2006). Molecular cloning and characterization of ouabain-insensitive Na^{+}-ATPase in the parasitic protist, *Trypanosoma cruzi*. *Biochim. Biophys. Acta*, 1758, 738–746.

[147] Burlando, B., Marchi, B., Krüppel, T., Orunesu, M., & Viarengo, A. (1999). Occurrence of Na^{+}-Ca^{2+} exchange in the ciliate *Euplotes crassus* and its role in Ca^{2+} homeostasis. *Cell Calcium*, 25, 153–160.

[148] Docampo, R. & Moreno, S. N. (1999). Acidocalcisome: A novel Ca^{2+} storage compartment in trypanosomatids and apicomplexan parasites. *Parasitol.* Today, 15, 443–448.

[149] Rooney, E. K. & Gross, J. D. (1992). ATP-driven Ca^{2+}/H^{+} antiport in acid vesicles from *Dictyostelium*. *Proc. Natl. Acad. Sci. USA.* 89, 8025–8029.

[150] Luo, S., Vieira, M., Graves, J., Zhong, L., & Moreno, S. N. (2001). A plasma membrane type Ca^{2+}-ATPase colocalizes with a vacuolar H^{+}-pyrophosphatase to acidocalcisomes of *Toxoplasma gondii*. *EMBO J.*, 20, 55–64.

[151] Dunn, T., Gable, K., & Beeler, T. (1994). Regulation of cellular Ca^{2+} by yeast vacuoles. *J. Biol. Chem.*, 269, 7273–7278.

[152] Miseta, A., Kellermayer, R., Aiello, D. P., Fu, L.,& Bedwell, D. M. (1999). The vacuolar Ca^{2+}/H^{+} exchanger Vcx1p/Hum1p tightly controls cytosolic Ca^{2+} levels in *S. cerevisiae*. *FEBS Lett.*, 451, 132–136.

[153] Pittman, J. K., Edmond, C., Sunderland, P. A., & Bray, C. M. (2009). A cation-regulated and proton gradient-dependent cation transporter from *Chlamydomonas reinhardtii* has a role in calcium and sodium homeostasis. *J. Biol. Chem.*, 284, 525–533.

[154] Rotmann, A., Sanchez, C., Guiguemde, A., Rohrbach, P., Dave, A., Bakouh, N., Planelles, G., & Lanzer, M. (2010). PfCHA is a mitochondrial divalent cation/H^+ antiporter in *Plasmodium falciparum*. *Mol. Microbiol.*, 76, 1591–1606.

[155] Iwasaki, W., Sasaki, H., Nakamura, A., Kohama, K., & Tanokura, M. (2003). Metal-free and Ca^{2+}-bound structures of a multidomain EF-hand protein, CBP40, from the lower eukaryote *Physarum polycephalum*. Structure, 11, 75–85.

[156] Alafag, J. I., Moon, E. K., Hong, Y. C., Chung, D. I., & Kong, H. H. (2006). Molecular and biochemical characterization of a novel actin bundling protein in *Acanthamoeba*. Korean *J. Parasitol.*, 44, 331–341.

[157] Jain, R., Kumar, S., Gourinath, S., Bhattacharya, S., & Bhattacharya, A. (2009). N- and C-terminal domains of the calcium binding protein EhCaBP1 of the parasite *Entamoeba histolytica* display distinct functions. *PLoS One*, 4, e5269.

[158] Pinto, A. P., Campana, P. T., Beltramini, L. M., Silber, A. M., & Araújo, A. P. (2003). Structural characterization of a recombinant flagellar calcium-binding protein from *Trypanosoma cruzi*. *Biochim. Biophys. Acta*, 1652, 07–14.

[159] Wu, Y., Deford, J., Benjamin, R., Lee, M. G., & Ruben, L. (1994). The gene family of EF-hand calcium-binding proteins from the flagellum of *Trypanosoma brucei*. *Biochem. J.*, 304, 833–841.

[160] Chakrabarty, P., Sethi, D. K., Padhan, N., Kaur, K. J., Salunke, D. M., Bhattacharya, S., & Bhattacharya, A. (2004). Identification and characterization of EhCaBP2. A second member of the calcium-binding protein family of the protozoan parasite *Entamoeba histolytica*. *J. Biol. Chem.*, 279, 12898–12908.

[161] Allen, R. D., Aihara, M. S., & Fok, A. K. (1998). The striated bands of *Paramecium* are immunologically distinct from the centrin-specific infraciliary lattice and cytostomal cord. *J. Eukaryot. Microbiol.*, 45, 202–209.

[162] Friedberg, F. (2006). Centrin isoforms in mammals. Relation to calmodulin. *Mol. Biol. Rep.*, 33, 243–252.

[163] Gonda, K., Komatsu, M., & Numata, O. (2000). Calmodulin and Ca^{2+}/calmodulin-binding proteins are involved in *Tetrahymena thermophila* phagocytosis. Cell Struct. Funct., 25, 243–251.

[164] Creutz, C. E., Tomsig, J. L., Snyder, S. L., Gautier, M. C., Skouri, F., Beisson, J., & Cohen, J. (1998). The copines, a novel class of C2 domain-containing, calcium-dependent, phospholipid-binding proteins conserved from *Paramecium* to humans. *J. Biol. Chem.*, 273, 1393–1402.

[165] Fok, A. K., Aihara, M. S., Ishida, M., & Allen, R. D. (2008). Calmodulin localization and its effects on endocytic and phagocytic membrane trafficking in *Paramecium multimicronucleatum*. *J. Eukaryot. Microbiol.*, 55, 481–491.

[166] Wilson, M. A., & Brunger, A. T. (2000). The 1.0 A crystal structure of Ca^{2+}-bound calmodulin: an analysis of disorder and implications for functionally relevant plasticity. *J. Mol. Biol.*, 301, 1237–1256.

[167] López, N. C., Valck, C., Ramírez, G., Rodríguez, M., Ribeiro, C., Orellana, J., Maldonado, I., Albini, A., Anacona, D., Lemus, D., Aguilar, L., Schwaeble, W., & Ferreira, A. (2010). Antiangiogenic and antitumor effects of *Trypanosoma cruzi* calreticulin. *PLoS Negl. Trop. Dis.*, 4, e730.

[168] Oladiran, A., & Belosevic, M. (2010). *Trypanosoma carassii* calreticulin binds host complement component C1q and inhibits classical complement pathway-mediated lysis. *Dev. Comp. Immunol.*, 34, 396–405.

[169] Garcia-Marchan, Y., Sojo, F., Rodriguez, E., Zerpa, N., Malave, C., Galindo-Castro, I., Salerno, M., & Benaim, G. (2009). *Trypanosoma cruzi* calmodulin: cloning, expression and characterization. *Exp. Parasitol.*, 123, 326–333.

[170] Wingard, J. N., Ladner, J., Vanarotti, M., Fisher, A. J., Robinson, H., Buchanan, K. T., Engman, D. M., & Ames, J. B. (2008). Structural insights into membrane targeting by the flagellar calcium-binding protein (FCaBP), a myristoylated and palmitoylated calcium sensor in *Trypanosoma cruzi*. *J. Biol. Chem.*, 283, 23388–23396.

[171] el-Sayed, N. M., Harkins, P. C., Fox, R. O., Anderson, K., & Patton, C. L. (1995). Crystallization and preliminary X-ray investigation of the recombinant *Trypanosoma brucei rhodesiense* calmodulin. *Proteins*, 21, 354–357.

[172] Benaim, G., Szabo, V., & Cornivelli, L. (1987). Isolation and characterization of calmodulin from *Leishmania braziliensis* and *Leishmania mexicana*. *Acta Cient. Venez.*, 38, 289–291.

[173] Fuertes, M. A., Pérez, J. M., Soto, M., López, M. C., & Alonso, C. (2001). Calcium-induced conformational changes in *Leishmania infantum* kinetoplastid membrane protein-11. *J. Biol. Inorg. Chem.*, 6, 107–117.

[174] Munoz, M. L., Weinbach, E. C., Wieder, S. C., Claggett, C. E., & Levenbook, L. (1987). *Giardia lamblia*: detection and characterization of calmodulin. *Exp. Parasitol.*, 63, 42–48.

[175] Brugerolle, G., Bricheux, G., & Coffe, G. (2000). Centrin protein and genes in *Trichomonas vaginalis* and close relatives. *J. Eukaryot. Microbiol.*, 47, 129–138.

[176] Keeling, P. J., Doherty-Kirby, A. L., Teh, E. M., & Doolittle, W. F. (1996). Linked genes for calmodulin and E2 ubiquitin-conjugating enzyme in *Trichomonas vaginalis*. *J. Eukaryot. Microbiol.*, 43, 468–474.

[177] Polson, H. E., Blackman, M. J. (2005). A role for poly(dA)poly(dT) tracts in directing activity of the *Plasmodium falciparum* calmodulin gene promoter. *Mol. Biochem. Parasitol.*, 141, 179–189.

[178] Pezzella-D'Alessandro, N., Le Moal, H., Bonhomme, A., Valere, A., Klein, C., Gomez-Marin, J., & Pinon, J. M. (2001). Calmodulin distribution and the actomyosin cytoskeleton in *Toxoplasma gondii*. *J. Histochem. Cytochem.*, 49, 445–454.

[179] . Dunn, P. P., Bumstead, J. M., & Tomley, F. M. (1996). Sequence, expression and localization of calmodulin-domain protein kinases in *Eimeria tenella* and *Eimeria maxima*. *Parasitology*, 113, 439–448.

[180] Kink, J. A., Maley, M. E., Preston, R. R., Ling, K. Y., Wallen-Friedman, M. A., Saimi, Y., & Kung C. (1990). Mutations in paramecium calmodulin indicate functional differences between the C-terminal and N-terminal lobes *in vivo*. *Cell*, 62, 165–174.

[181] Luo, S., Rohloff, P., Cox, J., Uyemura, S. A., & Docampo, R. (2004). *Trypanosoma brucei* plasma membrane-type Ca^{2+}-ATPase 1 (TbPMC1) and 2 (TbPMC2) genes encode functional Ca^{2+}-ATPases localized to the acidocalcisomes and plasma membrane, and essential for Ca^{2+} homeostasis and growth. *J. Biol. Chem.*, 279, 14427–14439.

[182] Luo, S., Vieira, M., Graves, J., Zhong, L., & Moreno, S. N. (2001). A plasma membrane-type Ca^{2+} ATPase co-localizes with a vacuolar H^+-pyrophosphatase to acidocalcisomes of *Toxoplasma gondii. EMBO J.*, 20, 55–64.

[183] Luo, S., Ruiz, F. A., Moreno, S. N. (2005). The acidocalcisome Ca^{2+} ATPase (TgA1) of *Toxoplasma gondii* is required for polyphosphate storage, intracellular calcium homeostasis and virulence. *Mol. Microbiol.*, 55, 1034–1045.

[184] Ghosh, S. K., Rosenthal, B., Rogers, R., & Samuelson, J. (2000). Vacuolar localization of an *Entamoeba histolytica* homologue of the plasma membrane ATPase (PMCA). *Mol. Biochem.Parasitol.*, 108, 125–130.

[185] Hamasaki-Katagiri, N., Molchanova, T., Takeda, K., & Ames, J. B. (2004). Fission yeast homolog of neuronal calcium sensor-1 (Ncs1p) regulates sporulation and confers calcium tolerance. *J. Biol. Chem.*, 279, 12744–12754.

In: Calcium Signaling
Editor: Masayoshi Yamaguchi

ISBN: 978-1-61324-313-8
©2012 Nova Science Publishers, Inc.

Chapter 2

THE STORE-OPERATED CALCIUM ENTRY PATHWAY

Juan A. Rosado[*]

Department of Physiology, Cell Physiology Group,
University of Extremadura, Cáceres, Spain.

ABSTRACT

Store-operated calcium entry is a major mechanism for calcium influx in non-excitable, but also in excitable, cells. This event is regulated by the filling state of the intracellular calcium stores, which, upon depletion, evoke the opening of calcium permeable channels in the plasma membrane. Recent studies have revealed that the protein STIM1 (Stromal Interaction Molecule-1) acts as a calcium sensor that communicates information about the amount of calcium stored to the plasma membrane.

STIM1 has also been located in the plasma membrane where it can regulate calcium channel gating. A number of studies have revealed that the calcium channels in the plasma membrane involve the protein Orai1, which shows a high selectivity for calcium. Since certain store-operated currents are not calcium selective, other channels, such as those of the TRP family, have been presented as candidates to conduct calcium entry. Store-operated calcium entry plays an important role in the regulation of a number of cellular functions and dysregulation of this mechanism leads to a number of dysfunctions.

Keywords: Transient receptor potential (TRP) channels, Orai, STIM, CRAC, store-operated Ca^{2+} entry.

[*] Correspondence to: Dr. J. A. Rosado, Department of Physiology, Faculty of Veterinary Sciences, University of Extremadura, Cáceres 10071, SPAIN; Tel: +34 927 257139; Fax: +34 927257110; E-mail: jarosado@unex.es

ABBREVIATIONS

2-APB	2-aminoethoxydiphenyl borate;
$[Ca^{2+}]_c$	cytosolic free Ca^{2+} concentration;
CAD	CRAC-activating domain;
CMD	CRAC modulatory domain;
CRAC	Ca^{2+} release-activated Ca^{2+} channel;
ER	endoplasmic reticulum;
OASF	Orai-activating small fragment;
ROCE	receptor-operated Ca^{2+} entry;
SERCA	sarcoplasmic/endoplasmic-reticulum Ca^{2+}-ATPase;
SMOC	second messenger-operated Ca^{2+} entry;
SOC	store-operated channel;
SOAR	STIM1 Orai-activating region;
SOCE	store-operated calcium entry;
STIM1	stromal interaction molecule 1;
TBHQ	2,5-di-(t-butyl)-1,4-hydroquinone;
TG	thapsigargin;
TRP	transient receptor potential.

1. INTRODUCTION

In a number of cell types regulation of cytosolic Ca^{2+} concentration modulates a variety of cellular functions ranging from short-term responses, such as muscle contraction and secretion to long-term processes like cell growth [1]. Changes in cytosolic free Ca^{2+} concentration ($[Ca^{2+}]_c$) are characterized by transient increases in the concentration of free calcium ions [2]. Agonists increase $[Ca^{2+}]_c$ either by releasing compartmentalised Ca^{2+} from intracellular stores or by facilitating the entry of extracellular Ca^{2+} across the plasma membrane. Due to the finite amount of Ca^{2+} accumulated in the intracellular stores Ca^{2+} release from the internal stores is usually transient; thus, many cellular events, as well as the refilling of the intracellular stores, require a sustained increase in $[Ca^{2+}]_c$, and, therefore, Ca^{2+} entry from the extracellular medium plays an important role.

Ca^{2+} entry might be activated by different mechanisms, ranging from membrane depolarization to receptor occupation or discharge of the intracellular stores. Voltage-operated Ca^{2+} entry requires the presence of voltage-sensitive Ca^{2+} channels and is characteristic of electrically excitable cells. Receptor occupation leads to the activation of Ca^{2+} entry, which might take different forms. Receptor-operated Ca^{2+} influx occurs through Ca^{2+} channels that form part of the receptor (true receptor-operated Ca^{2+} entry (ROCE)) or through channels activated by second messengers, such as diacylglycerol, a mechanism referred to as second messenger-operated Ca^{2+} entry (SMOC) [3]. In addition, occupation of G-protein coupled receptors by agonists leads to phospholipase C-dependent Ca^{2+} release from the intracellular Ca^{2+} stores, mediated by the inositol 1,4,5-trisphosphate (IP_3), which results in subsequent activation of Ca^{2+} entry across the plasma membrane. This Ca^{2+} influx pathway involves signalling from depleted intracellular Ca^{2+} stores to plasma membrane Ca^{2+}

permeable channels, independently or rises in $[Ca^{2+}]_c$, a process referred to as capacitative Ca^{2+} entry or store-operated Ca^{2+} entry (SOCE) [4-5].

SOCE is a major mechanism for Ca^{2+} entry in non-electrically excitable cells [4, 6-9], where, in addition to allow endoplasmic reticulum (ER) Ca^{2+} refilling, essential for protein synthesis and folding [10-11], it plays an important physiological role in Ca^{2+} signalling, including the support of Ca^{2+} oscillations [9] and a number of cellular processes, including cell proliferation and platelet aggregation and secretion [12-13].

SOCE is a mechanism regulated by the filling state of the intracellular Ca^{2+} stores. Since the identification of SOCE two decades ago the mechanisms underlying the communication between the Ca^{2+} stores and the plasma membrane channels, as well as the nature of the latter, have been a matter of intense investigation and debate. SOCE has been widely investigated by means of inhibitors of the sarco/endoplasmic reticulum Ca^{2+} ATPase (SERCA), including thapsigargin, 2,5-di-(t-butyl)-1,4-hydroquinone (TBHQ) or cyclopiazonic acid. Cell treatment with SERCA inhibitors results in passive discharge of intracellular Ca^{2+} stores, due to Ca^{2+} leak across the membrane of the stores, thereby gating the store-operated channels. The use of SERCA inhibitors provides a number of advantages. First, depletion of intracellular Ca^{2+} stores can be achieved independently of the activation of G protein-coupled receptors and phospholipase C. Second, since SERCA inhibitors activate SOCE to a similar extent as phospholipase C generated IP_3, the rises in the cellular concentrations of inositol phosphates are not essential for the activation of SOCE and movement of Ca^{2+} into the ER is not required for Ca^{2+} entry [14-15]. However, a number of concerns should be taken into account when interpreting the data obtained by the use of SERCA inhibitors. First, a number of experiments have reported that, at least in human platelets, $[Ca^{2+}]_c$ elevation induced by thapsigargin results in secretion of autocrine factors, such as ADP and 5-hydroxytriptamine, that activate phospholipase C-linked receptors as well as ROCE [16]. Second, SERCA inhibitors might result in the activation of monovalent cation flux across non selective capacitative cation channels, which, in the case of human platelets, results in the activation of the Na^+/Ca^{2+} exchanger working in reverse mode, thus introducing Ca^{2+} into the cell by a non-capacitative mechanism dependent on SOCE [17]. Finally, in excitable cells, membrane depolarization induced by SOCE, after treatment with SERCA inhibitors, might result in the activation of secondary voltage-activated Ca^{2+} influx.

In this chapter I review the current knowledge concerning the components of SOCE and the functional interactions between elements in the plasma membrane and in the membrane of the intracellular stores to allow the activation of Ca^{2+} entry upon Ca^{2+} store depletion.

2. CALCIUM CURRENTS

Electrophysiological studies of cells upon discharge of the intracellular Ca^{2+} stores have revealed ionic currents at the plasma membrane that show diverse biophysical properties, which reflects that different cells types might express distinct store-operated channels. The first identified and best characterized capacitative current is the Ca^{2+} release-activated Ca^{2+} current, I_{CRAC}, a non-voltage-gated current that was initially identified in electrophysiological recordings from rat peritoneal mast cells and human Jurkat T cells [18-19], and has since been extensively described in other cells [20-23].

Classically, I_{CRAC} was identified as a Ca^{2+} current activated in T cells and mast cells by physiological agonists, such as the occupation of the T-cell receptor or immunoglobulin E receptor, as well as by Ca^{2+} store depletion [24-26]. A number of studies have demonstrated that I_{CRAC} is a capacitative current. I_{CRAC} can be activated by mimicking physiological Ca^{2+} release through intracellular application of IP_3, as well as by inducing Ca^{2+} store depletion either by treatment with the Ca^{2+} ionophore ionomycin, or preventing reuptake of Ca^{2+} leaking from the stores either by treatment with the SERCA inhibitor thapsigargin or by loading the cells with cytoplasmic Ca^{2+} chelators, such as BAPTA [24-25, 27-28]. Finally, I_{CRAC} can be triggered by rapid chelation of free Ca^{2+} within the Ca^{2+} stores by treatment with the membrane-permeant chelator TPEN [N,N,N′,N′-tetrakis(2-pyridylmethyl) ethylenediamine] [21]. TPEN induces in situ chelation of free Ca^{2+} into the Ca^{2+} stores and provides unequivocal evidence that I_{CRAC} is activated by a decrease in the Ca^{2+} concentration in the lumen of the Ca^{2+} stores, independently of rises in $[Ca^{2+}]_c$ or the activation of phospholipase C.

I_{CRAC} channels show a low conductance calculated in 0.02 pS [29]. The estimated current carried by an individual CRAC channel at both physiological membrane potentials and extracellular Ca^{2+} concentration is about 10^4 Ca^{2+} ions per second [24-25, 27, 30]. In addition, I_{CRAC} is highly selective for Ca^{2+} over monovalent cations, including Na^+, at physiological conditions (P_{Ca}:P_{Na}=1000); as a result, there is no appreciable fraction of the inward current through CRAC channels carried out by Na^+ [30-31]. In support of the high Ca^{2+} selectivity of the CRAC channels, I_{CRAC} approaches zero at very positive voltages (approximately +60 mV), [32]. In addition, at physiological Ca^{2+} levels, extracellular Na^+ replacement by the large organic cation N-methyl-D-glucamine ($NMDG^+$) does not modify either the extent or the apparent reversal potential of I_{CRAC} [30], and removal of extracellular Ca^{2+} in the presence of external Na^+ and Mg^{2+} abolishes I_{CRAC} [24-25, 33]. However, CRAC channels lose their selectivity in divalent free external solution. In divalent-free bath solutions, the Na^+ current develops with a similar time course to that of Ca^{2+} following depletion of the intracellular Ca^{2+} stores [27, 30, 34-35].

While I_{CRAC} was the first capacitative Ca^{2+} current identified, it is not the only one, and other store-operated cation currents, referred to as I_{SOC}, have been identified in different cell types. The channels conducting I_{SOC} currents will be named SOC channels to differentiate them from the CRAC channels that mediate I_{CRAC}. I_{SOC} are non-voltage activated, non-selective currents, of small, but resolvable 0.7-11 pS conductance [29]. SOC channels have been identified at single channel and whole-cell current levels in different cell types, including endothelial cells, vascular smooth muscle cells, skeletal muscle cells, neurons and neuroendocine cells [36-39]. In mouse and rabbit aorta, a store-operated 3-pS channel has been described. This SOC channel is activated by thapsigargin in cells loaded with the intracellular Ca^{2+} chelator BAPTA; therefore, channel gating cannot be attributed to thapsigargin-induced rise in $[Ca^{2+}]_c$ but to store depletion [40]. In contrast to CRAC channels, the SOC channels in vascular smooth muscle cells were cation selective but did not discriminate between Na^+, K^+, Cs^+, Ca^{2+}, Ba^{2+}, or Sr^{2+} and the P_{Ca}:P_{Na} was estimated to be 1 [29, 37, 40]. I_{SOC} currents have also been reported in several different types of endothelial cells, neurons and chromaffin cells, showing different conductance and distinct P_{Ca}:P_{Na} ratios.

Under physiological conditions the I_{SOC} currents are carried by both Ca^{2+} and Na^+ and their role might be to provide a pathway for Ca^{2+} to enter the cell, to conduct Na^+ enter that might activate other secondary mechanisms such a Ca^{2+} entry through reverse Na^+/Ca^{2+}

exchange [17] or to provide the depolarization that is necessary for the more Ca^{2+}-selective voltage-gated Ca^{2+} channels to open in excitable cells [29].

Since I_{SOC} currents lack the Ca^{2+} selectivity characteristic of I_{CRAC} and monomeric Orai1 channels, the SOC channels might involve other non-selective channel subunits, such as the mammalian homologues of the *Drosophila* Transient Receptor Potential (TRP) channels, which were initially presented as candidates for the conduction of SOCE as described below.

3. STIM1

Almost two decades after the identification of SOCE by Putney [4] probably one of the most significant advances to understand the mechanism underlying the communication between the Ca^{2+} stores and the plasma membrane channels occurred: the identification of STIM1 as the intraluminal Ca^{2+} sensor. In 2005, Roos and co-workers, using an RNAi-based screen to identify genes that alter thapsigargin-evoked Ca^{2+} entry, and Zhang and co-workers, generating STIM1 mutants that lack the ability to bind Ca^{2+}, demonstrated that STIM1 is the ER sensor that report its Ca^{2+} filling state to the Ca^{2+} channels in the plasma membrane, essential for Ca^{2+} store depletion-triggered Ca^{2+} influx across de plasma membrane [41-42].

Reduction of the STIM1 expression by siRNA or shRNA or functional knockdown of STIM1 by electrotransjection of neutralizing antibodies reduces SOCE in different mammalian cell types, including HeLa cells, HEK293 cells and human platelets [41, 43-44] and I_{CRAC} in Jurkat T cells [45]. In addition, attenuation of dStim expression by RNAi reduced SOCE in *Drosophila* S2 cells [41, 43].

STIM1 proteins are mostly located in the ER, and a minor pool of STIM1 has also been located in the plasma membrane [42-44, 46-47]. STIM1 is a Ca^{2+}-binding protein with the N-terminal region within the ER lumen, or in the extracellular space, according to the protein location. STIM1 structure includes a N-terminal sterile-alpha motif (SAM) and an EF-hand, Ca^{2+}-binding, domain, a single transmembrane region, two coiled-coil regions, several ezrin–radixin–moesin-like domains, and proline rich domains in the C terminus, located in the cytosol in all cases [48-49]. The SAM domain has been reported to facilitate protein interactions with a number of signalling molecules [50]. The conserved EF-hand motif is composed of the 12-residue Ca^{2+}-binding loop and two surrounding α-helices [51] and shows a low Ca^{2+}-binding affinity (Kd $\approx$ 0.5 mM) [52].

In resting cells, STIM1 is diffusely distributed in the ER membrane [43], and redistributes upon store depletion in regions of close apposition between ER and plasma membrane [42-43, 53]. The relocalization of STIM1 in clusters close to the plasma membrane has been demonstrated both for overexpressed green fluorescent protein (GFP)–STIM1 by total internal reflection fluorescence (TIRF) microscopy, which places STIM1 within 200 nm of the plasma membrane, and for endogenous STIM1 and overexpressed horseradish peroxidase (HRP)–STIM1 by electron microscopy, which puts the closest approach within 10–25 nm of the plasma membrane [28, 54]. Translocation of STIM1 units and reorganization into clusters occurs by a lateral movement within the ER [43, 53, 55-56]. Therefore, it has been hypothesized that store depletion causes aggregation and translocation of STIM1 into puncta in close apposition to the plasma membrane, which results in the recruitment of Ca^{2+}

permeable channel protein Orai1 in the plasma membrane to the sites of STIM1 aggregates in order to assemble functional units of CRAC channels in a stoichiometric manner [56].

Although STIM1 has also been suggested to interact with the cation-permeable protein TRPC1 [57], an event that is probably associated to the activation of I_{SOC}, as discussed below, the association of STIM1 with Orai1 has been reported to mediate the highly selective Ca^{2+} current I_{CRAC} in a number of cell models [58]. Overexpression of both STIM1 and Orai1 proteins has been demonstrated to potentiate I_{CRAC}; however, individual overexpression of either protein fails to amplify this capacitative current, thus suggesting that STIM1 and Orai1 mutually limit store-operated currents [59]. In RBL cells and HEK293 cells, expression of Orai1 alone has been found to strongly reduce SOCE and I_{CRAC}; however, when Orai1 was co-expressed along with STIM1 SOCE was substantially enhanced up to 103-fold [46].

Four independent studies have identified a cytoplasmic region in STIM1, composed of an ezrin-radixin-moesin domain, which is essential for the activation of Orai1. This region has been named SOAR (STIM1 Orai-activating region) [60], OASF (Orai-activating small fragment) [61], CAD (CRAC-activating domain) [62] and CCb9 [63]. These studies provide similar sequences for the identified regions: SOAR (amino acids 344-442), OASF (amino acids 233-450/474), CAD (amino acids 342-448) and CCb9 (amino acids 339-444). To avoid confusion this region will be named SOAR throughout the manuscript. SOAR is located within STIM1 C-terminus and comprise two coiled-coil domains and an amino acid sequence that enhances interaction with Orai1 and enhances Ca^{2+} currents as discussed in the last section of the chapter.

In addition to the role of STIM1 as an ER Ca^{2+} sensor, STIM1 has been found in the plasma membrane in a number of cells. The EF-hand motif of plasma membrane-resident STIM1 faces the extracellular medium, which suggests that STIM1 might act also as a sensor of the extracellular free Ca^{2+} concentration. Recent studies have pointed out that STIM1 located at the plasma membrane might regulate SOCE through the interaction with plasma membrane Ca^{2+}-permeable channels. External application of an antibody addressed towards the STIM1 EF-hand motif has been reported to block both CRAC channels in hematopoietic cells and SOC channels in HEK293 cells [45]. Furthermore, external application of mouse anti-STIM1 (EF-hand) antibody abolishes the inhibition of SOCE induced by increasing extracellular Ca^{2+} concentrations in human platelets probably through its interaction with Orai1 but not with TRPC1 [47], an effect that was not achieved by cell treatment with the same amount of a non-specific mouse IgG (Lopez et al, personal communication). STIM1 at the plasma membrane has also been reported to participate in non-capacitative Ca^{2+} entry mechanisms, such as arachidonic acid-activated, ARC channels [64].

The analog of STIM1, STIM2, has a structure close to that of STIM1. Although the function of STIM2 in SOCE has been less extensively investigated, it has been reported to activate Ca^{2+} entry either in a store-operated or in a store-independent mode mediated by close interaction of STIM2 and Orai1 [65]. STIM2 has also been shown to participate in the modulation of Ca^{2+} mobilization as a feedback regulator that stabilizes resting cytosolic and ER Ca^{2+} concentrations, where STIM2 knockdown reduces resting and ER Ca^{2+} levels while STIM2 over-expression increases resting Ca^{2+} levels [52].

4. CA^{2+} CHANNELS

The nature of the channels that conduct SOCE has been a matter of investigation since the identification of this mechanism. One of the first hypotheses for a capacitative channel was postulated by Irvine and Moor in sea urchin eggs in the late 80's [66]. In these cells, the inositol 1,3,4,5-tetrakisphosphate (IP$_4$) located in the plasma membrane was proposed to couple with the IP$_3$ receptor in the ER to facilitate the entry of extracellular Ca^{2+} [67]. However, the role of the IP$_4$ receptor in SOCE was not free from controversy, with conflicting data and hypotheses arising from different tissues and experimental protocols in the early 90's.

Orai

Despite intense investigation, the nature of the channel that conducts I_{CRAC} remained elusive until 2006, with the identification of Orai1 (initially termed Ca^{2+} release-activated Ca^{2+} channel protein 1 (CRACM1)) as the CRAC channel subunit. The role of Orai1 in the conduction of I_{CRAC} was identified through the convergence of a whole-genome screening of *Drosophila* S2 cells and by gene mapping in patients with a hereditary severe combined immune deficiency (SCID) syndrome attributed to loss of I_{CRAC} [68-70]. In SCID patients, a point mutation in the DNA encoding Orai1 (R91W) renders the expressed protein non-functional, leading to impaired I_{CRAC}. Reconstitution of SCID patient T cells with wild-type human Orai1 restores I_{CRAC} [68]. Knockdown of dOrai in *Drosophila* cells, or of Orai1 in mammalian cells, by RNA interference significantly attenuates I_{CRAC} [68-70].

The Orai family includes three human homologs, Orai1, Orai2 and Orai3. All three isoforms contain cytosolic N- and C-terminus and four transmembrane segments, a putative coiled-coil domain in the C-terminus, a domain that is a common protein interaction motif [68-69, 71]. Orai1 is a 301 amino acids protein with four transmembrane domains and cytosolic N- and C-terminal tails. Within the N-terminus, Orai1 exclusively includes proline/arginine-rich regions. In addition, there is a membrane-proximal N-terminal domain of Orai1 (between residues 68 and 91) that binds calmodulin in a Ca^{2+}-dependent way and is involved in calmodulin-mediated Ca^{2+}-dependent inactivation [72]. The C-terminus contains a putative coiled-coil sequence that is essential for the association with STIM1 [73]. All three Orai isoforms show a flexible outer vestibule formed by residues in the first loop (including acidic or polar amino-acids at 3 positions), which are variable among the different Orais, that leads to a narrow pore [74]. All three Orais contain a putative coiled-coil sequence, a common protein interaction domain that plays a key role for dynamic coupling to STIM1 [75-76]. An Orai1 C terminus deletion mutant as well as a single point mutation within the putative coiled coil domain of Orai1 (L273S) disrupted the association and functional communication of Orai1 with STIM1 and failed to generate Ca^{2+} inward currents [75]. The three Orai isoforms can be activated by store depletion when co-expressed with STIM1 although the amplitude of the currents generated are smaller for Orai2 and Orai3, which might reflect that they interact with STIM1 with less efficiency [76-77]. Orai isoforms show slightly different selectivity for Na$^+$ (being Orai3 more permeable for Na$^+$ than Orai1 and Orai2). In addition, Orai proteins differ in their sensitivity to the pharmacological agent 2-

aminoethoxydiphenyl borate (2-APB) [78]. Orai3 currents can be substantially stimulated by 75 μM 2-APB independent of STIM1, whereas 2-APB at similar concentrations inhibited store-operated Orai1 currents [78]. Stimulation of Orai3 currents by 2-APB is mediated by an alteration of the permeation pathway and it is associated with an increase in Orai3 minimum pore size from about 3.8Å to more than 5.34 Å. Consistent with this, the Orai3 pore mutant E165Q particularly resembled in its permeation properties those of 2-APB stimulated Orai3 and exhibited a reduced response to 2-APB [78]. Orai1 and Orai2 currents have been reported to be stimulated by low concentrations of 2-APB when coexpressed with STIM1[79]. In addition, single-point mutations of critical amino acids in the selectivity filter of Orai1 (E106D and E190A) enable 2-APB to gate Orai1 in a STIM1-independent manner [79]. Therefore, current knowledge indicates that 2-APB has agonistic and antagonistic modes of action on Orai1-forming channels. 2-APB acts as a direct, store-independent, agonist for Orai3 and as a potentiating agonist for Orai1 and Orai2 upon store-operated and STIM1-dependent activation. In contrast, at high concentrations, 2-APB acts as an Orai1 antagonist, which might involve a direct block at the channel level and an additional uncoupling of STIM1 and Orai1. In addition, 2-APB has been reported to reverse the store-dependent multimerization of STIM1 at high concentrations [79].

The Ca^{2+} selectivity of Orai1 has been reported to be achieved by two conserved negatively charged glutamate located in the first and the third transmembrane segments (E106 and E190) and [78, 80-82] as demonstrated by means of point mutations. The Orai1 E106Q mutation prevents the reconstitution of large I_{CRAC} currents. More interestingly, a charge conserving mutation of glutamate to aspartate (E106D) resulted in a change in Ca^{2+} selectivity and smaller currents [81]. Mutations of the glutamate residue at position 190 reduced the Ca^{2+} selectivity of Orai1 characteristic of I_{CRAC} [81-82]. In addition to Orai1, Orai2 and Orai3 are also highly selective for Ca^{2+} over monovalent ions [71, 73, 78, 82-83]. Removal of extracellular Ca^{2+} in the presence of Mg^{2+} and Na^+ in the extracellular medium fully abolishes current influx [68]. However, in a medium free of divalent cations a prominent Na^+ influx has been reported, while permeability to other monovalent cations, such as Cs^+, is unusually low. The Na^+ currents through Orai1, Orai2, and Orai3, as well as the endogenous store-operated Na^+ currents in HEK293 cells, have been reported to be all inhibited by extracellular Ca^{2+} with a half-maximal concentration of approximately 20 μM [71]. In divalent free external solutions, Orai3 currents were considerably more stable than Orai1 or Orai2, indicating that Orai3 channels undergo a lesser degree of depotentiation [71].

A feature of I_{CRAC} is its fast Ca^{2+}-dependent inactivation within the first 100 ms upon CRAC channel gating [84-85], which limits Ca^{2+} influx to protect against cellular damage. This fast inactivation is induced by a local increase in cytosolic Ca^{2+} levels close to the channel mouth [84-85]. Concerning this issue, a decrease in intracellular Ca^{2+} buffering has been reported to result in Orai1 current fast inactivation, while Ba^{2+}-substitution for extracellular Ca^{2+} completely abolished fast Ca^{2+}-dependent inactivation [86]. Orai channels have been shown to exhibit different fast inactivation profiles [77]. Orai3 currents show a pronounced fast inactivation within the first 100ms, while that of Orai2 or Orai1 show less robust feedback regulation [77]. Recent studies have provided evidence that several domains within Orai proteins and STIM1 are relevant for fast Ca^{2+}-dependent inactivation. In Orai1 structure, the N and C-terminal tails, the second loop and the pore region have been reported to be important for this mechanism. A mutation in the calmodulin-binding site, located in the conserved membrane-proximal N-terminal domain between residues 68-91, has been reported to abrogate fast Ca^{2+}-dependent inactivation [72]. Furthermore, substitutions of alanine located at a central position within the second loop of Orai1 abolish fast inactivation [87]. Expression of a peptide comprising the second loop of Orai1 inhibited Orai1 channel gating, thus suggesting that the second loop might play an important role in the inactivation of Orai1 channels [87]. In the case of Orai2 and Orai3 forming channels, fast Ca^{2+}-dependent inactivation has been attributed to three conserved glutamates located at the C-termini [88]. Alanine substitutions in these glutamate residues reduce fast Ca^{2+}-dependent inactivation in Orai3 and only attenuate the time constant but not the extent of fast inactivation in Orai2.

Fast Ca^{2+}-dependent inactivation also involves the pore region itself. Mutations of negatively charged residues within the pore of Orai1 result in an increase in pore size, attenuation of Ca^{2+} selectivity, with increased Cs^+ permeability, and a decrease in Ca^{2+}-dependent inactivation [83]. A role for the pore region of Orai in Ca^{2+}-dependent inactivation of CRAC channels is consistent with the finding that the fast Ca^{2+} buffer BAPTA is more effective attenuating fast Ca^{2+}-dependent inactivation than the slow Ca^{2+} buffer EGTA, thus suggesting that domains that control fast inactivation must to be close to the channel pore [88]. Fast Ca^{2+}-dependent inactivation of the Orai channels has also been reported to require a domain of STIM1 that includes an anionic sequence that is C-terminal to the SOAR [88]. This region has recently been identified as the CRAC modulatory domain (CMD). Studies generating STIM1 mutants with either C-terminal deletions including CMD, and regions that are C-terminal to the CMD as control, or with 7 alanines replacing the negative amino acids within CMD, results in Orai1 currents that displayed attenuated or even abolished Ca^{2+}-dependent inactivation compared to STIM1 mutants with preserved CMD [86]. Similar observations were found with cytosolic C-terminal fragments of STIM1 in RBL-2H3 mast cells expressing CRAC channels endogenously; however, Ca^{2+}-dependent inactivation was much more pronounced than that of Orai1, suggesting that, in addition to Orai1, other channel subunits might be present in the endogenous CRAC channels. Co-expression of STIM1 constructs missing CMD with Orai3 also significantly reduced the extent of Ca^{2+}-dependent inactivation of this channel. Therefore, CMD within STIM1 C-terminal region provides a negative feedback signal to Ca^{2+} entry by inducing fast Ca^{2+}-dependent inactivation of ORAI/CRAC channels. Since some extent of Orai3 and CRAC inactivation was insensitive to CMD removal, additional Orai3 or CRAC subunit domains or factors might contribute to these inactivation mechanism [86].

Transient Receptor Potential (TRP) proteins

TRP proteins form ion channel non-selective for monovalent and divalent cations, including Na^+ and Ca^{2+}, that were initially identified in 1989 in the *trp* mutant of *Drosophila* [89]. The light-sensitive current in *Drosophila* photoreceptors is conducted by two Ca^{2+}-permeable channels encoded by the *trp* and *trpl* genes [90-91]. *Drosophila* TRP and TRPL channels conduct sustained light-sensitive depolarization due to Na^+ and Ca^{2+} influx; however, the *trp* mutant shows transient, rather than sustained, light-sensitive depolarization and receptor potential, which gave the name to these channels [92]. *Drosophila* TRP channels have been shown to be gated by diacylglycerol (DAG) or a metabolic byproduct, as well as by phosphatidylinositol 4,5-bisphosphate (PIP_2) depletion [93].

The first mammalian TRP protein, named canonical transient receptor potential protein-1 (TRPC1), was identified in 1995 in human [94-95] and mouse [96]. TRPC1 is 40% identical to *Drosophila* TRP [94]. Since the identification of TRPC1, a number of TRP proteins have been described, which are grouped into seven major subfamilies: four are closely related to *Drosophila* TRP (TRPC, TRPV, TRPA and TRPM), two more distantly related subfamilies (TRPP and TRPML), and the TRPN group expressed only in fish, flies and worms [97]. The TRPC subfamily comprises seven members (TRPC1-TRPC7, which, in turn, can be divided into four groups: TRPC1, TRPC2, TRPC3/6/7 and TRPC4/5), the vanilloid TRP subfamily (TRPV) consists of six members (TRPV1–TRPV6), the TRPA (ankyrin) subfamily includes only one mammalian protein, TRPA1, and the melastatin TRP subfamily (TRPM) comprises eight different proteins (TRPM1-TRPM8). The TRPP (polycystin) and the TRPML (mucolipin) subfamilies group three members each, and finally, the TRPN (no mechanoreceptor potential C, or NOMPC) are not expressed in mammals [98].

All TRP family members contain six transmembrane domains, with non-conserved cytosolic N- and C-termini depending on the subfamily, and a pore loop region between the transmembrane domains 5 and 6 [99]. Certain TRP families possess long N-terminal regions with several protein-protein interaction domains known as ankyrin repeats, such as TRPA, coiled coil regions, and a putative caveolin-binding domain. The C-terminal region of TRPs includes the TRP signature motif (EWKFAR) of unknown function, a kinase domain found in TRPM, a proline-rich motif and different functional regions that facilitate their interaction with calmodulin or inositol 1,4,5-trisphosphate (IP$_3$) receptor [100-102].

The initial experiments in *Drosophila* revealed that TRP channels were not store-operated but, instead, these channels were activated by downstream lipid products of phospholipase C [93, 103-104]. In fact, the light-sensitive current through TRP channels in *Drosophila* is mediated downstream of phospholipase C. *Drosophila* TRP channels have been shown to be activated by diacylglycerol or its metabolites (polyunsaturated fatty acids) and were originally associated to ROCE [90, 92-93]. In addition to their involvement in ROCE, the mammalian homologues of the *Drosophila* TRP channels were proposed to account for the conduction of SOCE and there is now considerable evidence supporting a role for TRP proteins in the conduction of cation entry during SOCE. Particular attention has been paid to members of the TRPC subfamily, the first mammalian homologues of the Drosophila TRP channels identified. Using different approaches, from overexpression of specific TRP proteins to knockdown of endogenous TRPs and knockout studies, it has been suggested that TRPC proteins can be gated by Ca^{2+} store depletion in a variety of cell types endogenously or transiently expressing TRPC proteins, including human platelets, salivary gland cells and gingival keratinocytes and HEK-293 cells [105-112]. All the members of the TRPC family have been reported to be activated by store depletion or to be involved in SOCE, including TRPC1 [105-106, 113-114], TRPC2 [115], TRPC3 [116-117], TRPC4 [118-119], TRPC5 [119-120], TRPC6 [121-123] and TRPC7 [124]. The involvement of TRPCs in ROCE or SOCE has been reported to depend on distinct circumstances, such as the expression level. At low expression levels TRPCs can be activated by depletion of the intracellular Ca^{2+} stores, while at relatively high levels of expression TRPCs are not longer sensitive to store depletion but activated by receptor occupation in a phospholipase C-dependent manner [116]. Studies in cells expressing TRPCs have often delivered controversial findings that might be explained, in some cases, by the mode of expression. It has been reported that TRPC channels participate in SOCE or ROCE in the same cell type depending on their expression. This is the case of studies in HEK-293 cells, where TRPC7 is activated by phospholipase C-coupled receptors when transiently expressed; in contrast, these channels are sensitive to store depletion when stably expressed [124]. The reason of this discrepancy has not been further investigated but it might be explained by the association of TRPC proteins with proteins that confer store depletion or receptor sensitivity. Recent studies have revealed that the participation of TRPC proteins in SOCE or ROCE might depend on its association with STIM1 and/or Orai1, as reported below.

Among TRP proteins, the role of TRPC1 in SOCE has been extensively investigated in different cell types. The involvement of TRPC1 in SOCE has been demonstrated by antisense experiments in human salivary glands [105, 125-126] and vascular endothelial cells [127]. In support of this, antibodies directed to the pore region of TRPC1 have been shown to reduce SOCE in vascular smooth muscle cells and human platelets [106, 114]. The inhibition of SOCE by two independent anti-TRPC1 antibodies directed towards the pore region has been suggested to occur by the antibody preservatives, specially sodium azide, rather by the antibody itself [128]; however, control experiments were done with the highest concentration of the same antibody neutralized by incubation with the control antigen peptide, which displayed no effect on SOCE in human platelets, which clearly demonstrate that the actions of the anti-TRPC1 antibody on SOCE cannot be attributed to the preservatives themselves [106]. Further studies supporting the role of TRPC1 in SOCE were performed in myoblasts, where TRPC1 depletion shows a substantially attenuated SOCE [129].

TRP channels are mostly nonselective for monovalent and divalent cations with $P_{Ca:Na}$ <10 [130]. This characteristic suggests the participation of TRPs in I_{SOC} current rather than the more Ca^{2+} selective I_{CRAC} currents probably mediated by Orai proteins. There are a number of exceptions concerning the ion selectivity of TRP forming channels, such as TRPM4 and TRPM5, which are selective for monovalent cations, and TRPV5 and TRPV6, with PCa:Na over 100.

5. STIM1-Orai-TRPC Interaction

As reported above, current evidence suggests that several capacitative Ca^{2+} currents with different properties operate in distinct cell types, with different pathways being emphasized in different tissues. These currents might be conducted through homo or heteromeric combinations of the different channel subunits identified with the participation of the ER Ca^{2+} sensor STIM1.

STIM1 enhances SOCE when co-expressed either with Orai1 [46, 55, 59], Orai2 [55] or Orai3 [77-78], thus indicating that these proteins are sufficient to mediate capacitative currents. However, recent studies have pointed out that an adequate stoichiometrical relationship between STIM1 and Orai1 in regions of close apposition between the plasma membrane and the ER is necessary for the activation of I_{CRAC} [46, 56]. The stochiometry of the STIM1-Orai1 association has been shown to play an important role of the biophysical properties of the channel formed. In cells expressing Orai1 and STIM1 at different ratios (from 4:1 to 1:4) low Orai1:STIM1 ratios result in I_{CRAC} with strong fast Ca^{2+}-dependent inactivation, while high Orai1:STIM1 ratios produce I_{CRAC} with strong activation at negative potentials. Moreover, the Orai1:STIM1 expression ratio affects Ca^{2+}, Ba^{2+} and Sr^{2+} conductance [131].

The nature of the interaction between STIM1 and Orai1 has been a matter of intense investigation in the last few years. As reported above the association between these proteins involves the SOAR region in the C-terminus of STIM1 and a sequence in the C-terminus of Orai1. In agreement with the proposed SOAR sequence, the Orai1 R91W mutant linked to the SCID syndrome did not impair the interaction of Orai1 with STIM1 but altered the activation of capacitative currents [75]. The Orai1-STIM1 interacting region has been reported to be able to homomerize, and to activate all identified Orai isoforms although the conductance is greater for Orai1 [60-61]. Further studies have revealed the existence of a regulatory domain at amino acids 474-485 of STIM1, which contains 7 negatively charged residues, named CMD (CRAC modulatory domain)/CDI (Ca^{2+}-dependent inactivation). CMD promotes Orai/CRAC channel closure in a Ca^{2+} concentration-dependent manner and might be involved in the fast Ca^{2+}-dependent inactivation of CRAC channels [72, 86, 88].

STIM1has also been reported to be able to interact and gate TRPC1-forming channels [108]. The association between STIM1 and TRPC1 has been reported to involve electrostatic interaction between two conserved, negatively charged, aspartates in TRPC1 ((639)DD(640)) and two positively charged lysine residues in STIM1((684)KK(685)) located in the C-terminal polybasic region [132]. A similar gating mechanism has been reported for TRPC3, but in this case mediated by the aspartate residues ((697)DD(698)) [132]. Therefore, STIM1 operates Orai1- and TRPC-forming channels by independent mechanism, involving the SOAR and CMD domains or the polybasic rich region, respectively [132].

The association of STIM1 and TRPC1 has been reported in a number of cell types and models including rat basophilic leukaemia cells [133], HEK293 cells [134] and human platelets endogenously expressing TRPC1 and STIM1[44]. In addition, STIM1 has been reported to associate with other TRPC family members including TRPC2 [135], TRPC4, TRPC5 [57] and TRPC6 [123]. The association of STIM1 with TRPC proteins has been reported to recruit these channels in the SOCE pathway. STIM1 recruits TRPC1 into lipid raft domains, plasma membrane microdomains enriched in cholesterol and sphingolipids [136], where TRPC1 functions as a SOC channel [137]. In contrast, in the absence of STIM1, TRPC1 participates in the formation of receptor-operated Ca^{2+} channels, which further supports the role of STIM1 in the activation mode of TRPC1 channels [137]. Lipid rafts participates in the generation of signalling microdomains necessary for clustering of STIM1 at ER-plasma membrane junctions in cells with depleted Ca^{2+} stores, facilitating the interaction between STIM1 and Ca^{2+} channel subunits [138].

In addition to the participation of Orai and TRPC proteins in the formation of CRAC and SOC channels, respectively, evidence for functional interactions between Orai1 and different TRPCs under the influence of STIM1 has been presented in different cell types. In human salivary gland cells, where TRPC1 is a major SOC channel component [105, 125-126], knockdown of Orai1 significantly reduces I_{SOC} [133]. In HEK293 cells, the TRPC proteins TRPC3 and TRPC6 become sensitive to store depletion upon expression of exogenous Orai [139]. Furthermore, in human platelets naturally expressing STIM1, Orai1 and TRPC members, interference with the association between STIM1 and Orai1 that occurs after Ca^{2+} store depletion abolishes co-immunoprecipitation between STIM1 and TRPC1, and subsequently changes the mode of activation of TRPC1 being no longer involved in SOCE but in ROCE mediated by DAG [111]. Similarly, the mode of activation of TRPC6 in human platelets have been reported to be regulated by its interaction with the Orai1-STIM1 complex or hTRPC3, for store-dependent or –independent modes, respectively [123]. Therefore, current evidence suggest that the involvement of Orai proteins in SOCE might be well explained either by a model in which Orai1 are self-contained Ca^{2+} channels activated by STIM1, the CRAC channel hypothesis [59, 80, 82, 140], or a model in which SOC channels consist of a combination of TRPC and Orai proteins, where Orai might take part in the formation of the channel pore or might communicate the information or the filling state of the Ca^{2+} stores from STIM1 to the TRPC subunits [133, 139, 141] (Figure 1). Consistent with this, Orai and TRPC proteins have been reported to participate both in SOCE and ROCE. Expression of Orai1, at conditions that enhance SOCE, leads to the activation of ROCE. In addition, the SCID Orai1 mutant (R91W) impairs both SOCE and ROCE in cells that, stably or transiently, express TRPC3 [142]. The store dependent activation of Orai1 has been reported to be associated to the channel location into lipid raft domains, which facilitate clustering of Ca^{2+} channels and their modulators in signalling microdomains that provide an excellent spatiotemporal regulation of Ca^{2+}-mediated cellular functions [143-145].

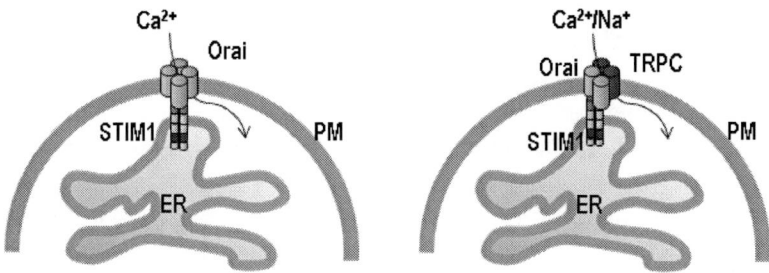

Figure 1. The current models for the participation of Orai1 in SOCE. A, STIM1, the ER Ca^{2+} sensor, is distributed into clusters immediately subjacent to the plasma membrane, where it activates Ca^{2+} influx through Ca^{2+} selective Orai1-formed CRAC channels. B, STIM1 interacts with Orai1 subunits taking part of the non-selective SOC channels together with different TRPC family members and located into lipid raft domains. PM, plasma membrane, ER, endoplasmic reticulum.

ACKNOWLEDGMENT

Supported by MEC-DGI grant BFU2010-21043-C02-01.

REFERENCES

[1] Berridge, MJ; Lipp, P; Bootman, MD. The versatility and universality of calcium signalling. *Nat. Rev. Mol. Cell Biol.* 2000 1, 11-21.
[2] Petersen, OH. Calcium signal compartmentalization. *Biol. Res.* 2002 35, 177-182.
[3] Rosado, JA; Sage, SO. Platelet signalling: calcium. In: Gresele P, Page CP, Fuster V, Vermylen J, eds. Platelets in Thrombotic and Non-Thrombotic Disorders Pathophysiology, Pharmacology and Therapeutics. Cambridge: Cambridge University Press; 2000; 260-271.

[4]   Putney, JW, Jr. A model for receptor-regulated calcium entry. *Cell Calcium.* 1986 7, 1-12.

[5]   Putney, JW, Jr. New molecular players in capacitative Ca^{2+} entry. *J. Cell Sci.* 2007 120, 1959-1965.

[6]   Abdullaev, IF; Bisaillon, JM; Potier, M; Gonzalez, JC; Motiani, RK; Trebak, M. Stim1 and Orai1 mediate CRAC currents and store-operated calcium entry important for endothelial cell proliferation. *Circ. Res.* 2008 103, 1289-1299.

[7]   Cioffi, DL; Stevens, T. Regulation of endothelial cell barrier function by store-operated calcium entry. *Microcirculation.* 2006 13, 709-723.

[8]   Leung, FP; Yung, LM; Yao, X; Laher, I; Huang, Y. Store-operated calcium entry in vascular smooth muscle. *Br. J. Pharmacol.* 2008 153, 846-857.

[9]   Wedel, B; Boyles, RR; Putney, JW, Jr.; Bird, GS. Role of the store-operated calcium entry proteins Stim1 and Orai1 in muscarinic cholinergic receptor-stimulated calcium oscillations in human embryonic kidney cells. *J. Physiol.* 2007 579, 679-689.

[10] Verkhratsky, A. Physiology and pathophysiology of the calcium store in the endoplasmic reticulum of neurons. *Physiol. Rev.* 2005 85, 201-279.

[11] Rao, RV; Ellerby, HM; Bredesen, DE. Coupling endoplasmic reticulum stress to the cell death program. *Cell Death Differ.* 2004 11, 372-380.

[12] Berridge, MJ. Capacitative calcium entry. *Biochem. J.* 1995 312 (Pt 1), 1-11.

[13] Redondo, PC; Harper, MT; Rosado, JA; Sage, SO. A role for cofilin in the activation of store-operated calcium entry by de novo conformational coupling in human platelets. *Blood.* 2006 107, 973-979.

[14] Kwan, CY; Takemura, H; Obie, JF; Thastrup, O; Putney, JWJ. Effects of MeCh, thapsigargin, and La^{3+} on plasmalemmal and intracellular Ca^{2+} transport in lacrimal acinar cells. *American Journal of Physiology-Cell Physiology.* 1990 258, 1006–1015.

[15] Rosado, JA. Discovering the mechanism of capacitative calcium entry. *Am. J. Physiol. Cell Physiol.* 2006 291, C1104-1106.

[16] Harper, AG; Mason, MJ; Sage, SO. A key role for dense granule secretion in potentiation of the Ca^{2+} signal arising from store-operated calcium entry in human platelets. *Cell Calcium.* 2009 45, 413-420.

[17] Harper, AG; Sage, SO. A key role for reverse Na^+/Ca^{2+} exchange influenced by the actin cytoskeleton in store-operated Ca^{2+} entry in human platelets: evidence against the de novo conformational coupling hypothesis. *Cell Calcium.* 2007 42, 606-617.

[18] Penner, R; Matthews, G; Neher, E. Regulation of calcium influx by second messengers in rat mast cells. *Nature.* 1988 334, 499-504.

[19] Lewis, RS; Cahalan, MD. Mitogen-induced oscillations of cytosolic Ca^{2+} and transmembrane Ca2+ current in human leukemic T cells. *Cell Regul.* 1989 1, 99-112.

[20] Artalejo, AR; Ellory, JC; Parekh, AB. Ca^{2+}-dependent capacitance increases in rat basophilic leukemia cells following activation of store-operated Ca^{2+} entry and dialysis with high-Ca2+-containing intracellular solution. *Pflugers Arch.* 1998 436, 934-939.

[21] Hofer, AM; Fasolato, C; Pozzan, T. Capacitative Ca2+ entry is closely linked to the filling state of internal Ca^{2+} stores: a study using simultaneous measurements of ICRAC and intraluminal $[Ca^{2+}]$. *J. Cell Biol.* 1998 140, 325-334.

[22] Bakowski, D; Parekh, AB. Voltage-dependent conductance changes in the store-operated Ca^{2+} current ICRAC in rat basophilic leukaemia cells. *J. Physiol.* 2000 529 Pt 2, 295-306.

[23] Fasolato, C; Nilius, B. Store depletion triggers the calcium release-activated calcium current (ICRAC) in macrovascular endothelial cells: a comparison with Jurkat and embryonic kidney cell lines. *Pflugers Arch.* 1998 436, 69-74.

[24] Hoth, M; Penner, R. Depletion of intracellular calcium stores activates a calcium current in mast cells. *Nature.* 1992 355, 353-356.

[25] Zweifach, A; Lewis, RS. Mitogen-regulated Ca^{2+} current of T lymphocytes is activated by depletion of intracellular Ca^{2+} stores. *Proc. Natl. Acad. Sci. U.S.A.* 1993 90, 6295-6299.

[26] Zhang, L; McCloskey, MA. Immunoglobulin E receptor-activated calcium conductance in rat mast cells. *J. Physiol.* 1995 483 (Pt 1), 59-66.

[27] Prakriya, M; Lewis, RS. Separation and characterization of currents through store-operated CRAC channels and Mg^{2+}- inhibited cation (MIC) channels. *Journal of General Physiology.* 2002 119, 487-507.

[28] Hogan, PG; Rao, A. Dissecting ICRAC, a store-operated calcium current. *TRENDS in Biochemical Sciences.* 2007 32, 235-245.

[29] Parekh, AB; Putney, JW, Jr. Store-operated calcium channels. *Physiol. Rev.* 2005 85, 757-810.

[30] Hoth, M; Penner, R. Calcium release-activated calcium current in rat mast cells. *Journal of Physiology.* 1993 465, 359-386.

[31] Matsuda, H; Noma, A. Isolation of calcium current and its sensitivity to monovalent cations in dialysed ventricular cells of guinea-pig. *Journal of Physiology.* 1984 357, 553-573.

[32] Parekh, AB; Penner, R. Store depletion and calcium influx. *Physiol. Rev.* 1997 77, 901-930.

[33] Fierro, L; Parekh, AB. Substantial depletion of the intracellular Ca^{2+} stores is required for macroscopic activation of the Ca^{2+} release-activated Ca^{2+} current in rat basophilic leukaemia cells. *Journal of Physiology.* 2000 522, 247-257.

[34] Bakowski, D; Parekh, AB. Monovalent cation permeability and Ca^{2+} block of the store-operated calcium current I_{CRAC} in rat basophilic leukaemia cells. *Pflugers Arch.* 2002 443, 892-902.

[35] Bakowski, D; Parekh, AB. Permeation through store-operated CRAC channels in divalent-free solution: potential problems and implications for putative CRAC channel genes. *Cell Calcium.* 2002 32, 379-391.

[36] Smani, T; Zakharov, SI; Csutora, P; Leno, E; Trepakova, ES; Bolotina, VM. A novel mechanism for the store-operated calcium influx pathway. *Nat. Cell Biol.* 2004 6, 113-120.

[37] Trepakova, ES; Gericke, M; Hirakawa, Y; Weisbrod, RM; Cohen, RA; Bolotina, VM. Properties of a native cation channel activated by Ca2+ store depletion in vascular smooth muscle cells. *J. Biol. Chem.* 2001 276, 7782-7790.

[38] Albert, AP; Saleh, SN; Peppiatt-Wildman, CM; Large, WA. Multiple activation mechanisms of store-operated TRPC channels in smooth muscle cells. *J. Physiol.* 2007 583, 25-36.

[39] Guibert, C; Ducret, T; Savineau, JP. Voltage-independent calcium influx in smooth muscle. *Prog. Biophys. Mol. Biol.* 2008 98, 10-23.

[40] Trepakova, ES; Csutora, P; Hunton, DL; Marchase, RB; Cohen, RA; Bolotina, VM. Calcium influx factor directly activates store-operated cation channels in vascular smooth muscle cells. *Journal of Biological Chemistry.* 2000 275, 26158-26163.

[41] Roos, J; DiGregorio, PJ; Yeromin, AV; Ohlsen, K; Lioudyno, M; Zhang, S; Safrina, O; Kozak, JA; Wagner, SL; Cahalan, MD; Velicelebi, G; Stauderman, KA. STIM1, an essential and conserved component of store-operated Ca^{2+} channel function. *J. Cell Biol.* 2005 169, 435-445.

[42] Zhang, SL; Yu, Y; Roos, J; Kozak, JA; Deerinck, TJ; Ellisman, MH; Stauderman, KA; Cahalan, MD. STIM1 is a Ca^{2+} sensor that activates CRAC channels and migrates from the Ca2+ store to the plasma membrane. *Nature.* 2005 437, 902-905.

[43] Liou, J; Kim, ML; Heo, WD; Jones, JT; Myers, JW; Ferrell, JE, Jr.; Meyer, T. STIM is a Ca^{2+} sensor essential for Ca^{2+}-store-depletion-triggered Ca^{2+} influx. *Curr. Biol.* 2005 15, 1235-1241.

[44] Lopez, JJ; Salido, GM; Pariente, JA; Rosado, JA. Interaction of STIM1 with endogenously expressed human canonical TRP1 upon depletion of intracellular Ca2+ stores. *J. Biol. Chem.* 2006 281, 28254-28264.

[45] Spassova, MA; Soboloff, J; He, LP; Xu, W; Dziadek, MA; Gill, DL. STIM1 has a plasma membrane role in the activation of store-operated Ca^{2+} channels. *Proc. Natl. Acad. Sci. U.S.A.* 2006 103, 4040-4045.

[46] Soboloff, J; Spassova, MA; Tang, XD; Hewavitharana, T; Xu, W; Gill, DL. Orai1 and STIM reconstitute store-operated calcium channel function. *J. Biol. Chem.* 2006 281, 20661-20665.

[47] Jardin, I; Lopez, JJ; Redondo, PC; Salido, GM; Rosado, JA. Store-operated Ca^{2+} entry is sensitive to the extracellular Ca^{2+} concentration through plasma membrane STIM1. *Biochim. Biophys. Acta.* 2009 1793, 1614-1622.

[48] Soboloff, J; Spassova, MA; Dziadek, MA; Gill, DL. Calcium signals mediated by STIM and Orai proteins--a new paradigm in inter-organelle communication. *Biochim. Biophys. Acta.* 2006 1763, 1161-1168.

[49] Schindl, R; Muik, M; Fahrner, M; Derler, I; Fritsch, R; Bergsmann, J; Romanin, C. Recent progress on STIM1 domains controlling Orai activation. *Cell Calcium.* 2009 46, 227-232.

[50] Schultz, J; Ponting, CP; Hofmann, K; Bork, P. SAM as a protein interaction domain involved in developmental regulation. *Protein Sci.* 1997 6, 249-253.

[51] Grabarek, Z. Structural basis for diversity of the EF-hand calcium-binding proteins. *J. Mol. Biol.* 2006 359, 509-525.

[52] Brandman, O; Liou, J; Park, WS; Meyer, T. STIM2 is a feedback regulator that stabilizes basal cytosolic and endoplasmic reticulum Ca^{2+} levels. *Cell.* 2007 131, 1327-1339.

[53] Wu, MM; Buchanan, J; Luik, RM; Lewis, RS. Ca^{2+} store depletion causes STIM1 to accumulate in ER regions closely associated with the plasma membrane. *J. Cell Biol.* 2006 174, 803-813.

[54] Hogan, PG; Rao, A. Dissecting ICRAC, a store-operated calcium current. *Trends Biochem. Sci.* 2007 32, 235-245.

[55] Mercer, JC; Dehaven, WI; Smyth, JT; Wedel, B; Boyles, RR; Bird, GS; Putney, JW, Jr. Large store-operated calcium selective currents due to co-expression of Orai1 or Orai2 with the intracellular calcium sensor, Stim1. *J. Biol. Chem.* 2006 281, 24979-24990.

[56] Xu, P; Lu, J; Li, Z; Yu, X; Chen, L; Xu, T. Aggregation of STIM1 underneath the plasma membrane induces clustering of Orai1. *Biochem. Biophys. Res. Commun.* 2006 350, 969-976.

[57] Yuan, JP; Zeng, W; Huang, GN; Worley, PF; Muallem, S. STIM1 heteromultimerizes TRPC channels to determine their function as store-operated channels. *Nat. Cell Biol.* 2007 9, 636-645.

[58] Luik, RM; Wu, MM; Buchanan, J; Lewis, RS. The elementary unit of store-operated Ca^{2+} entry: local activation of CRAC channels by STIM1 at ER-plasma membrane junctions. *J. Cell Biol.* 2006 174, 815-825.

[59] Peinelt, C; Vig, M; Koomoa, DL; Beck, A; Nadler, MJ; Koblan-Huberson, M; Lis, A; Fleig, A; Penner, R; Kinet, JP. Amplification of CRAC current by STIM1 and CRACM1 (Orai1). *Nat. Cell Biol.* 2006 8, 771-773.

[60] Yuan, JP; Zeng, W; Dorwart, MR; Choi, YJ; Worley, PF; Muallem, S. SOAR and the polybasic STIM1 domains gate and regulate Orai channels. *Nat. Cell Biol.* 2009 11, 337-343.

[61] Muik, M; Fahrner, M; Derler, I; Schindl, R; Bergsmann, J; Frischauf, I; Groschner, K; Romanin, C. A Cytosolic Homomerization and a Modulatory Domain within STIM1 C Terminus Determine Coupling to ORAI1 Channels. *J. Biol. Chem.* 2009 284, 8421-8426.

[62] Park, CY; Hoover, PJ; Mullins, FM; Bachhawat, P; Covington, ED; Raunser, S; Walz, T; Garcia, KC; Dolmetsch, RE; Lewis, RS. STIM1 clusters and activates CRAC channels via direct binding of a cytosolic domain to Orai1. *Cell.* 2009 136, 876-890.

[63] Kawasaki, T; Lange, I; Feske, S. A minimal regulatory domain in the C terminus of STIM1 binds to and activates ORAI1 CRAC channels. *Biochem. Biophys. Res. Commun.* 2009 385, 49-54.

[64] Mignen, O; Thompson, JL; Shuttleworth, TJ. The molecular architecture of the arachidonate-regulated Ca^{2+}-selective ARC channel is a pentameric assembly of Orai1 and Orai3 subunits. *J. Physiol.* 2009 587, 4181-4197.

[65] Parvez, S; Beck, A; Peinelt, C; Soboloff, J; Lis, A; Monteilh-Zoller, M; Gill, DL; Fleig, A; Penner, R. STIM2 protein mediates distinct store-dependent and store-independent modes of CRAC channel activation. *FASEB J.* 2008 22, 752-761.

[66] Irvine, RF; Moor, RM. Inositol(1,3,4,5)tetrakisphosphate-induced activation of sea urchin eggs requires the presence of inositol trisphosphate. *Biochem. Biophys. Res. Commun.* 1987 146, 284-290.

[67] Irvine, RF. Inositol phosphates and Ca^{2+} entry: toward a proliferation or a simplification? *FASEB J.* 1992 6, 3085-3091.

[68] Feske, S; Gwack, Y; Prakriya, M; Srikanth, S; Puppel, SH; Tanasa, B; Hogan, PG; Lewis, RS; Daly, M; Rao, A. A mutation in Orai1 causes immune deficiency by abrogating CRAC channel function. *Nature.* 2006 441, 179-185.

[69] Vig, M; Peinelt, C; Beck, A; Koomoa, DL; Rabah, D; Koblan-Huberson, M; Kraft, S; Turner, H; Fleig, A; Penner, R; Kinet, JP. CRACM1 is a plasma membrane protein essential for store-operated Ca^{2+} entry. *Science.* 2006 312, 1220-1223.

[70] Zhang, SL; Yeromin, AV; Zhang, XH; Yu, Y; Safrina, O; Penna, A; Roos, J; Stauderman, KA; Cahalan, MD. Genome-wide RNAi screen of Ca^{2+} influx identifies genes that regulate Ca^{2+} release-activated Ca^{2+} channel activity. *Proc. Natl. Acad. Sci. U.S.A.* 2006 103, 9357-9362.

[71] DeHaven, WI; Smyth, JT; Boyles, RR; Putney, JW, Jr. Calcium inhibition and calcium potentiation of Orai1, Orai2, and Orai3 calcium release-activated calcium channels. *J. Biol. Chem.* 2007 282, 17548-17556.

[72] Mullins, FM; Park, CY; Dolmetsch, RE; Lewis, RS. STIM1 and calmodulin interact with Orai1 to induce Ca^{2+}-dependent inactivation of CRAC channels. *Proc. Natl. Acad. Sci. U.S.A.* 2009 106, 15495-15500.

[73] Frischauf, I; Schindl, R; Derler, I; Bergsmann, J; Fahrner, M; Romanin, C. The STIM/Orai coupling machinery. *Channels.* (Austin). 2008 2, 261-268.

[74] McNally, BA; Yamashita, M; Engh, A; Prakriya, M. Structural determinants of ion permeation in CRAC channels. *Proc. Natl. Acad. Sci. U.S.A.* 2009 106, 22516-22521.

[75] Muik, M; Frischauf, I; Derler, I; Fahrner, M; Bergsmann, J; Eder, P; Schindl, R; Hesch, C; Polzinger, B; Fritsch, R; Kahr, H; Madl, J; Gruber, H; Groschner, K; Romanin, C. Dynamic coupling of the putative coiled-coil domain of ORAI1 with STIM1 mediates ORAI1 channel activation. *J. Biol. Chem.* 2008 283, 8014-8022.

[76] Frischauf, I; Muik, M; Derler, I; Bergsmann, J; Fahrner, M; Schindl, R; Groschner, K; Romanin, C. Molecular determinants of the coupling between STIM1 and Orai channels: differential activation of Orai1-3 channels by a STIM1 coiled-coil mutant. *J. Biol. Chem.* 2009 284, 21696-21706.

[77] Lis, A; Peinelt, C; Beck, A; Parvez, S; Monteilh-Zoller, M; Fleig, A; Penner, R. CRACM1, CRACM2, and CRACM3 are store-operated Ca^{2+} channels with distinct functional properties. *Curr. Biol.* 2007 17, 794-800.

[78] Schindl, R; Bergsmann, J; Frischauf, I; Derler, I; Fahrner, M; Muik, M; Fritsch, R; Groschner, K; Romanin, C. 2-aminoethoxydiphenyl borate alters selectivity of Orai3 channels by increasing their pore size. *J. Biol. Chem.* 2008 283, 20261-20267.

[79] Peinelt, C; Lis, A; Beck, A; Fleig, A; Penner, R. 2-Aminoethoxydiphenyl borate directly facilitates and indirectly inhibits STIM1-dependent gating of CRAC channels. *J. Physiol.* 2008 586, 3061-3073.

[80] Prakriya, M; Feske, S; Gwack, Y; Srikanth, S; Rao, A; Hogan, PG. Orai1 is an essential pore subunit of the CRAC channel. *Nature.* 2006 443, 230-233.

[81] Vig, M; Beck, A; Billingsley, JM; Lis, A; Parvez, S; Peinelt, C; Koomoa, DL; Soboloff, J; Gill, DL; Fleig, A; Kinet, JP; Penner, R. CRACM1 multimers form the ion-selective pore of the CRAC channel. *Curr. Biol.* 2006 16, 2073-2079.

[82] Yeromin, AV; Zhang, SL; Jiang, W; Yu, Y; Safrina, O; Cahalan, MD. Molecular identification of the CRAC channel by altered ion selectivity in a mutant of Orai. *Nature.* 2006 443, 226-229.

[83] Yamashita, M; Navarro-Borelly, L; McNally, BA; Prakriya, M. Orai1 mutations alter ion permeation and Ca^{2+}-dependent fast inactivation of CRAC channels: evidence for coupling of permeation and gating. *J. Gen. Physiol.* 2007 130, 525-540.

[84] Fierro, L; Lund, PE; Parekh, AB. Comparison of the activation of the Ca^{2+} release-activated Ca^{2+} current ICRAC to InsP$_3$ in Jurkat T-lymphocytes, pulmonary artery endothelia and RBL-1 cells. *Pflugers Arch.* 2000 440, 580-587.

[85] Zweifach, A; Lewis, RS. Rapid inactivation of depletion-activated calcium current (ICRAC) due to local calcium feedback. *J. Gen. Physiol.* 1995 105, 209-226.

[86] Derler, I; Fahrner, M; Muik, M; Lackner, B; Schindl, R; Groschner, K; Romanin, C. A CRAC modulatory domain (CMD) within STIM1 mediates fast Ca2+-dependent inactivation of ORAI1 channels. *J. Biol. Chem.* 2009.

[87] Srikanth, S; Jung, HJ; Ribalet, B; Gwack, Y. The intracellular loop of Orai1 plays a central role in fast inactivation of Ca^{2+} release-activated Ca^{2+} channels. *J. Biol. Chem.* 2010 285, 5066-5075.

[88] Lee, KP; Yuan, JP; Zeng, W; So, I; Worley, PF; Muallem, S. Molecular determinants of fast Ca^{2+}-dependent inactivation and gating of the Orai channels. *Proc. Natl. Acad. Sci. U.S.A.* 2009 106, 14687-14692.

[89] Montell, C; Rubin, GM. Molecular characterization of the Drosophila trp locus: a putative integral membrane protein required for phototransduction. *Neuron.* 1989 2, 1313-1323.

[90] Hardie, RC; Minke, B. The trp gene is essential for a light-activated Ca^{2+} channel in Drosophila photoreceptors. *Neuron.* 1992 8, 643-651.

[91] Phillips, AM; Bull, A; Kelly, LE. Identification of a Drosophila gene encoding a calmodulin-binding protein with homology to the trp phototransduction gene. *Neuron.* 1992 8, 631-642.

[92] Hardie, RC; Reuss, H; Lansdell, SJ; Millar, NS. Functional equivalence of native light-sensitive channels in the Drosophila trp301 mutant and TRPL cation channels expressed in a stably transfected Drosophila cell line. *Cell Calcium.* 1997 21, 431-440.

[93] Hardie, RC. Regulation of TRP channels via lipid second messengers. *Annu. Rev. Physiol.* 2003 65, 735-759.

[94] Wes, PD; Chevesich, J; Jeromin, A; Rosenberg, C; Stetten, G; Montell, C. TRPC1, a human homolog of a Drosophila store-operated channel. *Proc. Natl. Acad. Sci. U.S.A.* 1995 92, 9652-9656.

[95] Zhu, X; Chu, PB; Peyton, M; Birnbaumer, L. Molecular cloning of a widely expressed human homologue for the Drosophila trp gene. *FEBS Lett.* 1995 373, 193-198.

[96] Petersen, CC; Berridge, MJ; Borgese, MF; Bennett, DL. Putative capacitative calcium entry channels: expression of Drosophila trp and evidence for the existence of vertebrate homologues. *Biochem. J.* 1995 311 (Pt 1), 41-44.

[97] Montell, C; Birnbaumer, L; Flockerzi, V; Bindels, RJ; Bruford, EA; Caterina, MJ; Clapham, DE; Harteneck, C; Heller, S; Julius, D; Kojima, I; Mori, Y; Penner, R; Prawitt, D; Scharenberg, AM; Schultz, G; Shimizu, N; Zhu, MX. A unified nomenclature for the superfamily of TRP cation channels. *Mol. Cell.* 2002 9, 229-231.

[98] Pedersen, SF; Owsianik, G; Nilius, B. TRP channels: an overview. *Cell Calcium.* 2005 38, 233-252.

[99] Hoenderop, JG; Voets, T; Hoefs, S; Weidema, F; Prenen, J; Nilius, B; Bindels, RJ. Homo- and heterotetrameric architecture of the epithelial Ca^{2+} channels TRPV5 and TRPV6. *EMBO J.* 2003 22, 776-785.

[100] Vannier, B; Zhu, X; Brown, D; Birnbaumer, L. The membrane topology of human transient receptor potential 3 as inferred from glycosylation-scanning mutagenesis and epitope immunocytochemistry. *J. Biol. Chem.* 1998 273, 8675-8679.

[101] Vazquez, G; Wedel, BJ; Aziz, O; Trebak, M; Putney, JW, Jr. The mammalian TRPC cation channels. *Biochim. Biophys. Acta.* 2004 1742, 21-36.

[102] Montell, C; Birnbaumer, L; Flockerzi, V. The TRP channels, a remarkably functional family. *Cell.* 2002 108, 595-598.

[103] Montell, C. The venerable inveterate invertebrate TRP channels. *Cell Calcium.* 2003 33, 409-417.

[104] Montell, C. Drosophila TRP channels. *Pflugers Arch.* 2005 451, 19-28.

[105] Liu, X; Wang, W; Singh, BB; Lockwich, T; Jadlowiec, J; O'Connell, B; Wellner, R; Zhu, MX; Ambudkar, IS. Trp1, a candidate protein for the store-operated Ca^{2+} influx mechanism in salivary gland cells. *J. Biol. Chem.* 2000 275, 3403-3411.

[106] Rosado, JA; Brownlow, SL; Sage, SO. Endogenously expressed Trp1 is involved in store-mediated Ca^{2+} entry by conformational coupling in human platelets. *J. Biol. Chem.* 2002 277, 42157-42163.

[107] Zagranichnaya, TK; Wu, X; Villereal, ML. Endogenous TRPC1, TRPC3, and TRPC7 proteins combine to form native store-operated channels in HEK-293 cells. *J. Biol. Chem.* 2005 280, 29559-29569.

[108] Huang, GN; Zeng, W; Kim, JY; Yuan, JP; Han, L; Muallem, S; Worley, PF. STIM1 carboxyl-terminus activates native SOC, I(crac) and TRPC1 channels. *Nat. Cell Biol.* 2006 8, 1003-1010.

[109] Ambudkar, IS; Ong, HL; Liu, X; Bandyopadhyay, B; Cheng, KT. TRPC1: the link between functionally distinct store-operated calcium channels. *Cell Calcium.* 2007 42, 213-223.

[110] Fatherazi, S; Presland, RB; Belton, CM; Goodwin, P; Al-Qutub, M; Trbic, Z; Macdonald, G; Schubert, MM; Izutsu, KT. Evidence that TRPC4 supports the calcium selective I(CRAC)-like current in human gingival keratinocytes. *Pflugers Arch.* 2007 453, 879-889.

[111] Jardin, I; Lopez, JJ; Salido, GM; Rosado, JA. Orai1 mediates the interaction between STIM1 and hTRPC1 and regulates the mode of activation of hTRPC1-forming Ca^{2+} channels. *J. Biol. Chem.* 2008 283, 25296-25304.

[112] Salido, GM; Sage, SO; Rosado, JA. TRPC channels and store-operated Ca^{2+} entry. *Biochim. Biophys. Acta.* 2008.

[113] Brownlow, SL; Harper, AG; Harper, MT; Sage, SO. A role for hTRPC1 and lipid raft domains in store-mediated calcium entry in human platelets. *Cell Calcium.* 2004 35, 107-113.

[114] Xu, SZ; Beech, DJ. TrpC1 is a membrane-spanning subunit of store-operated Ca^{2+} channels in native vascular smooth muscle cells. *Circ. Res.* 2001 88, 84-87.

[115] Vannier, B; Peyton, M; Boulay, G; Brown, D; Qin, N; Jiang, M; Zhu, X; Birnbaumer, L. Mouse trp2, the homologue of the human trpc2 pseudogene, encodes mTrp2, a store depletion-activated capacitative Ca^{2+} entry channel. *Proc. Natl. Acad. Sci. U.S.A.* 1999 96, 2060-2064.

[116] Vazquez, G; Wedel, BJ; Trebak, M; St John Bird, G; Putney, JW, Jr. Expression level of the canonical transient receptor potential 3 (TRPC3) channel determines its mechanism of activation. *J. Biol. Chem.* 2003 278, 21649-21654.

[117] Yildirim, E; Kawasaki, BT; Birnbaumer, L. Molecular cloning of TRPC3a, an N-terminally extended, store-operated variant of the human C3 transient receptor potential channel. *Proc. Natl. Acad. Sci. U.S.A.* 2005 102, 3307-3311.

[118] Philipp, S; Cavalie, A; Freichel, M; Wissenbach, U; Zimmer, S; Trost, C; Marquart, A; Murakami, M; Flockerzi, V. A mammalian capacitative calcium entry channel homologous to Drosophila TRP and TRPL. *EMBO J.* 1996 15, 6166-6171.

[119] Brownlow, SL; Sage, SO. Transient receptor potential protein subunit assembly and membrane distribution in human platelets. *Thromb. Haemost.* 2005 94, 839-845.

[120] Philipp, S; Hambrecht, J; Braslavski, L; Schroth, G; Freichel, M; Murakami, M; Cavalie, A; Flockerzi, V. A novel capacitative calcium entry channel expressed in excitable cells. *EMBO J.* 1998 17, 4274-4282.

[121] Jardin, I; Redondo, PC; Salido, GM; Rosado, JA. Phosphatidylinositol 4,5-bisphosphate enhances store-operated calcium entry through hTRPC6 channel in human platelets. *Biochim. Biophys. Acta.* 2008 1783, 84-97.

[122] Brechard, S; Melchior, C; Plancon, S; Schenten, V; Tschirhart, EJ. Store-operated Ca^{2+} channels formed by TRPC1, TRPC6 and Orai1 and non-store-operated channels formed by TRPC3 are involved in the regulation of NADPH oxidase in HL-60 granulocytes. *Cell Calcium.* 2008.

[123] Jardin, I; Gomez, LJ; Salido, GM; Rosado, JA. Dynamic interaction of hTRPC6 with the Orai1-STIM1 complex or hTRPC3 mediates its role in capacitative or non-capacitative Ca^{2+} entry pathways. *Biochem. J.* 2009 420, 267-276.

[124] Lievremont, JP; Bird, GS; Putney, JW, Jr. Canonical transient receptor potential TRPC7 can function as both a receptor- and store-operated channel in HEK-293 cells. *Am. J. Physiol. Cell Physiol.* 2004 287, C1709-1716.

[125] Singh, BB; Liu, X; Ambudkar, IS. Expression of truncated transient receptor potential protein 1alpha (Trp1alpha): evidence that the Trp1 C terminus modulates store-operated Ca^{2+} entry. *J. Biol. Chem.* 2000 275, 36483-36486.

[126] Liu, X; Singh, BB; Ambudkar, IS. TRPC1 is required for functional store-operated Ca^{2+} channels. Role of acidic amino acid residues in the S5-S6 region. *J. Biol. Chem.* 2003 278, 11337-11343.

[127] Brough, GH; Wu, S; Cioffi, D; Moore, TM; Li, M; Dean, N; Stevens, T. Contribution of endogenously expressed Trp1 to a Ca^{2+}-selective, store-operated Ca^{2+} entry pathway. *FASEB J.* 2001 15, 1727-1738.

[128] Varga-Szabo, D; Authi, KS; Braun, A; Bender, M; Ambily, A; Hassock, SR; Gudermann, T; Dietrich, A; Nieswandt, B. Store-operated Ca^{2+} entry in platelets occurs independently of transient receptor potential (TRP) C1. *Pflugers Arch.* 2008 457, 377-387.

[129] Louis, M; Zanou, N; Van Schoor, M; Gailly, P. TRPC1 regulates skeletal myoblast migration and differentiation. *J. Cell Sci.* 2008 121, 3951-3959.

[130] Zhu, X; Jiang, M; Peyton, M; Boulay, G; Hurst, R; Stefani, E; Birnbaumer, L. trp, a novel mammalian gene family essential for agonist-activated capacitative Ca^{2+} entry. *Cell.* 1996 85, 661-671.

[131] Scrimgeour, N; Litjens, T; Ma, L; Barritt, GJ; Rychkov, GY. Properties of Orai1 mediated store-operated current depend on the expression levels of STIM1 and Orai1 proteins. *J. Physiol.* 2009 587, 2903-2918.

[132] Zeng, W; Yuan, JP; Kim, MS; Choi, YJ; Huang, GN; Worley, PF; Muallem, S. STIM1 gates TRPC channels, but not Orai1, by electrostatic interaction. *Mol. Cell.* 2008 32, 439-448.

[133] Ong, HL; Cheng, KT; Liu, X; Bandyopadhyay, BC; Paria, BC; Soboloff, J; Pani, B; Gwack, Y; Srikanth, S; Singh, BB; Gill, DL; Ambudkar, IS. Dynamic assembly of TRPC1-STIM1-Orai1 ternary complex is involved in store-operated calcium influx. Evidence for similarities in store-operated and calcium release-activated calcium channel components. *J. Biol. Chem.* 2007 282, 9105-9116.

[134] Cheng, KT; Liu, X; Ong, HL; Ambudkar, IS. Functional requirement for Orai1 in store-operated TRPC1-STIM1 channels. *J. Biol. Chem.* 2008 283, 12935-12940.

[135] Worley, PF; Zeng, W; Huang, GN; Yuan, JP; Kim, JY; Lee, MG; Muallem, S. TRPC channels as STIM1-regulated store-operated channels. *Cell Calcium.* 2007 42, 205-211.

[136] Simons, K; Toomre, D. Lipid rafts and signal transduction. *Nat. Rev. Mol. Cell Biol.* 2000 1, 31-39.

[137] Sampieri, A; Zepeda, A; Saldaña, C; Salgado, A; Vaca, L. STIM1 converts TRPC1 from a receptor-operated to a store-operated channel: Moving TRPC1 in and out of lipid rafts. *Cell Calcium.* 2008 44, 479-491.

[138] Pani, B; Ong, HL; Liu, X; Rauser, K; Ambudkar, IS; Singh, BB. Lipid rafts determine clustering of STIM1 in endoplasmic reticulum-plasma membrane junctions and regulation of store-operated Ca^{2+} entry (SOCE). *J. Biol. Chem.* 2008 283, 17333-17340.

[139] Liao, Y; Erxleben, C; Yildirim, E; Abramowitz, J; Armstrong, DL; Birnbaumer, L. Orai proteins interact with TRPC channels and confer responsiveness to store depletion. *Proc. Natl. Acad. Sci. U.S.A.* 2007 104, 4682-4687.

[140] Lorin-Nebel, C; Xing, J; Yan, X; Strange, K. CRAC channel activity in C. elegans is mediated by Orai1 and STIM1 homologues and is essential for ovulation and fertility. *J. Physiol.* 2007 580, 67-85.

[141] Liao, Y; Erxleben, C; Abramowitz, J; Flockerzi, V; Zhu, MX; Armstrong, DL; Birnbaumer, L. Functional interactions among Orai1, TRPCs, and STIM1 suggest a STIM-regulated heteromeric Orai/TRPC model for SOCE/Icrac channels. *Proc. Natl. Acad. Sci. U.S.A.* 2008 105, 2895-2900.

[142] Liao, Y; Plummer, NW; George, MD; Abramowitz, J; Zhu, MX; Birnbaumer, L. A role for Orai in TRPC-mediated Ca^{2+} entry suggests that a TRPC:Orai complex may mediate store and receptor operated Ca2+ entry. *Proc. Natl. Acad. Sc.i U.S.A.* 2009 106, 3202-3206.

[143] Sampieri, A; Angelica, Z; Carlos, S; Alfonso, S; Vaca, L. STIM1 converts TRPC1 from a receptor-operated to a store-operated channel: moving TRPC1 in and out of lipid rafts. *Cell Calcium.* 2008 44, 479-491.

[144] Jardin, I; Salido, GM; Rosado, JA. Role of lipid rafts in the interaction between hTRPC1, Orai1 and STIM1. *Channels* (Austin). 2008 2, 401-403.

[145] Pani, B; Singh, BB. Lipid rafts/caveolae as microdomains of calcium signaling. *Cell Calcium.* 2009 45, 625-633.

In: Calcium Signaling
Editor: Masayoshi Yamaguchi

ISBN: 978-1-61324-313-8
©2012 Nova Science Publishers, Inc.

Chapter 3

ROLE OF CALCIUM IN STRESS MEDIATED SIGNALING IN PLANTS

*Cheshta Sharma, Amita Pandey and Girdhar K. Pandey**

Department of Plant Molecular Biology,
University of Delhi South Campus, New Delhi, India

ABSTRACT

As the research is growing in the field of signal transduction, it is widely accepted that most of the signaling pathways exist as non-linear complex network. Calcium, a ubiquitous second messenger, plays regulatory role by positioning itself at the core of various signaling pathway in eukaryotes and act as "Hub" in multiple signaling pathways. In plants, this versatile cation is involved in environmental stresses, including both biotic and abiotic, and developmental processes. Recent studies suggest that in plants, perception of multiple stimuli, leads to generation of a 'calcium signature', which enables it to define the nature and level of response. However, each cell has to be well versed with the mechanism to decode this signature. The decoding process starts with sensing of fluctuation in intracellular calcium levels by Ca^{2+} sensors and responders, which result in alteration in their structural and functional properties. This leads to modulation of the expression and function of downstream targets. These targets ultimately modulate various cellular, biochemical and physiological processes. In this review, we are making an attempt to discuss the regulation of calcium homeostasis and various signaling pathways in plants and their role in encoding and generating a specific and overlapping response. Ultimately, this concerted interplay of calcium signaling enables the plant to adapt under different environmental stresses and developmental processes.

Keywords: Calcium, signal transduction, biotic and abiotic stress, calmodulin, CDPK, CBL, CIPK, calcium transporters and pumps

---

* Address for correspondence: gkpandey@south.du.ac.in

INTRODUCTION

Various environmental factors like drought, salinity, and low temperature have adverse effect on plant growth and development, which ultimately leads to an immense decrease in crop productivity. However, plants adapt themselves by sensing and transmitting these environmental and hormonal signals to complex cellular machinery by a process referred to as signal transduction. This process in general includes four main steps: a) signal perception by receptors, b) generation of second messengers, c) a cascade of events to control a set of stress regulated genes, d) stress tolerance, adaptation, death or any other phenotypic response which is the final outcome of gene expression and physiological changes inside the cell. Signal transduction requires proper spatial and temporal coordination of all signaling molecules, and several of these signaling molecules may also act as second messenger, such as calcium, inositol 1,4,5-phosphate (IP$_3$), and cyclic guanosine mono-phosphate (cGMP), nitric oxide (NO), reactive oxygen species (ROS) such as hydrogen peroxide (H$_2$O$_2$) etc.

Calcium: The Second Messenger. After extensive research in the field of signal transduction in eukaryotes, calcium, amongst all second messengers, has emerged as a versatile signaling ion. It acts as 'Hub' by positioning itself at the core of various signaling pathways. Calcium sensors or responders recognize any fluctuation in calcium level by binding calcium to domains such as EF-hand [Pandey *et al.* 2002; Luan *et al.* 2002; Sanders *et al.* 2002; Batistic and Kudla 2004; Pandey 2008; Dodd *et al.* 2010]. These calcium binding proteins then relay the information encoded in calcium signature to downstream targets like protein kinases, transcription factors, transporters etc [Reddy *et al.* 2004; Finkler *et al.* 2007; Luan 2008; Kudla and Batistic 2010; Kudla *et al.* 2010; Das and Pandey 2010; Pandey *et al.* 2010], thereby regulating the gene expression resulting in altered metabolism followed by phenotypic response of stress tolerance. The stress response could also be a growth inhibition or programmed cell death (apoptosis), which will depend upon up- or down-regulation of several different kinds of genes.

Calcium Uptake by Plants. Calcium being an essential plant nutrient is taken up by roots from soil solution and distributed to shoot through xylem system. It may traverse through the root either by cytoplasm of cells linked through plasmodesmata (the symplast) or via spaces between cells (apoplast) [Philip *et al.* 2003]. However, the movement of calcium through these pathways should be balanced to allow the root cells to signal using cytosolic calcium concentration, thereby controlling the rate of calcium delivery to the xylem and preventing the toxic cations from getting accumulated in shoot. Since a high [Ca^{2+}]$_{cyt}$ is cytotoxic, a submicromolar[Ca^{2+}]$_{cyt}$is maintained in non-stimulated cells by Ca^{2+}-ATPases and H$^+$/Ca^{2+}-antiporters [Sze *et al.* 2000; Hirschi *et al.* 2001; Dodd *et al.*, 2010]. Calcium deficiency is rare in nature occurring in soils with low base saturation and/or high levels of acidic deposition [McLaughlin and Wimmer 1999] when sufficient calcium is momentarily unavailable to the developing tissues. Deficiency symptoms are observed, like blossom end rot, tipburn in leafy vegetables, as calcium from older tissues is not mobilized and well distributed via phloem to developing tissues. The ability to extract Ca^{2+} from complex soil environments varies in different plants. In particular, Ca^{2+}requirement of dicots, when compared with monocots, is high for optimal growth [Tuteja and Mahajan 2007].

Cellular Functions of Calcium in Plants. Calcium plays a crucial role in determining cell wall structure. During cell wall formation, the carboxyl groups, released by de-esterification of methyl esters by pectin methyl esterase, binds calcium and forms calcium pectate. Low calcium concentration makes the cell wall more pliable and prone to rupture while its high concentrations rigidify the wall making it less plastic.

Variation in calcium concentration also has a profound effect on cell growth and development. Increase in calcium concentration lead to inhibition in shoot or coleoptile growth and its reduction results in cell and tissue elongation [Bennet-Clark 1956; Tagawa and Bonner 1957; Hepler 2005]. In pollen tubes, the existence of a tip-focused gradient of cytosolic free calcium $[Ca^{2+}]_{cyt}$ has been shown to be essential for growth [Hepler 2005; Obermeyer and Weisenseel 1991; Rathore *et al.* 1991; Miller *et al.* 1992; Malhó *et al.* 1994; Pierson *et al.* 1994] which is maintained by an asymmetric activity of Ca^{2+} channels [Malhó *et al.* 1995]. It was also shown earlier that increased $[Ca^{2+}]_{cyt}$ on one side of the pollen tube, reorients the growth axis towards that side only. Similarly decrease in $[Ca^{2+}]_{cyt}$ promoted bending of the pollen tube towards the opposite side. Thus, calcium is an essential ion required for growth, elongation and guidance of pollen tube [Miller *et al.* 1992; Malhó *et al.* 1996; Messerli *et al.* 2000; Sanders *et al.* 2002; Tuteja *et al.* 2007]. Pathogens such as fungi, bacteria invade and infect plant tissue by producing enzymes such as polyglacturonases and pectolytic enzymes that dissolve the middle lamella. It has been found that activity of these enzymes, from *Erwinia carotovorais*, surprisingly get reduced on increasing calcium content of the plant tissue [Easterwood 2002].

Abiotic stress, such as heat, results in generation of calcium waves, which have been detected using the calcium dependent luminescent protein, aequorin, in tobacco [Gong *et al.* 1998b]; suggesting the role of calcium in heat stress signaling. It has also been shown that calcium and calmodulin inhibitors limit the survival and the electrolyte leakage from membranes after heat treatment in maize [Gong *et al.* 1997a]. Heat stress induces the calcium uptake and some calmodulin related genes; however, the role of calcium is still not established for Hsp production in plants [Gong *et al.* 1997b]. This suggests that some other process is also required to induce Hsp for plant's survival after stress and calcium may be involved somewhere in between the perception of heat stress and induction of Hsp [Larkindale and Knight 2002].

On challenging *Arabidopsis* with hydrogen peroxide (H_2O_2), a ROS progenitor, a biphasic Ca^{2+} signature with first peak located in the cotyledons and the second peak in the root is triggered [Rentel and Knight 2004]. It is well established fact that $[Ca^{2+}]_{cyt}$ is involved in stomatal regulation via complex signal transduction pathway by acting as an elicitor that activate kinases related to protein phosphorylation [McAinsh *et al.* 2000; Allen *et al.* 2001; Hetherington 2001; Schroeder *et al.* 2001; Yang *et al.* 2003] or acting as a water channel blocker [Yang *et al.* 2006]. ABA, cold, H_2O_2 are some of the key players in regulating the opening and closing of the stomatal aperture in a calcium dependent manner. In *Arabidopsis*, these molecules usually induce the $[Ca^{2+}]_{cyt}$ oscillations in guard cells, finally resulting in steady state 'Ca^{2+} programmed' stomatal closure [Allen *et al.* 2000; 2001]. However, in *det3* mutants, the $[Ca^{2+}]_{cyt}$ oscillations and steady state stomatal closure occur only in response to ABA and cold whereas $H_2O_2$was not able to generate prolonged increase in $[Ca^{2+}]_{cyt}$ resulting in failure of stomatal closure [Allen *et al.* 2000]. The parameters for causing 'steady state' stomatal closure were extensively studied and defined as frequency, amplitude and duration of the oscillation [Allen *et al.* 2001]. It has been postulated that any oscillation above or below the defined level might lead to either short term or no closure of stomata [McAinsh and Pittman 2008].

Leguminous plants respond to rhizobia and arbuscular mycorrhizal fungi by changing the intracellular Ca^{2+}, suggesting role of calcium in these plants [Kosuta *et al.* 2008]. The Nodulation factor (Nod) secreted by nitrogen fixing bacteria in leguminous plants cause cellular internalization of rhizobia and root nodule development by increasing the perinuclear $[Ca^{2+}]_{cyt}$ [Oldroyd and Harrison 2009]. The nodulation defective mutants of *M. trucatula,* such as *dmi1* and *dmi2,* which are also defective in mycorrhizal infection [Catoira *et al.* 2000] fail to exhibit nod factor induced oscillations [Wais *et al.* 2000; Walker *et al,* 2000; Kanamouri *et al.* 2006; Miwa *et al.* 2006] further abolishing the mycorrhizal induced Ca^{2+} oscillations thereby implying a functional role of Ca^{2+}in symbiotic association and signaling [McAinsh and Pittman 2008].

Blue and red light (BL, RL) causes brief $[Ca^{2+}]_{cyt}$ transients without apparent oscillations. Calcium is also involved in the blue light (BL) signaling mediated by phototropin blue light photoreceptors (PHOTs) which causes calcium influx from plasma membrane voltage gated calcium channels [Baum *et al.* 1999; Harada *et al.*2003; Harada and Shimazaki 2009] and these BL-Ca^{2+} signals are also required for BL inhibition of seedling growth [Folta *et al.* 2003]. SUB1 (Short under blue light 1), an EF-hand containing protein is involved in light induced calcium signals relating to the decoding of cryptochrome and phytochrome mediated calcium alterations [Guo *et al.* 2001]. In Bryophyte, *Physcomitrella patens*, the BL induced $[Ca^{2+}]_{cyt}$ increases were mediated by both PHOTs and CRYs (Cryptochrome blue light photoreceptors) [Tucker *et al.* 2005]. However, the involvement of CRYs with BL-induced calcium signals in higher plants is still not clear [Dodd *et al.* 2010]

High calcium content in fruit flesh help in increasing its shelf life by delaying maturation, stabilizing cell membranes and strengthening cell wall [Penter *et al.* 2001], however, the fruit can become more prone to fungal attack [Bloksma and Jansonius 2000]. Calcium has emerged as an important mineral nutrient, which acts as second messenger in a variety of signaling pathways. Ca^{2+} signaling pathway can also regulate K^+ channel for low-K^+response in *Arabidopsis* [Xu *et al.* 2006; Li *et al.* 2006]. Calcium is also reported to be an essential component of the sucrose-signaling pathway that leads to the induction of fructan synthesis [Noel *et al.* 2006]. Calcium signaling is also involved in the regulation of cell cycle progression in response to abiotic stress [Tuteja *et al.* 2007].

In a few interesting reports, kinesin binding proteins, also regulated by calmodulin, have been shown to regulate the developmental processes linked to regulation of cell cycle and cell morphology [Oppenheimer *et al.* 1997; Vos *et al.* 2000; DeFalco *et al.* 2010].

SIGNAL-INDUCED CALCIUM CHANGES

The cytosolic calcium level that varies on its induction by different environmental and hormonal signals [Poovaiah and Reddy 1993] has to be regulated for the survival of the cell. Calcium stored in numerous vesicular compartments in plant cell can be released into the cytoplasm, when needed. The movement of calcium in and out of the cells and organelles is regulated by specific channels/transporters/pumps in plant cell [Sanders *et al.* 2002; Berridge *et al.* 2003; Mahajan *et al.* 2006a; Tuteja 2007; Miedema *et al.* 2008; Dodd *et al.* 2010]. The source of calcium resulting in its increase in the cytoplasm after stress has not been well studied. In general, Ca^{2+} can primarily be released from extracellular source (apoplastic space) as the addition of EGTA or BAPTA blocks calcium effects [Sanders *et al.* 2002; Tuteja 2009; Dodd *et al.* 2010; Kudla *et al.* 2010, Mehlmer *et al.* 2010]. Activation of phospholipase C leads to hydrolysis of PIP_2 to IP_3 and finally releasing Ca^{2+} from intracellular Ca^{2+} stores [Xiong *et al.* 2002; Pandey *et al.* 2000; Zhu 2002; Tuteja and Sopory 2008]. An additional level of regulation in calcium signaling is provided with the help of calcium binding proteins (calcium sensors). These sensor proteins recognize and decode the information present on calcium signature and impart it downstream to initiate phosphorylation cascade finally activating various stress inducible genes, which directly or indirectly provide the signal response which could be growth inhibition, cell death etc depending upon type of gene up-/or down-regulated [Kudla *et al.* 2010; Dodd *et al.* 2010; Luan *et al.* 2008; Mahajan *et al.* 2006a; Tuteja 2007]. Thus, the calcium-induced response could be a coordinated action of many genes. Abiotic and Biotic stimuli have also been reported to induce rapid and transient increase/ decrease in $[Ca^{2+}]_{cyt}$ [Poovaiah and Reddy 1993] e.g. light and gravity in maize coleoptiles and gibberellic acid in barley aleurone protoplast have been shown to be responsible for increasing $[Ca^{2+}]_{cyt}$ while ABA decreases the cytosolic Ca^{2+} of barley aleurone protoplast [Poovaiah and Reddy 1993].

As discussed earlier, perception of stimuli by plants result in generation of complex calcium signature that encode information related to nature and strength of stimuli. Upon transduction of the signal mediated by calcium, the cytosolic Ca^{2+} level is required to be maintained at extremely low concentration to prevent the damage caused by its higher concentration in the cytosol. Therefore, plants have evolved the calcium transport systems, which continuously regulate the homeostasis of calcium ion in the cell [Bothwell and Ng 2005]. In plants, this regulation is accomplished either through Ca^{2+} pumps powered by ATP hydrolysis or through Ca^{2+} antiporters powered by a proton motive force [Sanders *et al.* 2002]. These include Ca^{2+} influx channels, mediating calcium release in cytosol contributing to stimulus induced increase in $[Ca^{2+}]_{cyt}$ and Ca^{2+} efflux transporters which rapidly remove calcium from the cytosol, restoring $[Ca^{2+}]_{cyt}$ to resting values [McAinsh and Pittman 2008; Sanders *et al.* 2002; Hetherington and Brownlee 2004; Shigaki and Hirschi 2006; Boursiac and Harper 2007; Demidchik and Maathuis 2007; Pottosin and Schönknecht 2007].

COMPONENTS OF CALCIUM SIGNALING

Calcium Influx Channels

In plants, the calcium influx occurs by the channels that are permeable to cations including Ca^{2+} in contrast to the Ca^{2+} selective channels in animals [Schroeder and Hagiwara 1990; Thuleau *et al.* 1994; Pei *et al.* 2000; Very and Davies 2000; Zou *et al.* 2002]. These calcium influx channels are present in [McAinsh and Pittman 2008] plasma membrane and endomembranes that mediate Ca^{2+} release in the cytosol, contributing to stimulus-induced increase in $[Ca^{2+}]_{cyt}$.

Plasma Membrane Ca^{2+} Influx Channels

In plants, three main groups of Ca^{2+} permeable channel have been characterized in [McAinsh and Pittman 2008] plasma membrane. These are mechanosensitive Ca^{2+} channels (MCCs), the depolarization-activated Ca^{2+}channels (DACCs) and the hyperpolarization-activatedCa^{2+} channels (HACCs), which are all examples of nonselective cation channels (NSCCs) [Sanders *et al.* 2002; Hetherington and Brownlee 2004; Demidchik and Maathuis 2007; McAinsh and Pittman 2008].

Mechanosensitive Calcium Channels (MCCs)

MCCs are good candidates for shaping mechanically induced increase in $[Ca^{2+}]_{cyt}$ and found to be key components of signaling pathway. These have been recorded in various cell types including guard cells of *Vicia faba* [Cosgrove and Hedrich 1991], *Arabidopsis* mesophyll cells [Qi *et al.* 2004], pollen grains and tube tip protoplasts of *Lilium longiflorum* [Dutta and Robinson 2004]. It has also been shown through pharmacological studies that mechanical stimulation of chloroplast movements in ferns [Sato *et al.* 2001] and pollen tube germination and elongation [Dutta and Robinson, 2004] are both dependent on Ca^{2+} influx through MCC.

Depolarization Activated Calcium Channels (DACCs)

Biotic [Ehrhardt *et al.* 1992; Lhuissier *et al.* 2001] and abiotic [Okasaki *et al.* 2002] stimuli gives rise to elevated $[Ca^{2+}]_{cyt}$ which causes plasma membrane depolarization suggesting the involvement of non-selective, cation permeable DACCs [White 1998].However, the direct relation of DACC activity with *in situ* elevations in $[Ca^{2+}]_{cyt}$ still remains unexplored. The characterization of these channels is difficult due to small currents and activity rundown [Thion *et al.* 1998]. Their activity appears to be downregulated by microtubule cytoskeleton. Molecular identity of DACCs has not yet been verified, however, AtTPC1, a voltage dependent Ca^{2+} channel and ScCCH/MIDI gene in yeast have been proposed as the candidate genes [Furuichi *et al.* 2001; White *et al.* 2002].

Hyperpolarization Activated Calcium Channels (HACCs)

HACCs were first described in tomato cells [Gelli and Blumwald 1997] and were found to be activated in response to fungal elicitors [Gelli *et al.* 1997]. These have also been described in root hair apices, root epidermal, endodermal and cortical cells from the elongation zone; moreover, they were also reported in guard cell [Sanders *et al.* 2002]. Their activation in guard cells is stimulated by ABA [Pei *et al.* 2000] and reactive oxygen species [Pei *et al.* 2000; Murata *et al.* 2001], both of which lead to elevation of Ca^{2+} in guard cells [McAinsh *et al.* 1990; 1996]. These play an important role in sustained localization of $[Ca^{2+}]_{cyt}$ and Ca^{2+} influx [Kiegle *et al.* 2000; Very and Davies 2000; Hetherington and Brownlee 2004]. The activation of HACCs by ABA requires the presence of NADPH in the cytosol [Murata *et al.* 2001]. By genetic analysis of *Arabidopsis* double mutants, *atrbohD atrbohF* (NADPH oxidase double mutant), the ABA-induced $[Ca^{2+}]_{cyt}$ transients were not found to be abolished, however, they were significantly reduced, supporting a role of HACC and ROS in *Arabidopsis* guard cell ABA signaling [Kwak *et al.* 2003]. This suggested the significance of ROS for regulation of HACC by ABA and hence contributing to guard cell calcium signatures. Moreover, in guard cell HACC activity was also regulated by phosphorylation events [Köhler and Blatt 2002; Mori *et al.* 2006].In *Vicia*, ABA concentration of extracellular divalent cations and inhibitors of type 1 and 2A phosphatase increases the opening probability of HACC [Hamilton *et al.* 2000; Hamilton *et al.* 2001; Köhler *et al.* 2002].

Endomembrane Ca^{2+} Permeable Channels

Linking the activity of Ca^{2+} channels and elevated $[Ca^{2+}]_{cyt}$ in endomembrane is difficult as they cannot be monitored in intact cells [Hetherington and Brownlee 2004]. There are about four Ca^{2+} permeable channels identified in vacuolar membrane which may contribute to stimulus induced increases in $[Ca^{2+}]_{cyt}$. These include the inositol 1, 4, 5-trisphosphate ($InsP_3$) and cyclic ADP-ribose (cADPR)-gated channels, and the vacuolar voltage-gated Ca^{2+} channel (VVCa), and slow activating vacuolar (SV) channels [Sanders *et al.* 2002; Hetherington and Brownlee 2004; Pottosin and Schönknecht 2007]. Inositol 1,4,5-trisphosphate ($InsP_3$) and cyclic ADP-ribose (cADPR)-gated channels $InsP_3$ and cADPR, both have been shown to cause the release of Ca^{2+} from plant vacuoles leading to increase in $[Ca^{2+}]_{cyt}$ [Schumaker and Sze 1987; Gilroy *et al.* 1990; Wu *et al.* 1997; Leckie *et al.* 1998; MacRobbie 2000]. These have also been implicated in multiple Ca^{2+} mediated stress signaling pathways in plants [Knight *et al.* 1996; Wu *et al.* 1997; Drobak and Watkins 2000; Lecourieux *et al.* 2006]. The clear distribution of $InsP_3$ and cADPR-gated Ca^{2+} permeable channels on vacuolar membrane [Allen *et al.* 1995; Lommel and Felle 1997] or ER [Muir and Sanders 1997; Navazio *et al.* 2001], is still not known. The importance of ligand-gated endomembrane Ca^{2+} permeable channels to shape the calcium signature is shown by release of Ca^{2+} from the vacuole and ER in response to inositol hexakisphosphate ($InsP_6$) [Lemtiri-Chlieh *et al.* 2003] and nicotinic acid adenine dinucleotide phosphate (NAADP) [Navazio *et al.* 2000].

Vacuolar Voltage-Gated Ca^{2+} Channel (VVCa) and Slow Activating Vacuolar Channel (SV)

Vacuolar voltage-gated Ca^{2+} and SV channels are both voltage-dependent channels with a high affinity for Ca^{2+} [Pottosin and Schönknecht 2007]. The VVCa channel is gated by membrane hyperpolarization and activated by Ca^{2+} from the vacuolar side [Johannes *et al.* 1992], whereas the SV channel is gated by membrane depolarization and activated by $[Ca^{2+}]_{cyt}$ [Hedrich and Neher 1987], and has been suggested that these were the same channel but in opposite orientation [Pottosin and Schönknecht 2007]. The SV channel is the most abundant channel in the vacuolar membrane [Hedrich and Neher 1987] and is widely distributed among terrestrial plants [Hedrich *et al.* 1988] and is regulated by multiple factors including calmodulin (CaM) [Bethke and Jones 1994], phosphorylation [Allen and Sanders 1995] and 14-3-3 proteins [Wijngaard *et al.* 2001]. These features along with Ca^{2+} permeability and $[Ca^{2+}]_{cyt}$ dependent activation of the channels lead to the contribution of SV channel to increase $[Ca^{2+}]_{cyt}$ through the process of Ca^{2+} induced Ca^{2+} release (CICR) [Ward and Schroeder 1994; Bewell *et al.* 1999].

In *Arabidopsis*, SV channel is encoded by *AtTPC1* (Two-pore channel1) gene [Peiter *et al.* 2005] and its homologues in tobacco, rice and wheat suggest that they may encode Ca^{2+} permeable channels in the plasma membrane, entailing the species dependent targeting of

TPC1 channel proteins to different membranes [Demidchik and Maathuis 2007; Pottosin and Schönknecht 2007]. The demonstration of AtTPC1 encoding the *Arabidopsis* SV channel has permitted the first functional analysis of endomembrane Ca^{2+} permeable channel being involved in the generation of plant Ca^{2+} signatures at the molecular level [Peiter *et al.* 2005; Ranf *et al.* 2008].

In plants, plasma membrane has only two classes of ion channel homologous genes, which encode Ca^{2+} permeable channels, both of which are NSCCs (nonselective cation channel). These are cyclic nucleotide-gated channel (CNGC) [Mäser *et al.* 2001] and the glutamate receptor-like (GLR) genes [Lacombe *et al.* 2001]. The CNGCs were first identified in barley [Schuurink *et al.* 1998] and about 20 CNGCs are present in *Arabidopsis* genome [Mäser *et al.* 2001]. Ability of cAMP to stimulate Ca^{2+} influx in cultured carrot cells provides the indication of role of CNGCs in plant Ca^{2+} signaling [Kurosaki *et al.* 1994] and also that cAMP and cGMP both induce increase in $[Ca^{2+}]_{cyt}$ in aequorin-expressing tobacco protoplasts [Volotovski *et al.* 1998]. cAMP furthermore stimulates HACC activity in guard cells [Lemtiri-Chlieh and Berkowitz 2004]. However, the mechanisms by which the putative CNGC Ca^{2+} permeable channels are activated by their respective signaling pathways remains unknown. Similarly, 20 GLRs have been identified in *Arabidopsis* genome, with similarities to the animal ionotropic glutamate receptors [Lacombe *et al.* 2001; Davenport 2002]. Glutamate activated cation currents, present in the plasma membrane of *Arabidopsis* root protoplasts [Demidchik *et al.* 2004], may contribute to glutamate induced depolarization and increase in $[Ca^{2+}]_{cyt}$ by allowing calcium influx into cells and this increase in $[Ca^{2+}]_{cyt}$ is inhibited by antagonists of animal ionotropic glutamate receptors, thereby suggesting its functional role in generation of $[Ca^{2+}]_{cyt}$ [Dubos *et al.* 2003; Meyerhoff *et al.* 2005]. Disruptions of GLR3 1 in rice has been shown to disrupt calcium dependent processes, such as cell division, differentiation and programmed cell death in roots [Li *et al.* 2006a], while the overexpression of a radish GLR in *Arabidopsis* results in increased glutamate-induced Ca^{2+} influx and enhanced resistance to pathogen attack [Kang *et al.* 2006]. Another class of proteins called annexin, functions as Ca^{2+}-permeable channels when purified from maize and incorporated into planar lipid bilayers, can elicit elevations of $[Ca^{2+}]$ when added to protoplasts [Laohavisi *et al.* 2009]. The purified protein also has peroxidase activity, and it is suggested that annexin might form plasma membrane Ca^{2+} permeable channels during stress responses [Dodd *et al.* 2010].

Ca^{2+} Efflux Transporters

For Ca^{2+} to act as an effective signal, calcium transport systems have to maintain low $[Ca^{2+}]_{cyt}$ against a significant electrochemical potential difference. Plants have special pumps and exchangers to accomplish this task of rescuing themselves from dangers of increased $[Ca^{2+}]_{cyt}$ which might result in precipitation of cellular inorganic phosphate required for energy generation and electrochemical homeostasis across the cellular membranes [Sanders *et al.* 2002; Dodd *et al.* 2010].The presence of multiple Ca^{2+}-efflux transporters at various membrane locations help in refilling calcium stores, providing tolerance to increased $[Ca^{2+}]_{cyt}$ and also providing the potential to generate Ca^{2+} signatures in a sophisticated manner. Plants utilize two main pathways for $[Ca^{2+}]_{cyt}$ removal: High affinity Ca^{2+}-ATPases and low affinity Ca^{2+}-exchangers. Because of significant sequence conservation of these transporters throughout, their genetic basis has been known and studied in details over the past years, widening our knowledge about their kinetics, membrane localizations, expression pattern and physiological functions, most notably in *Arabidopsis* [Sze *et al.* 2000; Pittman and Hirschi 2003; Shigaki and Hirschi 2006; Boursiac and Harper 2007; McAinsh and Pittman 2009].

High-Affinity Ca^{2+} Efflux: Ca^{2+}-ATPases

A subgroup of the P-type ATPases (the P_2-ATPases) encompasses the Ca^{2+} pumps, [Baxter *et al.* 2003] which may be Ca^{2+}-ATPases belonging to high-affinity pumps or H^+/Ca^{2+} exchangers belonging to lower affinity pumps [Geisler *et al.* 2000; Pittman and Hirschi 2003; Sze *et al.* 2000; Hetherington and Brownlee 2004]. In plants, these are further subdivided into the type IIA or ER-type Ca^{2+}-ATPases (ECA) and type IIB or the autoinhibited Ca^{2+}-ATPases (ACA) [East 2000] based on their homologies with animal counterparts, namely, sarcoplasmic/ endoplasmic reticulum Ca^{2+}-ATPase (SERCA) and the CaM regulated plasma membrane Ca^{2+}-ATPase (PMCA) respectively. There exists two major points of differences between ECA and ACAs. ACA are targeted to endomembrane [Hong *et al.* 1999; Geisler *et al.* 2000] and can be directly activated by binding Ca^{2+}/CaM [Harper *et al.* 1998] and also contains an N-terminal autoinhibitory domain. However, its activity can be inhibited through calcium-dependent protein kinase (CDPK) binding and phosphorylation [Hwang *et al.* 2000]. Secondly, there are differences in membrane-located residues that are thought to be involved in Ca^{2+} binding. The insertional mutants in two of the four *Arabidopsis* ECA genes having Mn^{2+} as well as Ca^{2+} related phenotypes might be because of the difference in the membrane located residues. Thus, ECAs may be involved in delivery of cations to intracellular compartments where they are required for enzymatic activity etc. There is no direct regulation of type IIA ATPases in plants; thus, they may play a constitutive role in maintaining cytosolic Ca^{2+} level. There are 4 ECA and 10 ACA in *Arabidopsis*, of which type IIA Ca^{2+}-ATPases are either ER localized [Liang *et al.* 1997; Liang and Sze 1998] or else localized in plasma, vacuole or Golgi membrane [Ferrol and Bennett 1996].

Low-Affinity, High-Capacity Ca^{2+} Efflux: Ca^{2+} Exchangers

The Ca^{2+} exchangers usually have lower Ca^{2+} affinity than the Ca^{2+} pumps but transport of Ca^{2+} from cytosol occurs rapidly at high capacity. These Ca^{2+} exchangers are energized by counter exchange of another ion. The Na^+/Ca^{2+} exchangers, in animals, couple Na^+ influx to Ca^{2+} efflux, however, in plants, there occurs a structurally related family of cation exchanger (CAX) genes encoding H^+/Ca^{2+} exchangers [Cai and Lytton 2004; Shigaki and Hirschi 2006]. CAX transporters are members of the Major facilitator superfamily and predicted to have 11 transmembrane domains (TMDs). The first plant H^+/Ca^{2+} antiporter to be cloned and functionally expressed was *CAX1* (calcium exchanger 1) [Hirschi *et al.* 1996; Hirschi 2001; Sanders *et al.* 2002; Hirschi *et al.* 2006; Dodd *et al.* 2010]. *Arabidopsis* has six CAX genes (*AtCAX1-AtCAX6*) plus five related genes, designated cation-Ca^{2+} exchanger (CCX) (originally named *AtCAX7* to *AtCAX11*) that are more similar to an animal Na^+/Ca^{2+}-exchanger isoform [Cai and Lytton 2004; Shigaki *et al.* 2006]. The H^+/Ca^{2+}-antiporters present in the plasma membrane and tonoplast have been characterized biochemically [Sanders *et al.* 2002]. In *Arabidopsis*, eleven genes encoding putative H^+/Ca^{2+} antiporters have been identified, of which *AtCAX1*, *AtCAX2*, *AtCAX4* are localized at tonoplast [Hirschi 2001]. *AtCAX1* activity is regulated either by phosphorylation [Pittman *et al.* 2002a] or by

CAX interacting proteins (CXIPs), like CXIP4 or by N-terminal autoinhibitory domain like ACA pumps; however it does not bind CaM [Pittman and Hirschi 2001; Pittman *et al.* 2002a; Mei *et al.* 2007]. Under some developmental conditions, *AtCAX1* and *AtCAX3* may interact by forming a complex by altering kinetics of transport activity and thus, *AtCAX3* can be called as putative regulator of *AtCAX1* [Cheng *et al.* 2005].

Calcium Signal Propagation

Specificity in calcium signaling can be dealt with spatial localization of calcium signal. In addition to cytosol, the calcium signal can also be localized in different organelles with varying levels. The major reservoir for the calcium release center is vacuole, ER, apoplast, but in many cases organelles like chloroplasts, mitochondria can also function as calcium release sites [Weinl *et al.* 2008].

Mitochondrial Ca^{2+}

Mitochondrial calcium and its transport mechanism in animals is highly explored as $[Ca^{2+}]_{mit}$ plays an important role in modulation of $[Ca^{2+}]_{cyt}$ and apoptotic cell death regulation [Giacomello *et al.* 2007]. However, in plants the role of mitochondria in signaling is still not much explored, inspite of determining Ca^{2+} concentrations and Ca^{2+} oscillations using $[Ca^{2+}]_{mit}$ indicators [Logan and Knight 2003]. To compare the kinetics of $[Ca^{2+}]_{mit}$ and $[Ca^{2+}]_{cyt}$, aequorin were targeted to the mitochondrial matrix. It was found that average cellular "resting" $[Ca^{2+}]_{mit}$ was approximately double the $[Ca^{2+}]_{cyt}$ [Logan and Knight 2003]. Touch stimulus in *Arabidopsis* induces elevation in $[Ca^{2+}]_{mit}$ which is different from $[Ca^{2+}]_{cyt}$ signature and $[Ca^{2+}]_{cyt}$ recovers resting levels much more rapidly than $[Ca^{2+}]_{mit}$ [Logan and Knight 2003].However, the kinetics of both $[Ca^{2+}]_{mit}$ and $[Ca^{2+}]_{cyt}$ were comparable in response to cold and osmotic stress. Response of both $[Ca^{2+}]_{mit}$ and $[Ca^{2+}]_{cyt}$ were elevated to similar amplitude on exposure to H_2O_2. All these results suggests the existence of mitochondrial specific Ca^{2+} signals and different regulating mechanism for $[Ca^{2+}]_{mit}$ and $[Ca^{2+}]_{cyt}$.

Nuclear Ca^{2+}

In an interesting observation, it has been noted that calcium signaling is not only restricted to cytosol but similar changes in intracellular calcium are observed as Ca^{2+} oscillations that are localized to the nuclear region which were induced by nod-factor [Ehrhardt *et al.* 1996; Walker *et al.* 2000; Sun *et al.* 2007]. Functional analysis of *DMI1*, a gene encoding putative cation channel in *M. truncatula* was localized to nuclear periphery [Ané *et al.* 2004; Riely *et al.* 2007], it has been shown that *DMI1* is able to regulate Ca^{2+} release indirectly possibly by membrane potential modulation [Peiter *et al.* 2007]. It has been suggested that nucleoporin, a component of nuclear pore complex, may induce $[Ca^{2+}]_{nuc}$ oscillation [Kanamori *et al.* 2006; Saito *et al.* 2007] either by regulating the transport ofCa^{2+}from the cytosol into the nucleus or by generating another signal to activate calcium release into nuclear interior. In addition, the nuclear membrane also possesses various voltage-dependent and calcium-activated channels making it an obvious calcium store [Downie et al. 1998; Grygorczyk and Grygorczyk 1998; Bunney et al. 2000].

Chloroplast Ca^{2+}

Ca^{2+} oscillations can be independently generated in chloroplasts, [Ca^{2+}]$_{chl}$ because of its essential requirement, however excessive Ca^{2+} may be toxic as in case of cytosol. A significant increase in tobacco stromal Ca^{2+} oscillations following darkness [Johnson et al. 1995; Sai and Johnson 2002] suggests its possible role in generating circadian rhythm to ensure that photosynthetic processes being switched off at night [Sai and Johnson 2002], however, the source for generation of these [Ca^{2+}]$_{chl}$ is still not clear. During light conditions, Ca^{2+}traverses across inner envelope membrane in an ATP-dependent manner. In the inner envelope membrane of pea chloroplast, a membrane potential gradient-dependent, Ca^{2+}uniport mechanism has been detected in addition to the putative CaM regulated Ca^{2+}-ATPase, AtACA1 [Roh et al. 1998; Huang et al. 1993]. However, no ATPase activity has been detected at membrane [Roh et al. 1998], thus by proteomic analysis, AtACA1was found to be localized to the ER [Dunkley et al. 2006]. AtHMA1, a Cu^{+}-ATPase (heavy metal P-type ATPase), present at the envelope membrane is found to be essential for plants survival under high light conditions [Seigneurin-Berny et al. 2006]. H^{+}/Ca^{2+}exchanger loads the calcium into thylakoid membrane [Ettinger et al.1999], which further passes calcium to stroma by an unknown pathway. Overexpression of PPF1(a post-floral-specific gene expressed in short-day-grown green pea), a putative Ca^{2+} channel from pea in Arabidopsis guard cells reduces the [Ca^{2+}]$_{cyt}$ elevations as more calcium is retained in the chloroplast [Wang et al. 2003; Li et al. 2004].

SIGNAL PERCEPTION AND RELAY

The immediate decoding components which are responsible for detecting the changes in the concentration of calcium becomes essential player, designated as calcium sensor proteins, and are present in both cytosol and nucleus, which could exist as both free and attached to the membrane. The calcium sensor protein changes their conformation in calcium dependent manner and has specific Ca^{2+} binding characteristics, subcellular localizations and distribution. Calcium sensor proteins are classified as sensor-responders and sensor-relays on the basis of calcium perception and relay [Sanders et al. 2002]. Sensor-responder protein, like CDPKs, that combines within a single protein, has a response activity (like kinase activity) and a calcium-binding domain, which is responsible for Ca^{2+} induced conformational changes and function through bimolecular interactions [Sanders et al. 2002; Das and Pandey 2010; Pandey et al. 2010]. On the other hand, sensor relay protein, like calmodulin, also contains multiple calcium binding domains and undergo Ca^{2+} induced conformational changes but lack other effector domains. These undergo conformational changes in response to a signal and relay it further to an interacting partner, thereby regulating its activity [Luan et al. 2002; Das and Pandey 2009; Dodd *et al.* 2010; Pandey et al. 2010]. Binding of Ca^{2+} with high affinity to the calcium sensors occurs by helix-loop-helix motif termed the "EF hand" (named after the E and F regions of parvalbumin and discovered by Kretsinger and Nockolds in 1973) [Tuteja and Mahajan 2007].

Classification of CDPK-SnRK Superfamily

In *Arabidopsis*, 67 out of 1019 protein kinases have been implicated in Ca^{2+} signaling [Wang *et al.* 2003]. These Ca^{2+} regulated kinases belong to the CDPK-SnRK superfamily and have been reviewed extensively in last few years [Hrabak *et al.* 2003; Harmon *et al.* 2000; Assmann and Wang 2001; Evans *et al.* 2001; Harmon *et al.* 2001; Hetherington 2001; Cheng *et al.* 2002; Fasano *et al.* 2002; Xiong *et al.* 2002; Zhang and Lu 2003; Das and Pandey 2010; Pandey et al. 2010]. Characteristic feature of all of these protein kinases is the presence of catalytic domain of eukaryotic Ser-Thr kinases and this was used to classify these proteins into CDPK-SnRK superfamily [Hrabak *et al.* 2003; Assmann and Wang 2001]. The length and the sequence of the N-terminal domain are highly variable between subgroups and protein kinases. N-terminal domain has the myristoylation and palmitoylation sites in CDPKs contributing to membrane localization of these kinases [Hrabak *et al.* 2003]. Presence of an autoregulatory domain near C-terminal region of the kinase domain regulates the kinase activity and it also helps in mediating interactions with other proteins.

In an attempt to understand the functional role played by these proteins, we are trying to describe these briefly such as Calmodulin (CaM), CDPK, CRK, PPCK and PEPRK, CaMK and CCaMK, SnRKs in plants especially under abiotic stress conditions:

Calmodulin (CaM)
Calmodulin is one of the highly conserved proteins throughout eukaryotes. Babu et al. first solved CaM protein structure in 1985, suggesting saturation of all 4 EF hands by Ca^{2+}. This 17-kDa-calcium sensor protein is present in apoplast, cytosol, ER and nucleus of plant cell. The two globular domains in protein contain two EF-hands connected by a flexible, helical linker [Luan *et al.* 2002] as shown in Figure 1. CaM regulates gene expression by binding to specific transcription factors [Bouché *et al.* 2002] and also plays an important role in physiological processes like mechanical, heat stress, pathogens, osmotic stress etc. CaM works by binding and regulating the activity of a large number of target proteins called "CaM-binding proteins" in calcium dependent or independent manner [Reddy *et al.* 1996; Pooviah and Reddy 1993]. The first protein serine/ threonine phosphatase to interact with CaM in plants was PP7 and was reported to be involved in thermotolerance in *Arabidopsis* [Liu *et al.* 2007]. *Arabidopsis* has six CaM genes, encoding only three isoforms [McCormack and Braam 2003]. The CaM concentration in cytosol has been estimated to be about 5-40mM [Zielinski 1998; Rudd and Frankin-Tong 2001; McCormack and Braam 2003].

Calcium Dependent Protein Kinases (CDPKs)
Plant genome such as *Arabidopsis,* rice and wheat encodes a large number of calcium dependent protein kinase genes with 34 members in *Arabidopsis* and 31 genes in rice, 20 genes in wheat [Cheng *et al.* 2002; Das and Pandey 2010]. Presence of kinase catalytic

domain and an autoinhibitory domain, followed by calmodulin-like, calcium-binding regulatory domain fused to the C-terminal, as shown in Figure 1, makes it sometimes to be called as calmodulin-like domain protein kinase. These CDPKs are activated by calcium and help the plant in decoding calcium signals [Harper *et al.* 1993; Urao *et al.* 1994; Abo-El-Saad and Wu 1995; Hong *et al.* 1996; Hrabak *et al.* 1996; Saijo *et al.* 1997; Lee *et al.* 1998; Chico *et al.* 2002; Das and Pandey 2010]. Molecular masses of CDPK proteins range from 54.3 to 72.2-kDa and this variation in molecular mass is because of difference in variable domains. All *Arabidopsis* CDPKs (except *AtCPK25*) have four EF hands as predicted by SMART [Schultz *et al.* 2000]. Majority of CDPKs have myristoylation site at their N-terminal and a Cys residue that can serve as palmitoylation site. Most of these CDPKs are differentially regulated under different developmental stages and stress conditions.

In *Arabidopsis, AtCPK10, AtCPK11* play important role in osmotic stress as they are induced by drought and high salt [Urao *et al.* 1994]. *AtCPK32*, on the other hand regulates ABA mediated seed germination [Mori *et.al* 2006] whereas *AtCPK3, AtCPK6* control ABA mediated stomatal closure [Zhu *et al.* 2007]. Cold, drought and salt stress up- or down-regulate *OsCPK13* and *OsCPK17* in rice whereas *OsCPK6* and *OsCPK25* are up regulated by drought and heat stress respectively [Wan *et al.* 2007; Ray *et al.* 2007]. CDPKs from wheat, *TaCPK3, TaCPK6, TaCPK9, TaCPK12, TaCPK14, TaCPK15* were shown to be expressed in roots, stems, leaves, young spikes, however, *TaCPK1, TaCPK2, TaCPK5, TaCPK16* were shown to be expressed in all tissues in wheat [Li *et al.* 2008]. In case of *Vicia fava,* one of the CDPK, *VfCPK1* was shown to accumulate on encountering drought and ABA treatment [Liu *et al.* 2006] whereas in tobacco, *NtCDPK1* was found to be expressing in rapidly proliferating tissues like roots, stems, flowers and up-regulated by salt, wounding, calcium [Zhang *et al.* 2005; Das and Pandey 2010].

CDPK- Related Protein Kinases (CRKs)

Arabidopsis genome has eight CRK genes and there are five CRKs in rice genome [Takayuki *et al.* 2005]. These are located on branch close to subgroup containing 3 CDPKs, suggesting high similarity between their catalytic domains. However, the C-terminal domains of CRKs are poorly conserved and contain degenerate or truncated EF-hands, as shown in Figure 1, that are unable to bind Ca^{2+}. Maize CRK was equally active in presence of calcium or EGTA suggesting the absence of functional autoinhibitory domain [Furumoto *et al.* 1996]. The enzymatic activity of CRK is usually independent of calcium and calmodulin but in rice, CRK binds calmodulin in a calcium dependent manner [Zhang *et al.* 2002]. The CRKs are predicted to be myristoylated and palmitoylated, just like CDPKs, but are not acylated [Das and Pandey 2010]. CRKs have been studied in various plant species including maize and *Arabidopsis*. The CRKs possess calcium and CaM-independent kinase activity. *AtCRK3* activity is, however, inhibited by Ca^{2+} but is induced by ABA and thus seems to be phosphorylating the substrates in ABA signaling. Interestingly, the autophosphorylation and substrate phosphorylation of *AtCRK3* like *ZmCRK* is Ca^{2+} and CaM independent [Furumoto *et al.* 1996; Du *et al.* 2004].

Phosphoenolpyruvate Carboxylase Kinases and PPCK Related Kinases (PPCK and PEPRK)

PPCKs have been known to phosphorylate PEP carboxylase in calcium independent manner and have molecular masses ranging from 30-39-kDa [Vidal and Chollet 1997]. These have extensively been studied in crassulacean acid metabolism (CAM) and C4 plants (such as *K. fedtschenkoi* and maize, respectively) [Vidal and Chollet 1997; Nimmo 2000; Nimmo *et al.* 2001]. These enzymes are the smallest ATP-dependent protein kinases and are constitutively active. These kinases usually do not have N- or C-terminal extensions (Figure 1). There are two PPCK genes in *Arabidopsis* genome whose function has been confirmed *in vitro* [Hartwell *et al.* 1999a; Fontaine *et al.* 2002] and contain one intron close to the end of the catalytic domain. *Arabidopsis* also contain two more protein kinases with catalytic domains related to those of the PPCKs, out of which one was known to be expressed. These proteins, however, are constitutively active as they contain both N- and C-terminal extensions without any regulatory domains and bear no similarity to any other kinases in this superfamily. Molecular masses of these enzymes are 57.5 and 52.1 kDa and are termed as PEPRKs until their functions were clearly understood [Hrabak *et al.* 2003]. The *PPCK1* gene in *Arabidopsis* is highly expressed in rosette leaves and is also detected in roots, flowers, however, *PPCK2*, on the other hand is highly expressed in roots, flowers. The up regulation of *PPCK1* was detected by increased cytosolic pH; however, *PPCK2* remains unaffected [Nimmo 2003].

CaM Kinases and Ca²⁺/CaM-dependent Protein Kinases (CaMKs and CCaMKs)

CaMKs are highly expressed in rapidly growing cells and tissues of root and flower [Zhang and Lu 2003] and no representative of CaMKs have been identified in *Arabidopsis*. Activation of these proteins is brought about by calmodulin binding to autoinhibitory domain, located close to C-terminal kinase domain (Figure 1) [Hrabak *et al.* 2003]. Only one potential CaMK has been identified from apple, in case of plants [Watillon *et al.* 1995]. CCaMKs, like CaMK, are absent from *Arabidopsis* [Harper *et al.* 2004], but have been cloned from tobacco and lily [Patil *et al.* 1995; Liu *et al.* 1998]. CCaMKs require calcium in addition to CaM for their activation and carry out autophosphorylation in response to calcium binding to the EF-hands in C-terminal visinin like domain. This binding further increases the enzyme affinity for calcium/calmodulin complex thus enhancing substrate phosphorylation [Takezawa *et al.* 1996; Pandey and Sopory 1998; Sathyanarayanan *et al.*2000; Das and Pandey 2010]. Sequence analysis of CCaMKs revealed the presence of N-terminal catalytic domain, centrally located CaM binding domain overlapping autoregulatory domain [Ramachandiran *et al.* 1997] and a C-terminal visinin like domain having three EF-hands. The CCaMKs play a critical role in Nod factor signaling and gene regulation in leguminous plants [Gleason *et al.* 2006]. Involvement of CCaMKs in stress response has been reported in pea only in which on encountering the abiotic stress like cold and salinity, the protein level of *PsCCaMK* was found to increase and localize in nucleus [Pandey *et al.* 2002].

SNF1 Related Kinases (SnRKs)

Snf1 refers to a kinase mutation in yeast resulting in "Sucrose non-fermentation". 38 SnRKs in *Arabidopsis* are divided into three distinct families [Hrabak *et al.* 2003] based on domain structure and sequence similarity. SnRK1 are the largest proteins with molecular masses between 56.7 and 58.7 kDa and are closely related to SNF1 from yeast and to AMP-activated protein kinases (AMPK) from animals. AMP rather than allosterically regulating SnRK1, modulates it indirectly via regulation of the phosphorylation/dephosphorylation of a residue in the enzyme's activation loop (conserved subdomain VIII) [Sugden *et al.* 1999a].

The substrate specificity of SnRK1 overlaps with that of CDPKs [Bachmann *et al.* 1996; Huber and Huber 1996; Sugden *et al.* 1999b; Huang and Huber 2001].

Other two groups, SnRK2 and SnRK3 are unique to plants, with SnRK2 being shorter in size than SnRK1s. The presence of a patch of acidic amino acid in their C-terminal domain makes them different from other two SnRKs [Halford *et al.* 2000](Figure 1). Out of 10 SnRK2s in *Arabidopsis* genome, *SnRK2.6* has been studied in detail [Mustilli *et al.*2002; Yoshida *et al.* 2002]. It has been shown to be stimulated and regulated by ABA and is expressed in guard cell and vascular system. Extensively studied other SnRK2s includes *PKABA1* of wheat and [Anderberg and Walker-Simmons 1992] *REK* of rice [Hotta *et al.* 1998]. The transcription of *PKABA1* is induced by ABA. On the other hand, calcium stimulates the autophosphorylation activity of *REK* when expressed in bacterial cells [Hotta *et al.* 1998].

SnRK3 group in *Arabidopsis* is represented by protein kinases, also known as CBL-Interacting Protein Kinases and is comprised of 26 family member [Kudla *et al.* 1999; Shi *et al.* 1999; Kim *et al.* 2000; Albrecht *et al.* 2001; Luan et al. 2002; Kudla and Batistic 2008; Pandey 2008]. As the name suggests, these kinases interact with calcium binding proteins, (also known as calcineurin B-like protein (CBL) or (salt overly sensitive3 (SOS3), SOS2-like calcium binding protein (SCaBP) related to animal neuronal calcium sensor visinin and regulatory B-subunit of the protein phosphatase calcineurin, CNB [Liu and Zhu 1998; Kudla *et al.* 1999; Kim *et al.* 2000; Albrecht *et al.* 2001; Guo *et al.* 2001; Luan *et al.* 2002; Pandey 2008]. These kinases are involved in ABA, osmotic, salinity, K^+, NO^{3-} and sugar signaling in plants [Kim *et al.* 2000; Guo *et al.* 2001; Guo *et al.* 2002; Gong *et al.* 2002a; Gong *et al.* 2002c; Kim *et al.* 2002; Li *et al.* 2006; Cheong *et al.* 2007; Pandey *et al.* 2007; Pandey 2008; Luan *et al.* 2009]. Presence of Thr residue in activation loop (conserved subdomain VIII) of SnRKs distinguishes them from CDPK/CRK/PPCK/PEPRK, which has Asp and Glu at this position. Replacement of Thr residue in SOS2 and other SnRK3s by Asp residue in activation loop result in enhanced enzyme activity without any requirement for calcium [Guo *et al.* 2001; Gong *et al.* 2002b]. Calcium binding proteins interact with 21-24-residue region called NAF or FISL domain in SnRK3s and thus stimulate their activity, upon expression in bacteria [Shi *et al.* 1999; Albrecht *et al.* 2001; Guo *et al.* 2001]. This domain also serves as an autoinhibitory domain for kinase activity [Guo *et al.* 2001; Gong *et al.* 2002b]. Absence of putative N-terminal myristoylation site in SnRK3s makes them unlikely to be membrane localized. However, *SnRK3.11* (also called as *SOS2*) may be membrane localized through its interaction with *SOS3* [Halfter *et al.* 2000]. This pair was identified as *CIPK24* and *CBL4* respectively and was found to be interacting *in-vitro* and *in-vivo* during salinity stress through the regulation of plasma membrane localized Na^+/H^+-exchanger, *SOS1* [Li *et al.* 2003]. Another member of CBL family, *CBL10* was also shown to be involved in salinity stress and postulated to regulate vacuolar *NHX1* by pairing and targeting *CIPK24/SOS2* to vacuolar membrane in response to the salinity stress [Apse *et al.* 1999; Kim *et al.* 2007].

Mutant analysis has rendered a lot of information about individual members of CIPK gene family. *CIPK1* is targeted to the plasma membrane after interacting with *CBL1* and *CBL9* [D'Angelo *et al.* 2006]. Mutation of *CIPK1* makes plants hypersensitive to osmotic stress. *cbl1/cipk1* double mutant analysis showed that *CIPK1* is a crucial cross-talk junction that integrates different stress response pathways like ABA-dependent and ABA-independent abiotic stress signaling pathways [D'Angelo *et al.* 2006].

Reverse genetic approach has shown that *CIPK3*, a serine/threonine protein kinase is involved in regulation of ABA response during seed germination and ABA and other stress conditions like cold, wounding, drought and salinity. In *cipk3* mutant, expression of drought responsive genes was not altered showing that it is involved in specific pathways in response to abiotic stress and ABA. *CBL9-CIPK3* acts as negative regulator of ABA response during seed germination in *Arabidopsis* [Pandey *et al.* 2008]. In another study by Guo *et al.*, transgenic with RNA silenced *CBL1-CIPK15* showed hypersensitivity to ABA. Double mutants of *CBL1* and *CIPK15* did not produce cumulative phenotype and so were proved to act in the same pathway. Another protein known to interact with *CBL1-CIPK15* is *ABI1/2*. Recently, *AtERF7* has been known to interact with *CBL1-CIPK15*. The latter phosphorylates *AtERF7* in ABA hypersensitive responses during seed germination and seedling growth [Guo *et al.* 2003]. *AtERF7* together with *AtSIN3* and *HDA19* act as a negative regulator of ABA responsive stress signaling pathway.

By genetic analysis several members of SnRK3/CIPKs were functionally implicated in salt stress and root development such as *CIPK6* [Tripathi *et al.* 2009]; *CIPK23,* in low-K^+ uptake and nutrition by interacting with *CBL1* and *CBL9* and a voltage gated K^+ channel, *AKT1*, as the downstream target [Li *et al.* 2006; Cheong *et al.*2007]; *CIPK8* and *CIPK23* in nitrate uptake sensing and signaling [Hu *et al.* 2009]; *CIPK9*, another regulator of low potassium response in *Arabidopsis*, was shown to be highly induced in roots and shoots on potassium deprivation. Plants become hypersensitive to low K^+ media in *cipk9* mutants, however, there was no effect on the level of K^+ uptake [Pandey *et al.* 2007].

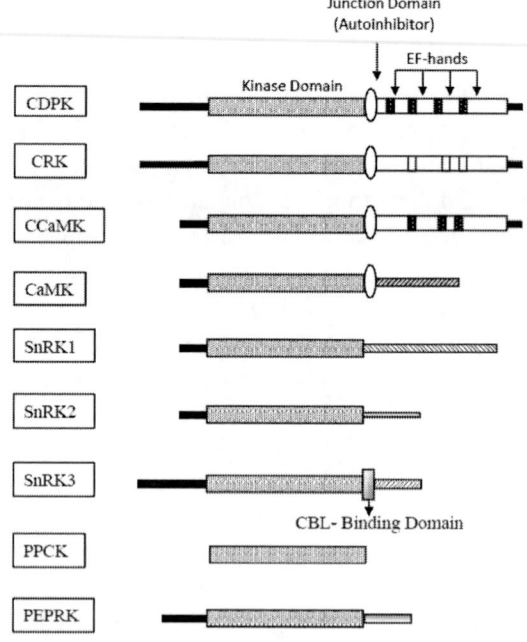

Figure 1. Schematic diagram showing the structures of protein kinase members of CDPK-SnRK superfamily. Kinase domain is present in all the members with variability only at the N-terminal domain. The kinase activity is regulated by the presence of autoregulatory domain near the C-terminal region, which also mediates the interaction with other proteins. EF hands helps in high affinity binding of Ca^{2+} to the calcium sensors.

CALCIUM AND STRESS SIGNALING

Stress may be defined as an organism's total response to environmental demands or pressures; however, it is very difficult to measure stress in a biological conditions. Moreover, stress experienced by one organism, such as in plant, may acts as an optimum condition for the other. Thus, definition of stress may be modified as an adverse condition, which restricts the normal functioning of a biological system [Jones and Jones 1989]. Most of the environmental stresses are hypothesized to be sensed by the plasma membrane with the combination of several components that includes the proteins or lipids present in the plasma membrane or any extracellular material, ligand, together with the plasma membrane proteins interact with a receptor molecule and act as an elicitor to generate a particular cellular response. Both biotic and abiotic signals acts as elicitors for plant cell response.

On perceiving a stress signal at the receptor, the plant transduces this signal further downstream to generate second messengers such as calcium, ROS, cyclic nucleotides (cGMP, cAMP), inositol phosphates etc. These second messengers further cause perturbations in intracellular calcium levels and lead to produce specific "calcium signature" which is further sensed by calcium sensors and on interacting with their interacting partners, they initiate a phosphorylation cascade and target some of the stress inducible genes [Luan *et al.* 2002; Kudla and Batistic 2004; Mahajan and Tuteja 2005; Pandey 2008]. The product of these stress genes is classified into two groups [Seki *et al.* 2004]. The first group includes the enzymes responsible for production of various osmoprotectants, like LEA, antifreeze proteins, and chaperones etc, which directly protect the plant against stresses by protecting cells from dehydration. The second group consists of transcription factors, protein kinases that regulate the gene expression and thus, the signal transduction pathways. Overall the major signal transduction pathways upon abiotic stress perception are divided into three types: (i) osmotic/ oxidative stress signaling that uses MAPK modules; (ii) Ca^{2+}-dependent signaling leading to the activation of several genes which provide tolerance such as LEA genes; (iii) Ca^{2+} dependent salt overly sensitive (SOS) signaling resulting in ion homeostasis [Xiong *et al.* 2002; Kaur *et al.* 2005; Mahajan *et al.* 2008; Pandey 2008; Pandey *et al.* 2010].

CALCIUM AND COLD STRESS SIGNALING

There is a requirement of optimal temperature for proper growth and development of the plants. However, any deviation from this optimum temperature may result in adverse effect on the growth and ultimately on productivity. Temperature below the optimal may result in cold stress that includes chilling and/ or frosting. There is a direct inhibition of metabolic reactions, which prevents the expression of complete genetic potential of a plant. Phenotypic changes that occur in response to chilling or frosting stress in plants include wilting, chlorosis, reduced leaf expansion, which may eventually result in cell death. Exposure of the rice plants to the chilling temperature during anthesis (floral opening) results in flower sterility [Jiang *et al.* 2002]. However, the process of cold acclimation helps the plants to acquire freezing tolerance by their prior exposure to low non-freezing temperatures (e.g., temperate plants). The process results in an overall protection and stabilization of cellular integrity, enhancement of oxidative mechanisms, increased levels of cryoprotectants like proline [Chinnusamy *et al.* 2007], polyamines, sugar and alcohols and glycine betaine [Mahajan *et al.* 2005]. Plants like barley, oat, rye etc. have a vernalization requirement that prevent the plant to enter the reproductive phase prematurely. Crops like rice, maize, soybean, tomato are cold sensitive and are incapable of cold acclimation, but still their temperature threshold for chilling damage is lowered by their prior exposures to the low temperature conditions. Low temperatures reduce the fluidity or dynamics of the cellular membranes [Steponkus 1984], changes the nucleic acid conformation and/ or metabolic concentrations [Chinnusamy *et al.* 2007]. Plant cell tends to activate protective machinery to combat the low temperature. One of the most important outcomes of low temperature exposure as depicted in Figure 2 is the induction of COR (COLD RESPONSIVE) genes due to plasma membrane rigidification resulting in cold acclimation in alfalfa and in *Brassica napus* [Orvar *et al.* 2000]. Membrane rigidification results in increased $[Ca^{2+}]_{cyt}$ by activation of mechano-sensitive or ligand activated Ca^{2+} channels. Calcium signatures impact cold signaling may also be induced by secondary signals, like, ABA, ROS which is exemplified by *Arabidopsis fro1* (frosbite1) mutant which accumulate high levels of ROS and impairs COR genes expression making them hypersensitive to chilling and freezing. The regulatory element structure of these genes has several copies of the DEHYDRATION RESPONSIVE ELEMENT (DRE)/ C-repeat (CRT) *cis*-element which is bound by transcription factors (DREBs or CBFs) of AP2/EREBP family (APETALA2/ETHYLENE RESPONSE ELEMENT BINDING PROTEIN) for activating the downstream genes in their promoters [Shinozaki and Shinozaki 1994; Stockinger *et al.* 1997; Liu *et al.* 1998]. The transient expression of the CBF/DREB1 genes in response to low temperature initiates a transcription cascade leading to RD/COR/KIN/LTI genes expression finally enabling the plant tolerant to freezing temperature. In a genetic screen *HOS1*, a constitutively expressing RING finger protein was found to be a negative regulator of COR genes as the *hos1* mutation in *Arabidopsis* results in sustained and super induction of *CBF2*, *CBF3* genes and their target regulatory genes, thereby modulating the expression levels of CBFs [Zhu *et al.* 1998]. In *Arabidopsis,* the MYC recognition elements in CBF3 promoters are bound by a MYC-type basic helix-loop-helix transcription factor, *ICE1* (INDUCER OF CBF EXPRESSION1) [Chinnusamy *et al.* 2003]. *ICE1*, which is localized and constitutively expressed in nucleus, induces CBFs only under cold stress suggesting the importance of stress induced posttranslational modification for *ICE1* [Chinnusamy *et al.* 2003]. The cold stress also results in *ICE1* phosphorylation and sumoylation of lysine residue at 393[rd]position [Chinnusamy *et al.* 2007]. The *ice1* mutation results in impairment of approximately 40% of cold regulated genes which includes genes involved in calcium signaling, lipid signaling etc [Lee *et al.* 2005], which eventually render plant cold sensitive.

Upstream transcription factor of *Arabidopsis*, *MYB15*, expressed even in absence of cold stress, bind the MYB recognition elements in CBFs promoters, thereby negatively regulating them [Agarwal *et al.* 2006]. However, *MYB15* is negatively regulated by ICE1 as *ice1* mutation results in its increased transcript level in contrast to wild type plants under cold stress [Chinnusamy *et al.* 2003; Agarwal *et al.* 2006]. Stress responsive genes such as *RD22* are activated through a different mechanism of MYB or MYC transcription factors because of lack of DRE/CRT elements. *AtMYB2* activate *RD22* by binding to *cis*-element of the promoters [Abe *et al.* 2007]. Additional cofactors are also required for transcriptional activation and determination of gene expression levels. Soybean *SCOF-1* gene (zinc finger protein) activates COR gene expression and increase freezing tolerance in non-acclimated plants when overexpressed in *Arabidopsis* and tobacco. *SCOF-1* interacts with G-Box binding bZIP protein, *SGBF-1*, which activates ABRE-driven reporter gene expression in *Arabidopsis* protoplasts, thereby regulating the *SGBF-1* activity for inducing COR gene expression [Kim *et al.* 2001].

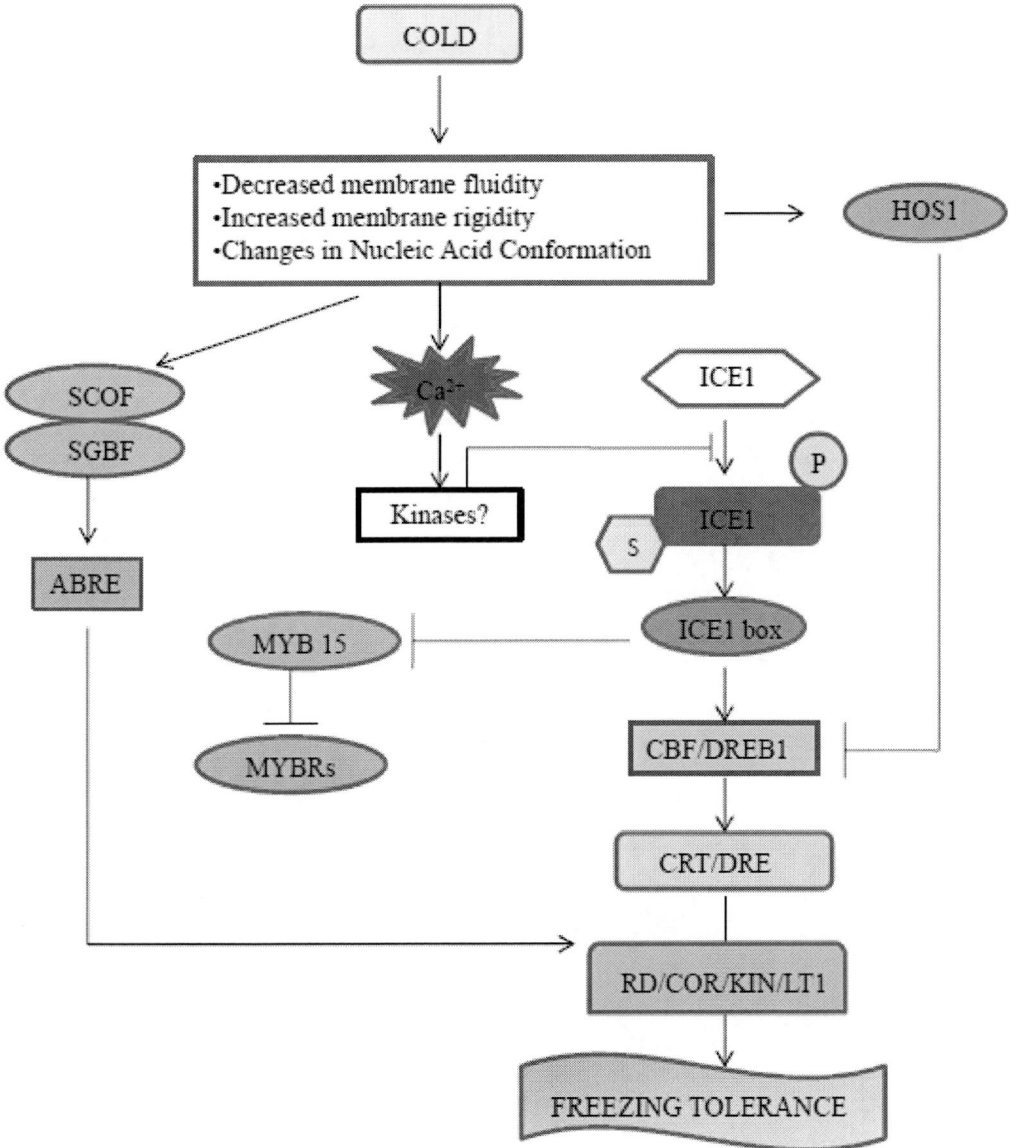

Figure 2. Diagram depicting the effect of cold stress in plants. Cold stress results in membrane rigidity and changes in nucleic acid conformation which lead to rise in cytosolic calcium level along with activation of various transcription factors and some unknown kinases. CBF/DREB1 and ABRE proteins bind to the *DRE/CRT cis*-element in the promoters of *RD/COR/KIN/LT1* genes, which is sufficient to activate the expression of these targets finally making the plant freezing tolerant. *HOS1* negatively regulates the cold signaling by targeting ICE or upstream signaling components. Here, P stands for Phosphorylation and S stands for Sumoylation.

IONIC STRESS SIGNALING AND TOLERANCE

Any increase in salt (NaCl) in land soil results in salinity stress which has devastating effect on plant's productivity by causing ionic and osmotic stress. Various factors are responsible for saline conditions in an area, like rate of evaporation, amount of precipitation, weathering of rocks etc. Under normal conditions, the plant uses its high osmotic pressure to draw water and other minerals from the soil solution. In contrast, the ability of plant to draw these nutrients diminishes as the osmotic pressure of soil exceeds that of the plant, resulting in entry of Na^+ and Cl^- inside the cell.

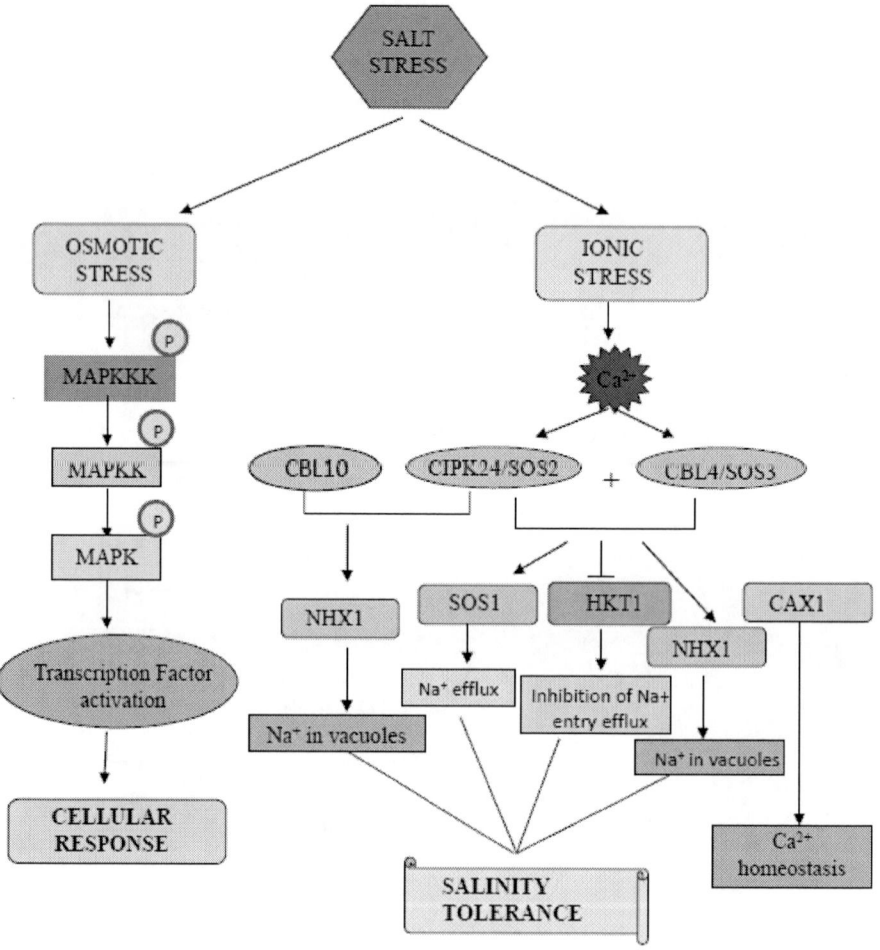

Figure 3. Representation of salt stress induced signaling pathway. Excess Na^+ in the growth medium result in ionic and osmotic stress induction which lead to increase in cytosolic calcium level. This increased calcium is perceived by calcium binding protein. *SOS3* which on interacting with its partner. *SOS3*, a kinase forms a complex to activate *SOS1*. Na^+/H^+ antiporter which helps in maintaining sodium homeostasis by throwing excess Na^+ outside the cell. The *SOS3-SOS2* complex also inhibits Na^+ entry by inhibiting the transporter, *HKT1*. Na^+/H^+ exchanger. *NHX1*, regulated by *C1PK24/CBL10*, also gets activated to compartmentalize excess sodium in the vacuoles. All these together help the plant in tolerating the pressure of high salinity. *SOS2* independently is also seen to maintain calcium homeostasis by activating H+/Ca2+ antiporter. The hyperosmotic stress, as already discussed, also result in activation MAP kinase cascade which on being activated by phosphorylation lead to activation of various transcription factors.

Presence of excessive Na^+ and Cl^- results in toxic effects on cellular membranes and metabolic activities in the cytosol [Greenway and Munns 1980; Hasegawa *et al.* 2000; Zhu 2001] which may eventually lead to effects like reduced cell expansion, assimilate production and membrane function and decreased cytosolic metabolism and finally production of reactive oxygen species (ROS). The salt-sensitive plants, glycophytes, have very limited capacity for sodium compartmentalization and osmolyte biosynthesis. Thus, plants with increased osmolyte production (proline, glycine, and betaine) are salt and drought tolerant [Halperin *et al.* 2003; Liang *et al.* 2009]. For a plant to become salt tolerant, three interconnected aspects are required, (i) damage to the cells must be prevented or alleviated (ii) homeostatic condition must be re-established (iii) growth of the plant must resume at least at a reduced rate.

SOS Regulatory Pathway

Various ion transporters are required to inspect the entry and exit of Na^+ and K^+ inside and outside of the cell and there is a need to regulate Na^+ compartmentalization in the vacuole and to selectively import K^+ over Na^+ in plant cells [Blumwald *et al.* 2000]. However, during salt stress some of these transporters gets activated or suppressed disrupting the ionic homeostasis by accumulating excess toxic Na^+ in cytoplasm and creating K^+ deficiency [Hasegawa *et al.* 2000].

Increased salinity results in release of calcium from apoplast and intracellular compartments thereby increasing $[Ca^{2+}]_{cyt}$ which ultimately leads to salt adaptation via signal transduction [Knight *et al.* 1997]. In order to identify the salt tolerance phenomenon, Liu and Zhu (1998) screened several mutants to identify genes for salt tolerance and finally identified SOS (*salt overly sensitive*) genes through positional cloning. In the salt tolerant pathway, the excess Na^+ is expelled out through plasma membrane Na^+/H^+-antiporter and the cellular homeostasis is retained. *SOS1* encodes plasma membrane localized Na^+/H^+-antiporter and thus function in exporting excess Na^+ from cytoplasm to the outside of the cell. *SOS1* seems to couple H^+ with H^+-K^+ co-transporter thereby indirectly acquiring K^+[Shi *et al.* 2000]. There is an upregulation in *SOS1* transcript level during salt stress in *Arabidopsis*. This upregulation during salt stress is partly under the control of *SOS2*, *SOS3* genes as the *SOS1* transcript level is substantially reduced in *sos2*, *sos3* mutants [Shi *et al.* 2000].

SOS2/CIPK24 gene encodes a serine/threonine protein kinase with self-interacting amino terminal catalytic domain and carboxy terminal regulatory domain [Liu *et al.* 2000; Guo *et al.* 2001]. This intramolecular interaction keeps the *SOS2* in an inactive state. A constitutively active protein can be produced either by deleting the regulatory domain or by mutating Thr residue at 168th position to Asp in the activation loop. *SOS3/CBL4* gene encodes a myristoylated calcium binding protein with low calcium affinity under normal conditions. However, during salt stress it is proposed to sense the perturbations in $[Ca^{2+}]_{cyt}$ [Liu and Zhu 1998]. During salt stress, increased level of $[Ca^{2+}]_{cyt}$ is sensed by *SOS3* which then physically interacts with *SOS2* through FISL motif present in *SOS2* regulatory domain [Halfter *et al.* 2000; Guo *et al.* 2001]. This SOS2-SOS3 complex then while regulating *SOS1* transcript level phosphorylates and activates the Na^+/H^+-antiporter [Qiu *et al.* 2002] (Figure 3), which expels out excess Na^+ ions thereby restoring ion homeostasis.

Other salt tolerance effectors regulated by SOS2-SOS3 complex includes *AtHKT1*, which is a low affinity Na^+ transporter. During salt stress, *sos3* mutation is suppressed with mutation in *Athkt1* mutation, which suggests the downregulation or inactivation of HKT1 gene expression [Rus *et al.* 2001]. Thus, *SOS2* and *SOS3* have been proved to be negative regulators of *AtHKT1* activity during salt stress [Zhu 2002] (Figure 3). The *SOS2* is shown to interact with vacuolar Na^+/H^+ antiporter and N-terminus of *CAX1* (H^+/Ca^{2+}) antiporter [Cheng *et al.* 2004]. However, activation of *CAX1* is independently activated and regulated by *SOS2* thereby maintaining Ca^{2+} homeostasis as shown in Figure 3.

Sodium homeostasis is maintained in the cell by exporting Na^+ out of the cell or by compartmentalizing it in the vacuoles as it act as an osmolyte in achieving the homeostasis. The H^+-ATPases act as primary pumps to provide necessary energy for these reactions. In *Arabidopsis*, NHX1 (Na^+/H^+ exchanger) encodes tonoplast Na^+/H^+ antiporter that compartmentalizes excess sodium in the vacuoles [Gaxiola *et al.* 1999] and its overexpression results in enhanced salt tolerance [Apse *et al.* 1999]. A member of SOS3 or calcineurin B-like protein family, *CBL10*, in *Arabidopsis*, was found to be involved in salt stress regulation [Kim *et al.* 2007; Quan *et al.* 2007]. It was usually expressed highly in green tissues but not in roots. Under high salt concentrations, *cbl10* mutants have been shown to suffer significant growth defect and encounter hypersensitive cell death in leaf tissues. To comment upon the role of *AtCBL10* in Ca^{2+} signaling pathway, the Na^+ content was recorded and was found to be lower than that of the wild type plant [Kim *et al.* 2007]. The *CBL10* protein was found to be physically interacting with *CIPK24/SOS2* and the *AtCBL10-AtCIPK24* complex was found to be associated with vacuolar compartments responsible for salt storage and detoxification in plant cells [Kim *et al.* 2007], thereby constituting an alternative pathway for salt tolerance thus regulating the sequestration and compartmentalization of Na^+ in plants.

MAPK Pathway

Under osmotic or dehydration stress conditions various compatible osmolytes such as proline, betaine glycine etc. are accumulated in the plant cell for osmotic adjustment. However, the osmolyte biosynthesis regulation in plant might be different than the well-studied yeast system. Several studies suggest that numerous protein kinases are activated on sensing osmotic imbalance. On encountering hyperosmotic stress, MAP kinases get activated and mediate signals from cell surface to the nucleus. These intracellular signaling modules phosphorylate serine/threonine residues and regulate different cellular activities of the target proteins [Zhu 2002]. The MAPK cascade encodes the genes with conserved sequences among the eukaryotes and comprises functionally interlinked protein kinases: *MAPKKK, MAPKK,* and *MAPK* [Mizoguchi *et al.* 1993; Mizoguchi *et al.* 2000; Agrawal *et al.* 2003]. Figure 3 illustrates that perception of stimuli activates *MAPK* and this activated MAPK is imported into the nucleus for phosphorylating specific downstream signaling components like transcription factors. A 46 kDa MAPK, namely, SALT-STRESS-INDCIBLE MAP KINASE (*SIMK*) in alfalfa cells and a 42 kDa MAPK, named, SALICYLIC ACID INDUCED PROTEIN KINASE (*SIPK*) in tobacco cells were activated upon hyperosmotic treatments [Munnik *et al.* 1999; Mikolajczyk *et al.* 2000]. Three MAPKs activated in response to cold, drought, wounding have been identified in *Arabidopsis* [Ichimura *et al.* 2000]. *AtMEKK1-MEK1/AtMKK2-AtMPK4* constitutes a stress signal transduction pathway in which *AtMPK4* is activated on encountering hyposmolarity, cold or wounding [Ichimura *et al.* 2000]. There occurs a crosstalk between CDPK and MAPK involving ethylene synthesis and perception, as shown by tobacco CDPK [Ludwig *et al.* 2005]. In an interesting report, Mehlmer *et al.* (2010) recently showed that *CPK3* and MAPKs act independently and parallel to each other in *Arabidopsis* on encountering salt stress, however, in microbe associated molecular patterns (MAMPs), the CDPK and MAPK are differentially regulated as stated by Boudsocq *et al.* 2010. Different form of cross talk is seen to be occurring between CDPK and MAPK in animal cells [Agell *et al.* 2002; Rozengurt 2007]

CALCIUM AND DROUGHT STRESS SIGNALING

Drought being one of the major abiotic stress results in tremendous loss of productivity and quality of the crop worldwide. The condition for drought prevails because of unavailability of water in optimum amounts which causes the disruption of membrane integrity, selectivity and also causes reduced or complete loss of cytosolic and organelle proteins. Drought stress along with salt stress eventually results in dehydration finally leading to osmotic imbalance. In response to dehydration or limited water availability, plant synthesizes ABA and triggers several cascades of signaling network. Calcium, regulated by ABA, in guard cells usually mediates the stomatal closure thereby reducing water loss and thus influencing leaf "water-use-efficiency".

Moreover, abscisic acid (ABA), also called as stress hormone, is synthesized in response to drought stress triggering a signaling cascade in guard cells causing stomatal closure, finally enabling plants in enhancing water-use-efficiency. In *Arabidopsis aba1, aba2, aba3*, ABA deficient mutants have been reported which grow normally in absence of stress like that of the wild type, however, these mutants wilt and finally die if the stress persists [Koornneef *et al.* 1998; Liotenberg *et al.* 1999]. The ABA-dependent and ABA-independent pathways are responsible for the osmotic stress generated by high salt and drought stress. However, there is not much difference in both the pathways and their components often cross-talk or converge at some point in the signaling network [Knight and Knight 2001; Xiong and Zhu 2001]. Calcium being one of the crucial second messengers serves the purpose of mediating such cross-talks. Presence of *cis*-acting elements called ABRE results in expression of stress responsive genes induced in response to ABA [Thomashow 1999; Shinozaki and Shinozaki 2000; Uno *et al.* 2000; Finkelstein *et al.* 2002]. Various transcription factors like bZIP, MYC/MYB, homeodomain Leu zipper, Zn-finger and ABI3/VP1 acts as mediators of ABA-induced gene expression as shown in Figure 4. Coupling element, *cis*-acting element required by bZIP transcription factors to binds as dimer to ABRE, may sometime be similar to DRE/CRT. Therefore, there occurs an interaction amongst ABRE bZIP and CRT/DRE binding AP2s to control ABA regulated gene expression [Shen and Ho 1995; Narusaka *et al.* 2003]. Recruitment of transcription factor *ABI3* to ABRE-containing promoters requires complex formation between *ABI3*and *ABI5* [Nakamura *et al.* 2001]. The binding efficiency of *SGBF-1*(Soyabean G-box binding factor 1) to *ABRE* is enhanced by a C_2H_2 Zn-finger protein, *SCOF-1*(Soyabean Cold inducible zinc Finger protein) however; *SCOF-1* itself cannot bind to *ABRE* [Kim *et al.* 2001]. The cooperation between *AtMYC2* and *AtMYB2* activates the ABA inducible genes such as *RD22* [Abe *et al.* 2003]. CRT binding factor (CBFs or DREB1s), a class of AP2 transcription family, plays a dual role in both ABA-dependent and ABA-independent pathway. It is reported that as the level of ABA in the plant increases, a calcium influx across plasma membrane through calcium channels is ensued and its release from intracellular stores eventually lead to increased $[Ca^{2+}]_{cyt}$ [Ramanjulu and Bartels 2002; Batistic and Kudla 2004; Bolle 2004; Zhou 2007]. This increased cytosolic calcium results in reduction in stomatal aperture thereby closing the stomata to minimize the water loss [Goldgur *et al.* 2007; Liu *et al.* 2007]. The dehydration-responsive element (DRE), C-repeat (CRT) *cis*-acting elements are responsible for ABA-independent gene expression of stress responsive genes.

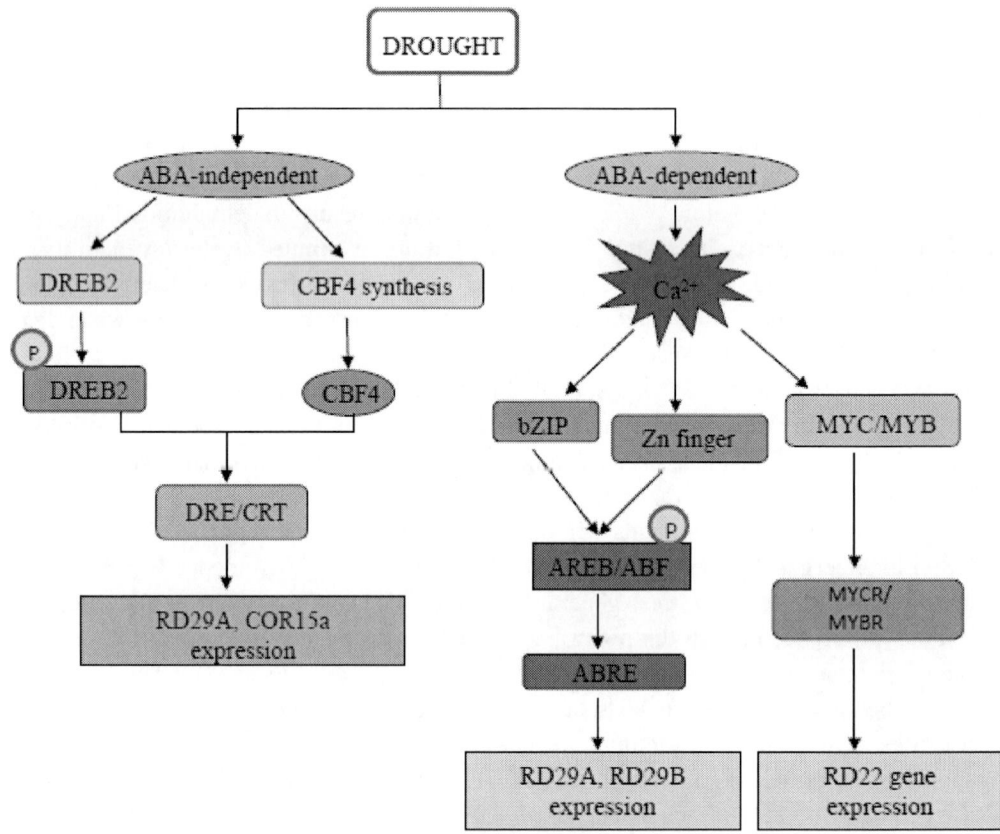

Figure 4. Hypothetical model depicting signal transduction mediated by drought stress. On perceiving dehydration conditions, plants generate two types of responses: ABA-dependent and ABA-independent. Increase in cytosolic calcium level leads to activation of bZIP, Zn-Finger, MYC/MYB transcription factors in the ABA-dependent pathway, which further binds to different coupling elements like ABRE etc. to induce the expression of stress responsive genes such as RD29A etc. However, the inactive DREB2 on activation by phosphorylation binds to DRE/CRT elements, thereby inducing the expression of stress responsive genes like COR15A in an ABA-independent pathway. Here, P stands for phosphorylation.

The CBF/ DREB2 mediates gene expression in response to drought or salinity stress. In *Arabidopsis*, *CBF4* is expressed within few minutes upon encountering drought stress [Haake *et al.* 2002]. Increased ABA level in the cell, as depicted in Figure 5, cause initial depolarization of the plasma membrane, which further results in the release of calcium from vacuoles, which changes the K^+ channel activity. Phosphatidic acid (PA) and inositol 1, 4, 5-triphosphate (IP_3) generated by PLC and PLD respectively acts as second messengers to inactivate the inward rectifying K^+ channel and stimulate calcium release from the vacuoles respectively [Blatt *et al.* 1990; Gilroy *et al.* 1990; Lee and Assmann 1991; Jacob *et al.* 1999; MacRobbie 2000; Munnik 2001] along with cyclic ADP ribose (cADPR) [MacRobbie 1998; Blatt 2000] (Figure 5). Increased calcium stimulates membrane bound NADPH oxidase, which results in NADPH dependent ROS production necessary for ABA activation of plasma membrane calcium channels [Murata *et al.* 2001; Sagi and Fluhr 2001]. Increased H_2O_2 in the guard cells cause stomatal closure, which is blocked by inhibition of NADPH oxidase with diphenylene iodonium chloride [Pie *et al.*2000]. H_2O_2 activates calcium currents in *abi1-1* mutant thereby confirming that *ABI1-1* encoded protein phosphatase is pH sensitive and is rapidly inactivated by H_2O_2 [Leube *et al.* 1998; Meinhard and Grill 2001]. ABA induces the cytoskeletal disruption resulting in inactivation of *AtRac1* finally leading to stomatal closure [Lemichez *et al.* 2001]. In another report, a farnesylated GTPase containing farnesylation motif at the -COOH terminal in *Arabidopsis* might function in stimulating NADPH activity by activating an intermediate protein kinase [Yalovsky *et al.* 2000]. *AtRac1* GTPase may acts as a putative target for ERA1-encoded farnesyl transferase suggesting occurrence of some cross-talk.

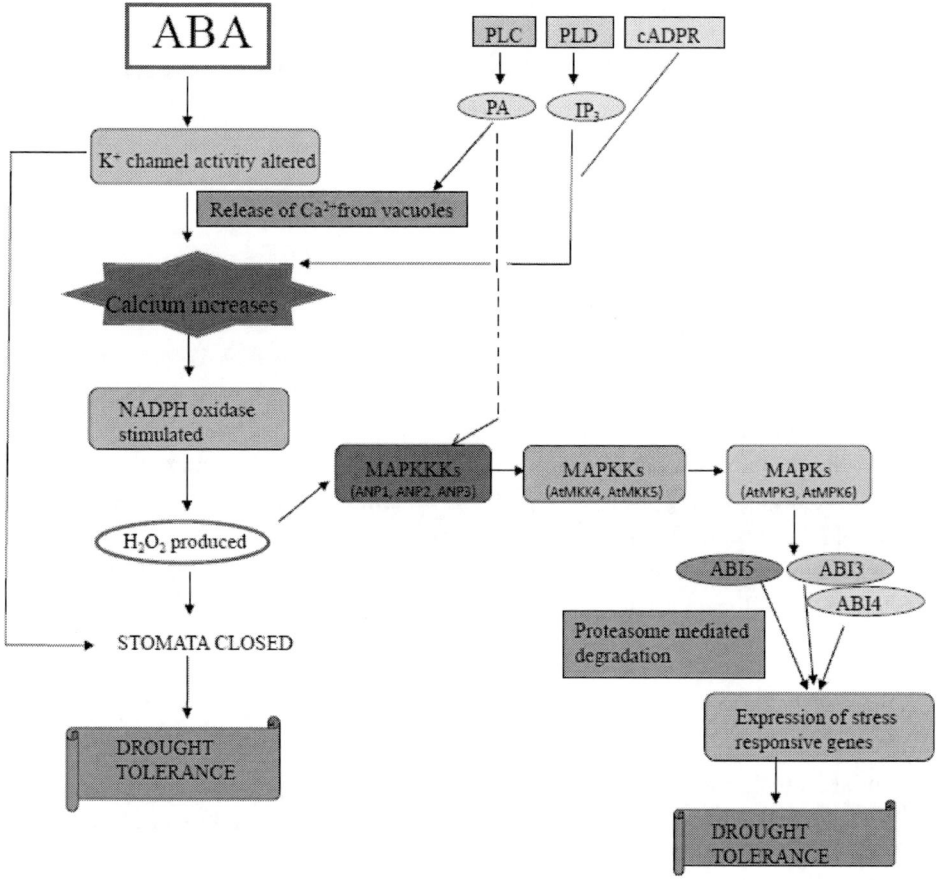

Figure 5. Model depicting the cross-talk between phospholipases, MAPKs, transcription factors, potassium channels in response to ABA, mimicking the effect of Drought Stress in Plants.

MAPK cascade comprising three consecutive protein kinases provides one more route for signal transmission to transcriptional machinery. On treating *Arabidopsis* with excess ABA or H_2O_2, *AtMPK3* and *AtMPK6* are activated by receiving signal from *ANP1*, *ANP2*, *ANP3* (MAPKKKs) through *AtMKK4* and *AtMKK5* [Kovtun *et al.* 2000] which finally lead to expression and regulation of stress responsive genes by transcription factors like *ABI3*, *ABI4*, *ABI5*.

CALCIUM AND POTASSIUM NUTRIENT SIGNALING

Role of Potassium in Plants

Macronutrient potassium is one of the most abundant inorganic cation present in plants [Leigh and Jones 1984]. K^+ plays an important regulatory role in growth, development and many other cellular functions in plants like activating the enzymes, maintaining the turgor pressure, reducing the respiration rate [Kochian and Lucas 1988; Marschner 1995; Maathius and Sanders 1996]. K^+ also helps in improving the crop quality as it increases the grain filling period and strengthening straw [Pandey *et al.* 2010]. K^+ deficiency results in growth arrest as the sugar level goes down, nitrogen balance and long distance transport are impaired [Marschner 1995]. The agricultural land affected by high saline and low K^+ conditions is severely damaged with respect to crop quality and quantity. A detail molecular understanding of K^+ nutrition signaling and homeostasis is required to improve the quality of plants so that they can grow on soil with low-K^+ and salinity [Amtmann *et al.* 2004; Pandey *et al.* 2008; Pandey *et al.* 2010].

Calcium and Low-K^+ Nutrition

Calcium is also involved in the regulation of K^+/Na^+ homeostasis and ionic selectivity under saline condition and exogenous Ca^{2+} can improve the salt tolerance [LaHaye and Epstein 1969; Lauchli 1990; Liu and Zhu 1997]. More significantly, calcium influx towards vacuole of older tissue was observed under K^+ deficient conditions thereby freeing the K^+ for transport to young growing tissues suggesting calcium homeostasis to be an important component of plant low K^+ condition adaptation [Armengaud *et al.* 2004].

CDPK are the major players in several abiotic stress signal transduction networks [Harper *et al.* 1991; 2004; Klimecka and Muszynska; 2007]. When the $[Ca^{2+}]_{cyt}$ increases in guard cells, CDPKs gets activated, eventually leading to phosphorylation of KAT1 (Potassium transporter) [Li *et al.* 1998], as shown in Figure 6. This K^+ channel is an important component of K^+ nutrition and is involved in maintaining K^+ homeostasis by Ca^{2+} in guard cells.

In a major breakthrough in potassium uptake from root cell under low-K^+ containing media, Wu and colleagues [Xu *et al.* 2006] isolated the *lks1* (low-K^+-sensitive) mutant by genetic screening for plant sensitivity to low potassium. *LKS1* was shown to encode the CBL (calcineurin B-like protein) interacting protein kinase *CIPK23*, a critical component for K^+ uptake in plants. Interestingly, *CIPK23* physically interact with *AKT1*, a voltage gated high affinity potassium channel *in vitro* and *in vivo* and the *CIPK23* and *AKT1* proteins were found to co-localize to the plasma membrane. In a parallel reverse genetic screen by Luan and colleagues [Cheong *et al.* 2007] identified loss of-function alleles of *CIPK23* as exhibiting severe growth impairment on media with low concentrations of potassium. It has been observed that the mutants of *CIPK23* were impaired in uptake of potassium into the roots from the medium. As depicted in Figure 6, the upstream components interacting with *CIPK23* were *CBL1* and *CBL9*, which target *CIPK23* to the membrane *in vivo*. There was a clear correlation identified between expression and localization in diverse tissues where *CIPK23* was found to be up-regulated by low-potassium conditions and co-expression analysis of the *CIPK23*, *CBL1* and *CBL9* genes has been detected [Cheong *et al.* 2007; Xu *et al.* 2006]. This suggests that these genes may function together in diverse tissues, including guard cells and root hairs. By genetic and biochemical approaches, the physiological substrate of CIPK23 was identified as voltage dependent high affinity potassium transporter, *AKT1* [Cheong *et al.* 2007; Xu *et al.* 2006]. A clear connection was established between calcium regulating the low-potassium uptake and nutrition where the *CBL1/CBL9-CIPK23* pair was implicated in activation of *AKT1* by phosphorylation.

After the discovery of role of *CIPK23* and *CBL1/9* in low potassium uptake, a comprehensive systematic approach was adopted by Luan and colleagues to determine the role of other CBLs and CIPKs by genome-wide expression analysis under low potassium condition. Strikingly, another CIPK family member, *CIPK9* transcript was detected to be highly induced under low-K condition, as illustrated in Figure 6. To ascertain the function of *CIPK9*, two independent alleles of CIPK9 T-DNA insertion, *cipk9-1* and *cipk9-2* were analysed (TAIR, Arabidopsis Resource Collection) and found to show hypersensitivity and impaired growth under K$^+$ deficient conditions [Pandey *et al.* 2007]. Interestingly, in the *CIPK9* mutant alleles the potassium uptake was not affected in contrast to *CIPK23* loss of function mutant [Xu *et al.* 2006; Cheong *et al.* 2007]. The authors postulated that *CIPK9* might be regulating some other aspect of K$^+$ nutrition and signaling linked either the distribution or homeostasis of K$^+$ rather than uptake from root [Pandey *et al.* 2007; Amtmann and Armengaud 2007].

A quest has begun after the initial studies of Wu and colleagues [Xu *et al.* 2006] and Luan group [Li *et al.* 2006; Cheong *et al.* 2007; Pandey *et al.* 2007] on low-potassium uptake and signaling which includes the mechanisms underlying the regulation of *AKT1* channel by the CBL-CIPK complexes. Since both AKT1 and CIPKs are multi-domain proteins, it is very crucial to determine the domains that mediate the interaction of AKT1-CIPK for understanding the structural basis for specificity of the interaction. Moreover, Luan group [Li *et al.* 2006] has found an important finding that the AKT1 activity was not abolished completely in *cbl1cbl9* double or *cipk23* mutant, raising a striking possibility that multiple CBL-CIPK complexes might be regulating the AKT1 channels activity cooperatively.

To elucidate the mechanistic detail and identification of the role of CBLs and CIPKs in regulating the AKT1 activity cooperatively, yeast two-hybrid analysis and *Xenopus* oocytes model systems were used for protein-protein interaction and electrophysiological channels recording, respectively, instead of *in planta* study [Lee *et al.* 2007] where the action of CBL-CIPK pair in the regulation of AKT1, potassium channel was studied in detail [Lee *et al.* 2007].

By yeast two-hybrid analysis, two more CIPKs, CIPK6 and CIPK16 were found to be interacting with AKT1 in addition to CIPK23. Similarly, two more CBLs, *CBL2* and *CBL3* in addition to CBL1 and CBL9, which were reported in earlier studies [Li *et al.* 2006; Xu *et al.* 2006] were found to interact with all three CIPKs such as CIPK6, CIPK16 and CIPK23 [Lee *et al.* 2007]. Therefore, a few more CBLs and CIPKs such as 4 CBLs and 3 CIPKs were found to be interacting with AKT1, as illustrated in Figure 6, forming a multivalent interacting network. The electrophysiological analysis also led to identification of the finding that there was differential activation of AKT1channel activity in *Xenopus* oocytes between these four different CBLs and three CIPKs, for example interaction of CBL1-CIPK23 with *AKT1* produces the strongest channel activity whereas interaction combination of CBL2, 3 or 9- CIPK6 or CIPK16 withAKT1 produces weaker channel activity [Lee *et al.* 2007] and hence indicated the degree of overlap and complexity in the cellular regulation by CBL-CIPK signaling network. In addition to finding a detailed regulatory aspect of AKT1 channel by differential binding of different CBL-CIPK complexes and affecting the activity, the study by Lee *et al.* (2007) also found a protein phosphatase 2C, AIP1 which dephosphorylate the channelAKT1 [Lee et al. 2007]. This finding completely dissects the signaling pathway where calcium mediated phosphorylation and de-phosphorylation regulation was found out as important criteria for high affinity potassium uptake under low-K^+ conditions by root cell (Figure 6).

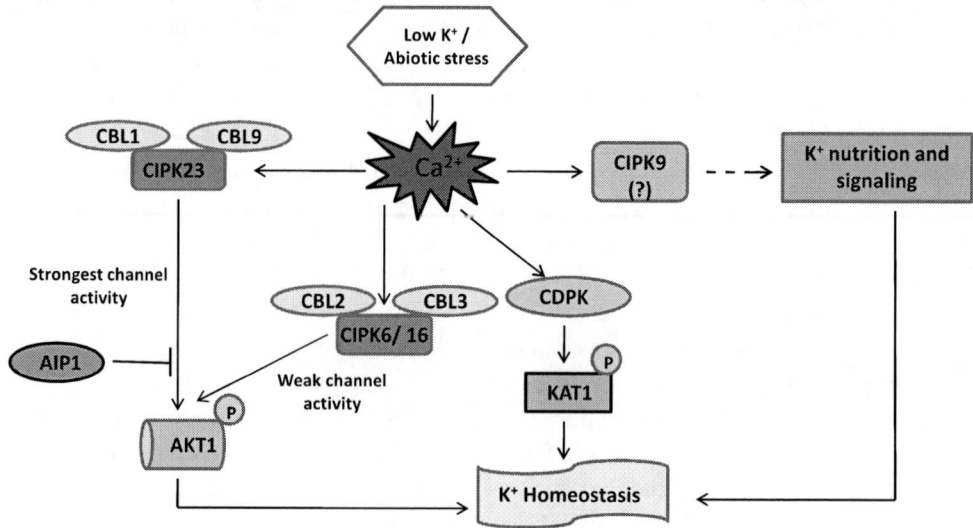

Figure 6. Hypothetical model depicting the role of calcium in potassium nutrient signaling. Here, increase in level of cytosolic calcium causes activation of CIPK23, CIPK6 and CIPK16 which further cause the activation of AKT1, a high affinity potassium channel. However AIP1, a PP2C, dephosphorylates AKT1, thereby helping in maintaining the potassium homeostasis. Another CIPK member, CIPK9 was found to be expressed at high level on encountering K^+ deficient conditions. Abiotic stress triggers the CDPK which then phosphorylates KAT1, a K^+ channel eventually helping in K^+ homeostasis maintenance. Strong arrow below CIPK23-CBL1 complex indicates the strongest channel activity of AKT1, while the weaker arrow below the CIPK6/CIPK16-CBL2/CBL3 complex indicates the weaker channel activity.

CONCLUSIONS AND PROSPECTIVE

Adverse environmental conditions are the major threat for agricultural practices all over the world. Calcium, a ubiquitous divalent cation, has been accepted as an important player in helping the plant to sense and cope up with these devastating conditions. It acts as one of the major second messenger in various signaling networks. Recently, the role-played by calcium in abiotic stress signaling and responses is appreciated by a large number of researchers throughout the world. Here, in this review, we have emphasized the role of calcium signaling in abiotic stress management, where a large of number of components are involved. A variety of calcium sensors decode the calcium signatures and relay the signal downstream by affecting kinases, phosphatases and other accessory proteins, which finally regulate either transcription factors or transporters/ channels to execute gene expression or physiological changes in response to the signal respectively. In addition, a plethora of transporters and channels are involved in regulating the ion homeostasis including the calcium concentration in the cytosol. All these components are integrated and their concerted interplay determines the generation of final response such as tolerance or adaptation against a particular stress condition.

Increasing human population demands an enormous increase in the crop productivity to feed the growing population. Because of limited agricultural land and unforeseen environmental stresses, plant biologists need to develop and generate the crop varieties, which can tolerate and grow in these adverse conditions without affecting the yield and eventually enhancing the crop productivity. Based on the previous research in calcium-mediated signaling pathway in *Arabidopsis*, a few genetic components have been identified, which could be the key player in genetically manipulating the crop to achieve higher yield and productivity. At the same time a detailed molecular understanding of these calcium mediated signaling cascades is needed to be investigated in detail, which will further enhance our understanding about how these abiotic stress signaling mechanism operate at physiological level.

ACKNOWLEDGMENT

We are thankful to Department of Biotechnology (DBT) and Department of Science and Technology (DST), Government of India for funding the research work in GKP's lab.

REFERENCES

Abe, H.; Urao, T.; Ito, T; Seki, M.; Shinozaki, K.; Yamaguchi-Shinozaki, K. Arabidopsis AtMYC2 (bHLH) and AtMYB2 (MYB) function as transcriptional activators in abscisic acid signaling. *Plant Cell* 2003, 15(1), 63-78.

Agrawal, G.K.; Iwahashi, H.; Rakwal, R. Rice MAPKs.*Biochemistry and Biophysics Research Communication* 2003, 302171–302180.

Agell, N.; Bachs, O.; Rocamora, N.; Villalonga, P. Modulation of the Ras/Raf/MEK/ERK pathway by Ca^{2+}, and calmodulin. *Cell Signal* 2002, *14* (8), 649- 654.

Albrecht, V.; Ritz, O.; Linder, S.; Harter, K.; Kudla, J. The NAF domain defines a novel protein-protein interaction module conserved in Ca^{2+}-regulated kinases. *EMBO J* 2001,*20* (5), 1051-1063.

Allen, G. J.; Chu, S. P.; Harrington, C. L.; Schumacher, K.; Hoffmann, T.; Tang, Y. Y.; Grill, E.; Schroeder, J. I. A defined range of guard cell calcium oscillation parameters encodes stomatal movements. *Nature* 2001,*411* (6841), 1053-1057.

Allen, G. J.; Muir, S. R.; Sanders, D. Release of Ca^{2+} from individual plant vacuoles by both InsP₃ and cyclic ADP-ribose. *Science* 1995,*268* (5211), 735-737.

Allen, G. J.; Sanders, D. Calcineurin, a Type 2B Protein Phosphatase, Modulates the Ca^{2+}-Permeable Slow Vacuolar Ion Channel of Stomatal Guard Cells. *Plant Cell* 1995,*7* (9), 1473-1483.

Amtmann, A.; Armengaud, P.; Volkov, V. Potassium nutrition and salt stress. In: Blatt MR (Ed) membrane transport in plants, Blackwell Publishing, Oxford, UK, pp293-339.

Anderberg, R. J.; Walker-Simmons, M. K. Isolation of a wheat cDNA clone for an abscisic acid-inducible transcript with homology to protein kinases.*Proc Natl Acad Sci U S A* 1992,*89* (21), 10183-10197.

Apse, M. P.; Aharon, G. S.; Snedden, W. A.; Blumwald, E. Salt tolerance conferred by overexpression of a vacuolar Na^+/H^+ antiport in *Arabidopsis*. *Science* 1999,*285* (5431), 1256-1268.

Asano, T.; Tanaka, N.; Yang, G.; Hayashi, N.; Komatsu, S. Genome-wide identification of the rice calcium-dependent protein kinase and its closely related kinase gene families: comprehensive analysis of the CDPKs gene family in rice. *Plant Cell Physiol* 2005,*46* (2), 356-366.

Assmann, S. M.; Wang, X. Q. From milliseconds to millions of years: guard cells and environmental responses. *Curr Opin Plant Biol* 2001,*4* (5), 421-428.

Batistic, O.; Kudla, J. Integration and channeling of calcium signaling through the CBL calcium sensor/CIPK protein kinase network. *Planta* 2004,*219* (6), 915-924.

Baum, G.; Long, J. C.; Jenkins, G. I.; Trewavas, A. J. Stimulation of the blue light phototropic receptor NPH1 causes a transient increase in cytosolic Ca^{2+}. *Proc Natl Acad Sci U S A* 1999,*96* (23), 13554-13569.

Baxter, I.; Tchieu, J.; Sussman, M. R.; Boutry, M.; Palmgren, M. G.; Gribskov, M.; Harper, J. F.; Axelsen, K. B. Genomic comparison of P-type ATPase ion pumps in *Arabidopsis* and rice. *Plant Physiol* 2003,*132* (2), 618-628.

Berridge, M. J.; Bootman, M. D.; Roderick, H. L. Calcium signalling: dynamics, homeostasis and remodelling. *Nat Rev Mol Cell Biol* 2003,*4* (7), 517-529.

Bethke, P. C.; Jones, R. L. Ca^{2+}-Calmodulin Modulates Ion Channel Activity in Storage Protein Vacuoles of Barley Aleurone Cells. *Plant Cell* 1994,*6* (2), 277-285.

Bewell, M. A.; Maathuis, F. J.; Allen, G. J.; Sanders, D. Calcium-induced calcium release mediated by a voltage-activated cation channel in vacuolar vesicles from red beet. *FEBS Lett* 1999,*458* (1), 41-44.

Blatt, M. R. Cellular signaling and volume control in stomatal movements in plants. *Annu Rev Cell Dev Biol* 2000,*16*, 221-241.

Blatt, M. R. Ca^{2+} signalling and control of guard-cell volume in stomatal movements. *Curr Opin Plant Biol* 2000,*3* (3), 196-204.

Blumwald, E.; Aharon, G. S.; Apse, M. P. Sodium transport in plant cells. *Biochim Biophys Acta* 2000, *1465* (1-2), 140-51.Bothwell, J. H.; Ng, C. K. The evolution of Ca^{2+} signaling in photosynthetic eukaryotes. *New Phytol* 2005, *166* (1), 21-38.

Bouche, N.; Scharlat, A.; Snedden, W.; Bouchez, D.; Fromm, H. A novel family of calmodulin-binding transcription activators in multicellular organisms. *J Biol Chem* 2002,*277* (24), 21851-21861.

Boudsocq, M.; Willmann, M. R.; McCormack, M.; Lee, H.; Shan, L.; He, P.; Bush, J.; Cheng, S. H.; Sheen, J. Differential innate immune signaling via Ca^{2+} sensor protein kinases. *Nature* 2010,*464* (7287), 418-422.

Boursiac, Y.; Harper, J. F. The origin and function of calmodulin regulated Ca^{2+} pumps in plants. *J Bioenerg Biomembr* 2007,*39* (5-6), 409-414.

Broadley, M. R.; Bowen, H. C.; Cotterill, H. L.; Hammond, J. P.; Meacham, M. C.; Mead, A.; White, P. J. Variation in the shoot calcium content of angiosperms. *J Exp Bot* 2003,*54* (386), 1431-1446.

Bunney, T. D.; Shaw, P. J.; Watkins, P. A.; Taylor, J. P.; Beven, A. F.; Wells, B.; Calder, G. M.; Drobak, B. K. ATP-dependent regulation of nuclear Ca^{2+} levels in plant cells. *FEBS Lett* 2000,*476* (3), 145-149.

Cai, X.; Lytton, J. The cation/Ca^{2+} exchanger superfamily: phylogenetic analysis and structural implications. *Mol Biol Evol* 2004,*21* (9), 1692-1703.

Carter, A. J.; Grauert, M.; Pschorn, U.; Bechtel, W. D.; Bartmann-Lindholm, C.; Qu, Y.; Scheuer, T.; Catterall, W. A.; Weiser, T. Potent blockade of sodium channels and protection of brain tissue from ischemia by BIII 890 CL. *Proc Natl Acad Sci U S A* 2000,*97* (9), 4944-4949.

Catoira, R.; Galera, C.; de Billy, F.; Penmetsa, R. V.; Journet, E. P.; Maillet, F.; Rosenberg, C.; Cook, D.; Gough, C.; Denarie, J.Four genes of *Medicago* controlling components of a nod factor transduction pathway.*Plant Cell* 2000,*12* (9), 1647-1666.

Cheng, N. H.; Liu, J. Z.; Nelson, R. S.; Hirschi, K. D. Characterization of CXIP4, a novel *Arabidopsis* protein that activates the H^+/Ca^{2+} antiporter, CAX1. *FEBS Lett* 2004,*559* (1-3), 99-106.

Cheng, N. H.; Pittman, J. K.; Shigaki, T.; Lachmansingh, J.; LeClere, S.; Lahner, B.; Salt, D. E.; Hirschi, K. D. Functional association of *Arabidopsis* CAX1and CAX3 is required for normal growth and ion homeostasis. *Plant Physiol* 2005,*138* (4), 2048-2060.

Cheng, N. H.; Pittman, J. K.; Zhu, J. K.; Hirschi, K. D. The protein kinaseSOS2 activates the *Arabidopsis* H^+/Ca^{2+} antiporter CAX1 to integrate calcium transport and salt tolerance. *J Biol Chem* 2004,*279* (4), 2922-2926.

Cheng, S. H.; Willmann, M. R.; Chen, H. C.; Sheen, J. Calcium signaling through protein kinases. The *Arabidopsis* calcium-dependent protein kinase gene family.*Plant Physiol* 2002,*129* (2), 469-485.

Cheong, Y. H.; Pandey, G. K.; Grant, J. J.; Batistic, O.; Li, L.; Kim, B. G.; Lee, S. C.; Kudla, J.; Luan, S. Two calcineurin B-like calcium sensors, interacting with protein kinase CIPK23, regulate leaf transpiration and root potassium uptake in *Arabidopsis*. *Plant J* 2007,*52* (2), 223-239.

Chico, J. M.; Raices, M.; Tellez-Inon, M. T.; Ulloa, R. M. A calcium-dependent protein kinase is systemically induced upon wounding in tomato plants. *Plant Physiol* 2002,*128* (1), 256-270.

Chinnusamy, V.; Ohta, M.; Kanrar, S.; Lee, B. H.; Hong, X.; Agarwal, M.; Zhu, J. K.*ICE1*: a regulator of cold-induced transcriptome and freezing tolerance in *Arabidopsis. Genes Dev* 2003,*17* (8), 1043-1054.

Chinnusamy, V.; Zhu, J.; Zhu, J. K. Cold stress regulation of gene expression in plants.*Trends Plant Sci* 2007,*12* (10), 444-451.

Cosgrove, D. J.; Hedrich, R. Stretch-activated chloride, potassium, and calcium channels coexisting in plasma membranes of guard cells of *Vicia faba* L. *Planta* 1991,*186* (1), 143-153.

D'Angelo, C.; Weinl, S.; Batistic, O.; Pandey, G. K.; Cheong, Y. H.; Schultke, S.; Albrecht, V.; Ehlert, B.; Schulz, B.; Harter, K.; Luan, S.; Bock, R.; Kudla, J. Alternative complex formation of the Ca^{2+}-regulated protein kinase CIPK1 controls abscisic acid-dependent and independent stress responses in *Arabidopsis. Plant J* 2006,*48* (6), 857-872.

Das, R.; Pandey, G. K. Expressional analysis and role of calcium regulated kinases in abiotic stress signaling. *Curr Genomics* 2010,*11* (1), 2-13.

Davenport, R. Glutamate receptors in plants.*Ann Bot* 2002,*90* (5), 549-557.

DeFalco, T.A., Bender, K.W., Snedden, W.A. Breaking the code: Ca^{2+} sensors in plant signaling. *Biochem. J*2010, 425, 27–40.

Demidchik, V.; Essah, P. A.; Tester, M. Glutamate activates cation currents in the plasma membrane of *Arabidopsis* root cells. *Planta* 2004,*219* (1), 167-175.

Demidchik, V.; Maathuis, F. J. Physiological roles of nonselective cation channels in plants: from salt stress to signalling and development. *New Phytol* 2007,*175* (3), 387-404.

den Hartog, M.; Musgrave, A.; Munnik, T. Nod factor-induced phosphatidic acid and diacylglycerol pyrophosphate formation: a role for phospholipase C and D in root hair deformation. *Plant J* 2001,*25* (1), 55-65.

Dodd, A. N.; Kudla, J.; Sanders, D.The language of calcium signaling.*Annu Rev Plant Biol* 2010,*61*, 593-620.

Drobak, B. K.; Watkins, P. A. Inositol(1,4,5)trisphosphate production in plant cells: an early response to salinity and hyperosmotic stress. *FEBS Lett* 2000,*481* (3), 240-244.

Dubos, C.; Huggins, D.; Grant, G. H.; Knight, M. R.; Campbell, M. M.A role for glycine in the gating of plant NMDA-like receptors.*Plant J* 2003,*35* (6), 800-810.

Dunkley, T. P.; Hester, S.; Shadforth, I. P.; Runions, J.; Weimar, T.; Hanton, S. L.; Griffin, J. L.; Bessant, C.; Brandizzi, F.; Hawes, C.; Watson, R. B.; Dupree, P.; Lilley, K. S. Mapping the *Arabidopsis* organelle proteome. *Proc Natl Acad Sci U S A* 2006,*103* (17), 6518-6523.

Lilley, K. S. Mapping the *Arabidopsis* organelle proteome. *Proc Natl Acad Sci U S A* 2006,*103* (17), 6518-6523.

Dutta, R.; Robinson, K. R. Identification and characterization of stretch-activated ion channels in pollen protoplasts.*Plant Physiol* 2004,*135* (3), 1398-1406.

East, J. M.Sarco(endo)plasmic reticulum calcium pumps: recent advances in our understanding of structure/function and biology (review). *Mol Membr Biol* 2000,*17* (4), 189-200.

Easterwood, G. Calcium's role in plant nutrition. Fluid Journal.(2002).

Ehrhardt, D. W.; Wais, R.; Long, S. R. Calcium spiking in plant root hairs responding to *Rhizobium* nodulation signals. *Cell* 1996,*85* (5), 673-681.

Ettinger, W. F.; Clear, A. M.; Fanning, K. J.; Peck, M. L. Identification of a Ca^{2+}/H^+ antiport in the plant chloroplast thylakoid membrane. *Plant Physiol* 1999,*119* (4), 1379- 1386.

Evans, N. H.; McAinsh, M. R.; Hetherington, A. M. Calcium oscillations in higher plants.*Curr Opin Plant Biol* 2001,*4* (5), 415-420.

Fasano, J. M.; Massa, G. D.; Gilroy, S. Ionic signaling in plant responses to gravity and touch.*J Plant Growth Regul* 2002,*21* (2), 71-88.

Finkelstein, R. R.; Gampala, S. S.; Rock, C. D. Abscisic acid signaling in seeds and seedlings. *Plant Cell* 2002,*14 Suppl*, S15-45.

Finkler, A.; Ashery-Padan, R.; Fromm, H. CAMTAs: calmodulin-binding transcription activators from plants to human. *FEBS Lett* 2007,*581* (21), 3893-3898.

Folta, K. M.; Lieg, E. J.; Durham, T.; Spalding, E. P. Primary inhibition of hypocotyl growth and phototropism depend differently on phototropin-mediated increases in cytoplasmic calcium induced by blue light. *Plant Physiol* 2003,*133* (4), 1464-1470.

Fontaine, V.; Hartwell, J.; Jenkins, G.I.; Nimmo, H.G. *Arabidopsis thaliana* contains two phospho*enol*pyruvate carboxylase kinase genes with different expression patterns. *Plant Cell Environ* 2002, 25: 115–122.

Furuichi, T; Cunningham, K.W.; Muto, S. A putative two-pore channel AtTPC1 mediates Ca^{2+} flux in *Arabidopsis* leaf cells. *Plant Cell Physiol.*2001, 42:900–905.

Furumoto, T.; Ogawa, N.; Hata, S.; Izui, K. Plant calcium-dependent protein kinase- related kinases (CRKs) do not require calcium for their activities. *FEBS Lett* 1996, *396* (2-3), 147-151.

Gampala, S. S.; Finkelstein, R. R.; Sun, S. S.; Rock, C. D.ABI5 interacts with abscisic acid signaling effectors in rice protoplasts. *J Biol Chem* 2002,*277* (3), 1689-1694.

Gaxiola, R. A.; Rao, R.; Sherman, A.; Grisafi, P.; Alper, S. L.; Fink, G. R. The *Arabidopsis thaliana* proton transporters, AtNhx1 and Avp1, can function in cation detoxification in yeast. *Proc Natl Acad Sci U S A* 1999,*96* (4), 1480-1485.

Geisler, M.; Axelsen, K. B.; Harper, J. F.; Palmgren, M. G. Molecular aspects of higher plant P-type Ca^{2+}-ATPases. *Biochim Biophys Acta* 2000,*1465* (1-2), 52-78.

Gelli, A.; Blumwald, E. Hyperpolarization-activated Ca^{2+}-permeable channels in the plasma membrane of tomato cells.*J Membr Biol* 1997,*155* (1), 35-45.

Gelli, A.; Higgins, V.J.; Blumwald, E. Activation of Plant Plasma Membrane Ca^{2+}- Permeable Channels by Race-Specific Fungal Elicitors.*Plant Physiol* 1997, 113(1):269- 279.

Giacomello, M.; Drago, I.; Pizzo, P.; Pozzan, T. Mitochondrial Ca^{2+} as a key regulator of cell life and death.*Cell Death Differ* 2007,*14* (7), 1267-1274.

Gilroy, S.; Read, N. D.; Trewavas, A. J. Elevation of cytoplasmic calcium by caged calcium or caged inositol triphosphate initiates stomatal closure. *Nature* 1990, *346* (6286), 769-771.

Gleason, C.; Chaudhuri, S.; Yang, T.; Munoz, A.; Poovaiah, B. W.; Oldroyd, G. E. Nodulation independent of rhizobia induced by a calcium-activated kinase lacking autoinhibition. *Nature* 2006,*441* (7097), 1149-1152.

Goldgur, Y.; Rom, S.; Ghirlando, R.; Shkolnik, D.; Shadrin, N.; Konrad, Z.; Bar-Zvi, D. Desiccation and zinc binding induce transition of tomato abscisic acid stress ripening 1, a water stress- and salt stress-regulated plant-specific protein, from unfolded to folded state. *Plant Physiol* 2007,*143* (2), 617-628.

Gong, D.; Gong, Z.; Guo, Y.; Chen, X.; Zhu, J. K. Biochemical and functional characterization of PKS11, a novel *Arabidopsis* protein kinase. *J Biol Chem* 2002, *277* (31), 28340-52830.

Gong, D.; Gong, Z.; Guo, Y.; Zhu, J. K. Expression, activation, and biochemical properties of a novel *Arabidopsis* protein kinase. *Plant Physiol* 2002,*129* (1), 225-234.

Gong, D.; Guo, Y.; Jagendorf, A. T.; Zhu, J. K. Biochemical characterization of the *Arabidopsis* protein kinase SOS2 that functions in salt tolerance. *Plant Physiol* 2002,*130* (1), 256-264.

Gong, D.; Zhang, C.; Chen, X.; Gong, Z.; Zhu, J. K. Constitutive activation and transgenic evaluation of the function of an *Arabidopsis*PKS protein kinase. *J Biol Chem* 2002,*277* (44), 42088-42096.

Gong, M; Chen, S.N.; Song, Y.Q.; Li, Z.G. Effect of calcium and calmodulin on intrinsic heat tolerance in relation to antioxidant systems in maize seedlings. *J Plant Physiology* 1997a, 24: 371–379.

Gong, M.; Li, Y.J.; Dai, X.; Tian, M.; Li, Z.G. Involvement of calcium and calmodulin in the acquisition of heat-shock induced thermotolerance in maize. *J Plant Physiology* 1997 b, 150: 615–621.

Gong, M.; Van der Luit A.H.; Knight, M.R. Trewavas, A.J. Heat-shock-induced changes in intracellular Ca^{2+} level in tobacco seedlings in relation to thermotolerance. *Plant Physiology* (1998b) 116, 429–437.

Guo, H.; Mockler, T.; Duong, H.; Lin, C., SUB1, an *Arabidopsis* Ca^{2+}-binding protein involved in cryptochrome and phytochrome coaction. *Science* 2001,*291* (5503), 187 190.

Guo, Y.; Halfter, U.; Ishitani, M.; Zhu, J. K. Molecular characterization of functional domains in the protein kinase SOS2that is required for plant salt tolerance. *Plant Cell* 2001,*13* (6), 1383-1400.

Haake, V.; Cook, D.; Riechmann, J. L.; Pineda, O.; Thomashow, M. F.; Zhang, J. Z. Transcription factor CBF4 is a regulator of drought adaptation in *Arabidopsis*. *Plant Physiol* 2002,*130* (2), 639-648.

Halfter, U.; Ishitani, M.; Zhu, J. K. The *Arabidopsis* SOS2 protein kinase physically interacts with and is activated by the calcium-binding protein SOS3. *Proc Natl Acad Sci U S A* 2000, *97* (7), 3735-40.

Hamilton, D. W.; Hills, A.; Kohler, B.; Blatt, M. R. Ca^{2+} channels at the plasma membrane of stomatal guard cells are activated by hyperpolarization and abscisic acid. *Proc Natl Acad Sci U S A* 2000,*97* (9), 4967-4972.

Hamilton, D.W.A.; Hills, A.; Blatt, M.R. Extracellular $Ba^{2+}$and voltage interact to gate Ca^{2+} channels at the plasma membrane of stomatal guard cells. *FEBS Lett.*2001,491: 99–103.

Harada, A.; Sakai, T.; Okada, K. Phot1 and Phot2mediate blue light-induced transient increases in cytosolic Ca^{2+} differently in *Arabidopsis* leaves. *Proc Natl Acad Sci U S A* 2003, *100* (14), 8583-8588.

Harada, A.; Shimazaki, K., Measurement of changes in cytosolic Ca^{2+} in *Arabidopsis* guard cells and mesophyll cells in response to blue light.*Plant Cell Physiol* 2009,*50* (2), 360-373.

Harmon, A.C.; Gribskov, M.; Gubrium, E.; Harper, J.F.The CDPK superfamily of protein kinases.*New Phytol*2001, 151:175– 183.

Harper, J. F.; Breton, G.; Harmon, A. Decoding Ca^{2+} signals through plant protein kinases.*Annu Rev Plant Biol* 2004,*55*, 263-288.

Harper, J. F.; Hong, B.; Hwang, I.; Guo, H. Q.; Stoddard, R.; Huang, J. F.; Palmgren, M. G.; Sze, H. A novel calmodulin-regulated Ca^{2+}-ATPase (ACA2) from Arabidopsis with an N-terminal autoinhibitory domain. *J Biol Chem* 1998,*273* (2), 1099-1106.

Harper, J.F.; Sussman, M.R.; Schaller, G.E.; Putnam-Evans, C.; Charbonneau, H.; Harmon, A.C.A novel calcium-dependent protein kinase with a regulatory domain similar to calmodulin.*Science* 1991, 252: 951-954.

Hepler, P. K. Calcium: a central regulator of plant growth and development. *Plant Cell* 2005,*17* (8), 2142-2155.

Hetherington, A. M.Guard cell signaling. *Cell* 2001,*107* (6), 711-714.

Hetherington, A.M. and Brownlee, C. The generation of calcium signals in plants. *Annu Rev Plant Biol* 2004, 55,401–427.

Hirschi, K. Vacuolar H^+/Ca^{2+} transport: who's directing the traffic? *Trends Plant Sci* 2001,*6* (3), 100-104.

Hirschi, K. D.; Miranda, M. L.; Wilganowski, N. L. Phenotypic changes in *Arabidopsis* caused by expression of a yeast vacuolar Ca^{2+}/H^+ antiporter. *Plant Mol Biol* 2001,*46* (1), 57-65.

Hong, Y.; Takano, M.; Liu, C. M.; Gasch, A.; Chye, M. L.; Chua, N. H. Expression of three members of the calcium-dependent protein kinase gene family in *Arabidopsis thaliana*. *Plant Mol Biol* 1996,*30* (6), 1259-1275.

Hotta, H.; Aoki, N.; Matsuda, T.; Adachi, T. Molecular analysis of a novel protein kinase in maturing rice seed. *Gene* 1998,*213* (1-2), 47-54.

Hrabak, E. M.; Chan, C. W.; Gribskov, M.; Harper, J. F.; Choi, J. H.; Halford, N.; Kudla, J.; Luan, S.; Nimmo, H. G.; Sussman, M. R.; Thomas, M.; Walker-Simmons, K.; Zhu, J. K.; Harmon, A. C.The *Arabidopsis* CDPK-SnRK superfamily of protein kinases.*Plant Physiol* 2003,*132* (2), 666-680.

Hrabak, E. M.; Dickmann, L. J.; Satterlee, J. S.; Sussman, M. R. Characterization of eight new members of the calmodulin-like domain protein kinase gene family from Hu, H. C.; Wang, Y. Y.; Tsay, Y. F.AtCIPK8, a CBL-interacting protein kinase, regulates the low-affinity phase of the primary nitrate response. *Plant J* 2009,*57* (2), 264-278.

Hu, HC; Wang, Y.Y.; Tsay, Y.F.; Huang, J-Z; Huber, S.C.Phosphorylation of synthetic peptides by a CDPK and plant SNF1-related protein kinase: influence of proline and basic amino acid residues at selected positions. *Plant Cell Physiol* 2001, 42, 1079–1087.

Hwang, I.; Sze, H.; Harper, J. F. A calcium-dependent protein kinase can inhibit a calmodulin-stimulated Ca^{2+} pump (ACA2) located in the endoplasmic reticulum of *Arabidopsis. Proc Natl Acad Sci U S A* 2000,*97* (11), 6224-6229.

Ichimura, K.; Mizoguchi, T.; Yoshida, R.; Yuasa, T.; Shinozaki, K. Various abiotic stresses rapidly activate *Arabidopsis* MAP kinases ATMPK4and ATMPK6. *Plant J* 2000, 24, 655-665.

Ishitani, M.; Xiong, L.; Lee, H.; Stevenson, B.; Zhu, J. K. HOS1, a genetic locus involved in cold-responsive gene expression in *Arabidopsis. Plant Cell* 1998,*10* (7), 1151- 1161.

Jang, I. C.; Pahk, Y. M.; Song, S. I.; Kwon, H. J.; Nahm, B. H.; Kim, J. K. Structure and expression of the rice class-I type histone deacetylase genes OsHDAC1-3:

OsHDAC1overexpression in transgenic plants leads to increased growth rate and altered architecture. *Plant J* 2003,*33* (3), 531-541.

Jiang, Q.W.; Kiyoharu, O.; Ryozo, I. Two novel mitogen-activated protein signaling components, OsMEK1 and OsMAP1, are involved in a moderate low-temperature signaling pathway in Rice1, *Plant Physiol* 2002, 129, 1880–1891.

Johannes, E.; Brosnan, J.M.; Sanders, D. Parallel pathways for intracellular Ca^{2+} release from the vacuole of higher-plants.*Plant Journal* 1992, 2, 97–102.

Kanamori, N.; Madsen, L. H.; Radutoiu, S.; Frantescu, M.; Quistgaard, E. M.; Miwa, H.; Downie, J. A.; James, E. K.; Felle, H. H.; Haaning, L. L.; Jensen, T. H.; Sato, S.; Nakamura, Y.; Tabata, S.; Sandal, N.; Stougaard, J. A nucleoporin is required for induction of Ca^{2+} spiking in legume nodule development and essential for rhizobial and fungal symbiosis. *Proc Natl Acad Sci U S A* 2006,*103* (2), 359-364.

Kang, S.; Kim, H. B.; Lee, H.; Choi, J. Y.; Heu, S.; Oh, C. J.; Kwon, S. I.; An, C. S. Overexpression in *Arabidopsis* of a plasma membrane-targeting glutamate receptor from small radish increases glutamate-mediated Ca^{2+} influx and delays fungal infection. *Mol Cells* 2006,*21* (3), 418-427.

Kaur, N., Gupta, A.K. Signal transduction pathways under abiotic stresses in plants. *Current Science*, 2005, 88, 1771- 1780.

Kiegle, E.; Gilliham, M.; Haseloff, J.; Tester, M. Hyperpolarisation-activated calcium currents found only in cells from the elongation zone of *Arabidopsis thaliana* roots. *Plant J* 2000,*21* (2), 225-229.

Kim, B. G.; Waadt, R.; Cheong, Y. H.; Pandey, G. K.; Dominguez-Solis, J. R.; Schultke, S.; Lee, S. C.; Kudla, J.; Luan, S. The calcium sensor CBL10 mediates salt tolerance by regulating ion homeostasis in Arabidopsis. *Plant J* 2007, *52* (3), 473-484.

Kim, J. A.; Agrawal, G. K.; Rakwal, R.; Han, K. S.; Kim, K. N.; Yun, C. H.; Heu, S.; Park, S. Y.; Lee, Y. H.; Jwa, N. S. Molecular cloning and mRNA expression analysis of a novel rice (Oryzasativa L.) MAPK kinase kinase, OsEDR1, an ortholog of *Arabidopsis* AtEDR1, reveal its role in defense/stress signalling pathways and development. *Biochem Biophys Res Commun* 2003, 300 (4), 868-876.Kovtun, Y.; Chiu, W. L.; Tena, G.; Sheen, J. Functional analysis of oxidative stress- activated mitogen-activated protein kinase cascade in plants. *Proc Natl Acad Sci U S A* 2000, *97* (6), 2940-2945.

Kudla, J.; Xu, Q.; Harter, K.; Gruissem, W.; Luan, S. Genes for calcineurin B-like proteins in *Arabidopsis* are differentially regulated by stress signals. *Proc Natl Acad Sci U S A* 1999,*96* (8), 4718-4723.

Kwak, J. M.; Mori, I. C.; Pei, Z. M.; Leonhardt, N.; Torres, M. A.; Dangl, J. L.; Bloom, R. E.; Bodde, S.; Jones, J. D.; Schroeder, J. I. NADPH oxidase AtrbohDand AtrbohF genes function in ROS-dependent ABA signaling in *Arabidopsis*. *EMBO J* 2003,*22* (11), 2623-2633.

Kwak, J. M.; Murata, Y.; Baizabal-Aguirre, V. M.; Merrill, J.; Wang, M.; Kemper, A.; Hawke, S. D.; Tallman, G.; Schroeder, J. I. Dominant negative guard cell K^+ channel mutants reduce inward-rectifying K^+ currents and light-induced stomatal opening in *Arabidopsis*. *Plant Physiol* 2001,*127* (2), 473-485.

Lacombe, B.; Becker, D.; Hedrich, R.; DeSalle, R.; Hollmann, M.; Kwak, J. M.; Schroeder, J. I.; Le Novere, N.; Nam, H. G.; Spalding, E. P.; Tester, M.; Turano, F. J.; Chiu, J.;

Coruzzi, G.The identity of plant glutamate receptors.*Science* 2001,*292* (5521), 1486-1497.

Laohavisit, A.; Mortimer, J. C.; Demidchik, V.; Coxon, K. M.; Stancombe, M. A.; Macpherson, N.; Brownlee, C.; Hofmann, A.; Webb, A. A.; Miedema, H.; Battey, N. H.; Davies, J. M.*Zea mays* annexins modulate cytosolic free Ca^{2+} and generate a Ca^{2+}-permeable conductance. *Plant Cell* 2009,*21* (2), 479-493.

Larkindale, J.; Knight, M. R. Protection against heat stress-induced oxidative damage in *Arabidopsis* involves calcium, abscisic acid, ethylene, and salicylic acid. *Plant Physiol* 2002,*128* (2), 682-695.

Lecourieux, D.; Ranjeva, R.; Pugin, A. Calcium in plant defence-signalling pathways.*New Phytol* 2006,*171* (2), 249-269.

Lee, B. H.; Henderson, D. A.; Zhu, J. K.The *Arabidopsis* cold-responsive transcriptome and its regulation by ICE1.*Plant Cell* 2005,*17* (11), 3155-3175.

Lee, H.; Xiong, L.; Gong, Z.; Ishitani, M.; Stevenson, B.; Zhu, J. K. The *Arabidopsis*HOS1 gene negatively regulates cold signal transduction and encodes a RING finger protein that displays cold-regulated nucleo-cytoplasmic partitioning. *Genes Dev* 2001,*15* (7), 912-924.

Lee, S.C.; Lan, W.Z.; Kim, B.G.; Li, L.; Choeng, Y.H.; Pandey, G.K.; Lu, G.; Buchanan, B.; Luan, S. A protein phosphorylation/dephosphorylation network regulates a plant potassium channel *Proceedings of the National Academy of Sciences USA,2007,*104, 15959-15964.

Leigh, R.A.; Jones, R.G.W.A hypothesis relating critical potassium concentrations for growth to the distribution and functions of this ion in the plant cell.*New Phytologist* 1984, 97, 1-13.

Lemichez, E.; Wu, Y.; Sanchez, J. P.; Mettouchi, A.; Mathur, J.; Chua, N. H. Inactivation of *AtRac1* by abscisic acid is essential for stomatal closure. *Genes Dev* 2001,*15* (14), 1808-1816.Lemtiri-Chlieh, F.; MacRobbie, E. A.; Webb, A. A.; Manison, N. F.; Brownlee, C.; Skepper, J. N.; Chen, J.; Prestwich, G. D.; Brearley, C. A. Inositol hexakisphosphate mobilizes an endomembrane store of calcium in guard cells. *Proc Natl Acad Sci U S A* 2003,*100* (17), 10091-10095.

Lhuissier, F.G.P.; de Ruijter, N.C.A.; Sieberer, B.J.; Esseling, J.J.; Emons, A.M.C. Time course of cell biological events evoked in legume root hairs by *Rhizobium* Nod factors: state of the art. *Ann. Bot.* 2001, 87:289–302.

Li, A. L.; Zhu, Y. F.; Tan, X. M.; Wang, X.; Wei, B.; Guo, H. Z.; Zhang, Z. L.; Chen, X. B.; Zhao, G. Y.; Kong, X. Y.; Jia, J. Z.; Mao, L. Evolutionary and functional study of the CDPK gene family in wheat (*Triticum aestivum L.*). *Plant Mol Biol* 2008,*66* (4), 429-443.

Li, J.; Kinoshita, T.; Pandey, S.; Ng, C. K.; Gygi, S. P.; Shimazaki, K.; Assmann, S. M. Modulation of an RNA-binding protein by abscisic-acid-activated protein kinase. *Nature* 2002,*418* (6899), 793-797.

Li, J.; Wang, D. Y.; Li, Q.; Xu, Y. J.; Cui, K. M.; Zhu, Y. X.*PPF1* inhibits programmed cell death in apical meristems of both G2 pea and transgenic *Arabidopsis* plants possibly by delaying cytosolic Ca^{2+} elevation. *Cell Calcium* 2004,*35* (1), 71-77.

Li, J.; Zhu, S.; Song, X.; Shen, Y.; Chen, H.; Yu, J.; Yi, K.; Liu, Y.; Karplus, V. J.; Wu, P.; Deng, X. W. A rice glutamate receptor-like gene is critical for the division and survival of individual cells in the root apical meristem. *Plant Cell* 2006,*18* (2), 340-349.

Li, L.; Kim, B. G.; Cheong, Y. H.; Pandey, G. K.; Luan, S. A Ca^{2+} signaling pathway regulates a K^+ channel for low-K response in *Arabidopsis*. *Proc Natl Acad Sci U S A* 2006, *103* (33), 12625-12630.

Li, Z. J.; Onodera, H.; Ugaki, M.; Tanaka, H. and Komatsu, S. Characterization of calreticulin as a phosphoprotein interacting with cold-induced protein kinase in rice. *Biol Pharm Bull* 2003, 26, 256–261.

Liang, C.; Zhang, X.Y.; Luo, Y.; Wang, G.P; Zou, Q.; Wang, W. Over-accumulation of glycine betaine alleviates the negative effects of salt stress in wheat. *Russian J Plant Physiol* 2009, 56, 370-376.

Liang, F.; Sze, H. A high-affinity Ca^{2+} pump, ECA1, from the endoplasmic reticulum is inhibited by cyclopiazonic acid but not by thapsigargin. *Plant Physiol* 1998, 118(3), 817-825.

Light, P. E.; Bladen, C.; Winkfein, R. J.; Walsh, M. P.; French, R. J. Molecular basis of protein kinase C-induced activation of ATP-sensitive potassium channels. *Proc Natl Acad Sci U S A* 2000, *97* (16), 9058-9063.

Lin, H.; Yang, Y.; Quan, R.; Mendoza, I.; Wu, Y.; Du, W.; Zhao, S.; Schumaker, K. S.; Pardo, J. M.; Guo, Y. Phosphorylation of SOS3-LIKE CALCIUM BINDING PROTEIN8 by SOS2protein kinase stabilizes their protein complex and regulates salt tolerance in *Arabidopsis*. *Plant Cell* 2009, *21* (5), 1607-1619.

Liotenberg, S.; North, H.; Poll, A.M. Molecular biology and regulation of abscisic acid biosynthesis in plants, Plant Physiol. Biochem. 1999, 37, 341–350.

Liu, G.; Chen, J.; Wang, X.VfCPK1, a gene encoding calcium-dependent protein kinase from *Vicia faba*, is induced by drought and abscisic acid. *Plant Cell Environ* 2006, *29* (11), 2091-2099.

Liu, H. T.; Li, G. L.; Chang, H.; Sun, D. Y.; Zhou, R. G.; Li, B. Calmodulin-binding protein phosphatase PP7is involved in thermotolerance in *Arabidopsis*. *Plant Cell Environ* 2007, *30* (2), 156-164.

Liu, J.; Ishitani, M.; Halfter, U.; Kim, C. S.; Zhu, J. K. The *Arabidopsis thaliana*SOS2gene encodes a protein kinase that is required for salt tolerance. *Proc Natl Acad Sci U S A* 2000, *97* (7), 3730-3734.

Liu, J.; Zhu, J. K. A calcium sensor homolog required for plant salt tolerance. *Science* 1998, *280* (5371), 1943-1945.

Liu, Q.; Kasuga, M.; Sakuma, Y.; Abe, H.; Miura, S.; Yamaguchi-Shinozaki, K.; Shinozaki, K. Two transcription factors, DREB1 and DREB2, with an EREBP/AP2 DNA binding domain separate two cellular signal transduction pathways in drought- and low-temperature-responsive gene expression, respectively, in *Arabidopsis*. *Plant Cell* 1998, *10* (8), 1391-1406.

Liu, Z.; Xia, M.; Poovaiah, B. W. Chimeric calcium/calmodulin-dependent protein kinase in tobacco: differential regulation by calmodulin isoforms. *Plant Mol Biol* 1998, *38* (5), 889-897.

Liu, K.M.; Wang, L.; Xue, Y.Y. et al. Overexpression of Os- COIN, a putative cold inducible zinc finger protein, increased tolerance to chilling, salt and drought, and enhanced praline level in rice, *Planta*2007, 226, 1007–1016.

Logan, D. C.; Knight, M. R. Mitochondrial and cytosolic calcium dynamics are differentially regulated in plants. *Plant Physiol* 2003, *133* (1), 21-24.

Logan, D. C.; Scott, I.; Tobin, A. K.The genetic control of plant mitochondrial morphology and dynamics.*Plant J* 2003,*36* (4), 500-509.

Luan, S.; Kudla, J.; Rodriguez-Concepcion, M.; Yalovsky, S.; Gruissem, W. Calmodulins and calcineurin B–like proteins: Calcium sensors for specific signal response coupling in plants. *Plant Cell* 2002, 14, S389–400.

Luan, S. Tyrosine phosphorylation in plant cell signaling. *Proc Natl Acad Sci U S A* 2002,*99* (18), 11567-11569.

Ludwig, A. A.; Saitoh, H.; Felix, G.; Freymark, G.; Miersch, O.; Wasternack, C.; Boller, T.; Jones, J. D.; Romeis, T. Ethylene-mediated cross-talk between calcium-dependent protein kinase and MAPK signaling controls stress responses in plants. *Proc Natl Acad Sci U S A* 2005,*102* (30), 10736-10741.

MacRobbie, E. A. Signal transduction and ion channels in guard cells. *Philos Trans R Soc Lond B Biol Sci* 1998,*353* (1374), 1475-1488.

MacRobbie, E. A. ABA activates multiple Ca^{2+} fluxes in stomatal guard cells, triggering vacuolar K^+(Rb^+release. *Proc Natl Acad Sci U S A* 2000,*97* (22), 12361-12368.

Mahajan, S.; Sopory, S. K.; Tuteja, N. Cloning and characterization of CBL-CIPK signalling components from a legume (*Pisum sativum*).*FEBS J* 2006,*273* (5), 907-925.

Mahajan, S.; Tuteja, N. Cold, salinity and drought stresses: an overview. *Arch Biochem Biophys* 2005,*444* (2), 139-158.

Mahajan, S.; Sopory, S. K.; and Tuteja, N. CBL-CIPK paradigm: Role in calcium and stress signaling in plants. *Proc Indian NatnSci Acad* 2006a, 72, 63–78.

Malho, R.; Read, N. D.; Trewavas, A. J.; Pais, M. S. Calcium Channel Activity during Pollen Tube Growth and Reorientation. *Plant Cell* 1995,*7* (8), 1173-1184.

Marschner, H. Mineral nutrition of higher plants, 2nd edn.Academic Press, London 1995.

Martinez, D. E.; Bartoli, C. G.; Grbic, V.; Guiamet, J. J. Vacuolar cysteine proteases of wheat (Triticum aestivum L.) are common to leaf senescence induced by different factors. *J Exp Bot* 2007,*58* (5), 1099-1107.

Martinez-Noel, G.; Tognetti, J.; Nagaraj, V.; Wiemken, A.; Pontis, H. Calcium is essential for fructan synthesis induction mediated by sucrose in wheat. *Planta* 2006,*225* (1), 183-191.

Maser, P.; Thomine, S.; Schroeder, J. I.; Ward, J. M.; Hirschi, K.; Sze, H.; Talke, I. N.; Amtmann, A.; Maathuis, F. J.; Sanders, D.; Harper, J. F.; Tchieu, J.; Gribskov, M.; Persans, M. W.; Salt, D. E.; Kim, S. A.; Guerinot, M. L. Phylogenetic relationships within cation transporter families of *Arabidopsis*. *Plant Physiol* 2001,*126* (4), 1646-1667.

McAinsh, M. R.; Clayton, H.; Mansfield, T. A.; Hetherington, A. M. Changes in Stomatal Behavior and Guard Cell Cytosolic Free Calcium in Response to Oxidative Stress. *Plant Physiol* 1996,*111* (4), 1031-1042.

McAinsh, M. R.; Pittman, J. K. Shaping the calcium signature. *New Phytol* 2009,*181* (2), 275-294.

McCormack, E; and Braam, J. Calmodulins and related potential calcium sensors of *Arabidopsis. New Phytologist* 2003, 159, 585–598.

McLaughlin S.B.; Wimmer R. Calcium physiology and terrestrial ecosystem processes. *New Phytologist* 1999, 142, 373–417.

Mehlmer, N.; Wurzinger, B.; Stael, S.; Hofmann-Rodrigues, D.; Csaszar, E.; Pfister, B.; Bayer, R.; Teige, M. The Ca^{2+}-dependent protein kinase *CPK3* is required for MAPK-independent salt-stress acclimation in *Arabidopsis. Plant J* 2010, 63 (3), 484-498.

Mei, H.; Zhao, J.; Pittman, J. K.; Lachmansingh, J.; Park, S.; Hirschi, K. D.In planta regulation of the *Arabidopsis* Ca^{2+}/H^+antiporter CAX1.*J Exp Bot* 2007,*58* (12), 3419-3427.

Meinhard, M.; Grill, E. Hydrogen peroxide is a regulator of ABI1, a protein phosphatase 2C from *Arabidopsis*. *FEBS Lett* 2001,*508* (3), 443-446.

Meinhard, M.; Rodriguez, P. L.; Grill, E. The sensitivity of ABI2 to hydrogen peroxide links the abscisic acid-response regulator to redox signalling. *Planta* 2002,*214* (5), 775-782.

Messerli, M. A.; Creton, R.; Jaffe, L. F.; Robinson, K. R. Periodic increases in elongation rate precede increases in cytosolic Ca^{2+} during pollen tube growth. *Dev Biol* 2000,*222* (1), 84-98.

Meyerhoff, O.; Muller, K.; Roelfsema, M. R.; Latz, A.; Lacombe, B.; Hedrich, R.; Dietrich, P.; Becker, D. AtGLR3.4, a glutamate receptor channel-like gene is sensitive to touch and cold. *Planta* 2005,*222* (3), 418-427.

Miedema, H.; Demidchik, V.; Very, A. A.; Bothwell, J. H.; Brownlee, C.; Davies, J. M. Two voltage-dependent calcium channels co-exist in the apical plasma membrane of *Arabidopsis thaliana* root hairs. *New Phytol* 2008,*179* (2), 378-385.

Mikolajczyk, M.; Awotunde, O. S.; Muszynska, G.; Klessig, D. F.; Dobrowolska, G. Osmotic stress induces rapid activation of a salicylic acid-induced protein kinase and a homolog of protein kinase ASK1 in tobacco cells. *Plant Cell* 2000,*12* (1), 165-178.

Miller, D.; Callaham, D.A.; Gross, D.J.; and Hepler, P.K. Free calcium gradient in growing pollen tubes of *Lilium*.J. *Cell Sci* 1992, 101, 7–12.

Miwa, H.; Sun, J.; Oldroyd, G. E.; Downie, J. A. Analysis of calcium spiking using a cameleon calcium sensor reveals that nodulation gene expression is regulated by calcium spike number and the developmental status of the cell. *Plant J* 2006,*18* (6), 883 891.

Miwa, H.; Sun, J.; Oldroyd, G.E.; Downie, J.A. Analysis of Nod-factor-induced calcium signaling in root hairs of symbiotically defective mutants of Lotus japonicus. *Mol Plant Microbe Interact* 2006, 19(8),914-923.

Mizoguchi, T.; Hayashida, N.; Yamaguchi-Shinozaki, K.; Kamada, H.; Shinozaki, K., ATMPKs: a gene family of plant MAP kinases in *Arabidopsis thaliana*. *FEBS Lett* 1993,*336* (3), 440-444.

Mizoguchi, T.; Ichimura, K.; Yoshida, R.; Shinozaki, K. MAP kinase cascades in *Arabidopsis*: their roles in stress and hormone responses. *Results Probl Cell Differ* 2000,*27*, 29-38.

Mori, I. C.; Murata, Y.; Yang, Y.; Munemasa, S.; Wang, Y. F.; Andreoli, S.; Tiriac, H.; Alonso, J. M.; Harper, J. F.; Ecker, J. R.; Kwak, J. M.; Schroeder, J. I. CDPKs CPK6 and CPK3function in ABA regulation of guard cell S-type anion- and Ca^{2+}- permeable channels and stomatal closure. *PLoS Biol* 2006,*4* (10), e327.

Munnik, T. Phosphatidic acid: an emerging plant lipid second messenger. *Trends Plant Sci* 2001,*6* (5), 227-233.Murata, Y.; Pei, Z. M.; Mori, I. C.; Schroeder, J. Abscisic acid activation of plasma membrane Ca^{2+} channels in guard cells requires cytosolic NAD(P)H and is

differentially disrupted upstream and downstream of reactive oxygen species production in *abi1-1* and *abi2-1* protein phosphatase 2C mutants. *Plant Cell* 2001,*13* (11), 2513-2523.

Mustilli, A. C.; Merlot, S.; Vavasseur, A.; Fenzi, F.; Giraudat, J.*Arabidopsis* OST1 protein kinase mediates the regulation of stomatal aperture by abscisic acid and acts upstream of reactive oxygen species production. *Plant Cell* 2002,*14* (12), 3089-3099.

Nakamura, S.; Lynch, T. J.; Finkelstein, R. R. Physical interactions between ABA response loci of *Arabidopsis.Plant J* 2001,*26* (6), 627-635.

Narusaka, Y.; Nakashima, K.; Shinwari, Z. K.; Sakuma, Y.; Furihata, T.; Abe, H.; Narusaka, M.; Shinozaki, K.; Yamaguchi-Shinozaki, K. Interaction between two cis- acting elements, ABRE and DRE, in ABA-dependent expression of *Arabidopsis* rd29A gene in response to dehydration and high-salinity stresses. *Plant J* 2003,*34* (2), 137-148.

Navazio, L.; Bewell, M. A.; Siddiqua, A.; Dickinson, G. D.; Galione, A.; Sanders, D. Calcium release from the endoplasmic reticulum of higher plants elicited by the NADP metabolite nicotinic acid adenine dinucleotide phosphate. *Proc Natl Acad Sci U S A* 2000,*97* (15), 8693-8698.

Navazio, L.; Mariani, P.; Sanders, D. Mobilization of Ca^{2+} by cyclic ADP-ribose from the endoplasmic reticulum of cauliflower florets.*Plant Physiol* 2001,*125* (4), 2129-2138.

Nimmo, H. G.The regulation of phosphoenolpyruvate carboxylase in CAM plants.*Trends Plant Sci* 2000,*5* (2), 75-80.

Nimmo, H. G. Control of the phosphorylation of phosphoenolpyruvate carboxylase in higher plants. *Arch Biochem Biophys* 2003,*414* (2), 189-196.

Oppenheimer, D. G.; Pollock, M. A.; Vacik, J.; Szymanski, D. B.; Ericson, B.; Feldmann, K.; Marks, M. D. Essential role of a kinesin-like protein in *Arabidopsis* trichome morphogenesis. *Proc Natl Acad Sci U S A* 1997,*94* (12), 6261-6266.

Ordenes, V. R.; Reyes, F. C.; Wolff, D.; Orellana, A. A thapsigargin-sensitive Ca^{2+} pump is present in the pea Golgi apparatus membrane. *Plant Physiol* 2002,*129* (4), 1820- 1828.

Orvar, B. L.; Sangwan, V.; Omann, F.; Dhindsa, R. S. Early steps in cold sensing by plant cells: the role of actin cytoskeleton and membrane fluidity. *Plant J* 2000,*23* (6), 785-794.

Pandey, G. K.; Cheong, Y. H.; Kim, B. G.; Grant, J. J.; Li, L.; Luan, S. CIPK9: a calcium sensor-interacting protein kinase required for low-potassium tolerance in *Arabidopsis*. *Cell Res* 2007,*17* (5), 411-421.

Pandey, G.K.; Yadav, A.K.; Kanwar, P.; Tuteja, N. Role of calcium in regulating Potassium-Sodium Homeostasis and Potassium as nutrient signal during abiotic stress conditions. Plant stress. Global Science Books; 2010.

Pandey, G.K. Emergence of a novel calcium signaling pathway in plants: CBL-CIPK signaling network. Physiology and Molecular Biology of Plants 2008, 14, 51-68.

Pandey, G. K.; Grant, J. J.; Cheong, Y. H.; Kim, B. G.; Li le, G.; Luan, S. Calcineurin-B-like protein *CBL9* interacts with target kinase CIPK3 in the regulation of ABA response in seed germination. *Mol Plant* 2008,*1* (2), 238-248.

Pandey, S.; Sopory, S. K. Biochemical evidence for a calmodulin-stimulated calcium-dependent protein kinase in maize. *Eur J Biochem* 1998,*255* (3), 718-726.

Pandey, S.; Tiwari, S. B.; Tyagi, W.; Reddy, M. K.; Upadhyaya, K. C.; Sopory, S. K. A Ca^{2+}/CaM-dependent kinase from pea is stress regulated and in vitro phosphorylates a protein that binds to AtCaM5promoter. *Eur J Biochem* 2002,*269* (13), 3193-3204.

Pandey, S.; Tiwari, S. B.; Upadhyaya, K.C. and Sopory, S.K. Calcium Signaling: Linking Environmental Signals to Cellular Functions.*Critical Reviews in Plant Sciences* 2000,19, 4, 291-318.

Patil, S.; Takezawa, D.; Poovaiah, B.W. Chimeric plant calcium/calmodulin-dependent protein kinase gene with a neural visinin-like calcium-binding domain.*Proc Natl Acad Sci U S A* 1995, 92(11), 4897-4901.

Pei, Z. M.; Murata, Y.; Benning, G.; Thomine, S.; Klusener, B.; Allen, G. J.; Grill, E.; Schroeder, J. I. Calcium channels activated by hydrogen peroxide mediate abscisic acid signalling in guard cells. *Nature* 2000,*406* (6797), 731-734.

Peiter, E.; Maathuis, F. J.; Mills, L. N.; Knight, H.; Pelloux, J.; Hetherington, A. M.; Sanders, D. The vacuolar Ca^{2+}-activated channel TPC1regulates germination and stomatal movement. *Nature* 2005,*434* (7031), 404-408.

Peiter, E.; Sun, J.; Heckmann, A. B.; Venkateshwaran, M.; Riely, B. K.; Otegui, M. S.; Edwards, A.; Freshour, G.; Hahn, M. G.; Cook, D. R.; Sanders, D.; Oldroyd, G. E.; Downie, J. A.; Ane, J. M. The *Medicago truncatula* DMI1 protein modulates cytosolic calcium signaling. *Plant Physiol* 2007,*145* (1), 192-203.

Pittman, J. K.; Hirschi, K. D. Regulation of CAX1, an *Arabidopsis* Ca^{2+}/H^+ antiporter.Identification of an N-terminal autoinhibitory domain.*Plant Physiol* 2001,*127* (3), 1020-1029.

Pittman, J. K.; Hirschi, K. D.Don't shoot the (second) messenger: endomembrane transporters and binding proteins modulate cytosolic Ca^{2+} levels. *Curr Opin Plant Biol* 2003,*6* (3), 257-262

Pittman, J. K.; Shigaki, T.; Cheng, N. H.; Hirschi, K. D. Mechanism of N-terminal autoinhibition in the Arabidopsis Ca^{2+}/H^+antiporter CAX1. *J Biol Chem* 2002,*277* (29), 26452-26459.

Poovaiah, B. W.; Reddy, A. S. Calcium and signal transduction in plants.*CRC Crit Rev Plant Sci* 1993,*12* (3), 185-211.

Pottosin, II; Schonknecht, G. Vacuolar calcium channels.*J Exp Bot* 2007,*58* (7), 1559- 1569.

Qi, Z.; Kishigami, A.; Nakagawa, Y.; Iida, H.; Sokabe, M.A mechanosensitive anion channel in Arabidopsis thaliana mesophyll cells.*Plant Cell Physiol* 2004,*45* (11), 1704-1708.

Qiu, Q. S.; Guo, Y.; Dietrich, M. A.; Schumaker, K. S.; Zhu, J. K. Regulation of *SOS1*, a plasma membrane Na^+/H^+ exchanger in *Arabidopsis thaliana*, by SOS2 and SOS3. *Proc Natl Acad Sci U S A* 2002,*99* (12), 8436-8441.

Quan, R.; Lin, H.; Mendoza, I.; Zhang, Y.; Cao, W.; Yang, Y.; Shang, M.; Chen, S.; Pardo, J. M.; Guo, Y.SCABP8/CBL10, a putative calcium sensor, interacts with the protein kinase *SOS2* to protect *Arabidopsis* shoots from salt stress. *Plant Cell* 2007,*19* (4), 1415-1431.

Ramachandiran, S.; Takezawa, D.; Wang, W.; Poovaiah, B. W. Functional domains of plant chimeric calcium/calmodulin-dependent protein kinase: regulation by autoinhibitory and visinin-like domains. *J Biochem* 1997,*121* (5), 984-990.

Ranf, S.; Wunnenberg, P.; Lee, J.; Becker, D.; Dunkel, M.; Hedrich, R.; Scheel, D.; Dietrich, P. Loss of the vacuolar cation channel, AtTPC1, does not impair Ca^{2+} signals induced by abiotic and biotic stresses. *Plant J* 2008,*53* (2), 287-299.

Ray, S.; Agarwal, P.; Arora, R.; Kapoor, S.; Tyagi, A. K. Expression analysis of calcium-dependent protein kinase gene family during reproductive development and 80 abiotic stress conditions in rice (*Oryza sativa* L. ssp. indica). *Mol Genet Genomics* 2007,*278* (5), 493-505.

Rentel, M.C.; and Knight, M.R. Oxidative Stress-Induced Calcium Signaling in *Arabidopsis Plant Physiology*, 2004, (135), 1471–1479.

Riely, B. K.; Lougnon, G.; Ane, J. M.; Cook, D. R. The symbiotic ion channel homolog DMI1is localized in the nuclear membrane of *Medicago truncatula* roots. *Plant J* 2007,*49* (2), 208-216.

Roderick, H.L.; Berridge, M.J.; Bootman, M.D. Calcium-induced calcium release.*Curr Biol* 2003, 13(11)R425.

Ritchie, J. M.; Black, J. A.; Waxman, S. G. Sodium channels in the cytoplasm of Schwann cells. *Proc Natl Acad Sci U S A* 2000,*97* (4), 1949.

Roh, M. H.; Shingles, R.; Cleveland, M. J.; McCarty, R. E. Direct measurement of calcium transport across chloroplast inner-envelope vesicles. *Plant Physiol* 1998,*118* (4), 1447-1454.

Rus, A.; Yokoi, S.; Sharkhuu, A.; Reddy, M.; Lee, B. H.; Matsumoto, T. K.; Koiwa, H.; Zhu, J. K.; Bressan, R. A.; Hasegawa, P. M. AtHKT1is a salt tolerance determinant that controls Na$^+$entry into plant roots. *Proc Natl Acad Sci U S A* 2001,*98* (24), 14150-14155.

Sagi, M.; Fluhr, R. Superoxide production by plant homologues of the *gp91*(phox) NADPH oxidase. Modulation of activity by calcium and by tobacco mosaic virus infection.*Plant Physiol* 2001,*126* (3), 1281-1290.

Saito, K.; Yoshikawa, M.; Yano, K.; Miwa, H.; Uchida, H.; Asamizu, E.; Sato, S.; Tabata, S.; Imaizumi-Anraku, H.; Umehara, Y.; Kouchi, H.; Murooka, Y.;Szczyglowski, K.; Downie, J. A.; Parniske, M.; Hayashi, M.; Kawaguchi, M.NUCLEOPORIN85 is required for calcium spiking, fungal and bacterial symbioses, and seed production in Lotus japonicus. *Plant Cell* 2007,*19* (2), 610-624.

Sanders, D.; Pelloux, J.; Brownlee, C.; Harper, J. F. Calcium at the crossroads of signaling.*Plant Cell* 2002,*14*, S401-417.

Sangwan, V.; Foulds, I.; Singh, J.; Dhindsa, R. S. Cold-activation of *Brassica napus*BN115promoter is mediated by structural changes in membranes and cytoskeleton, and requires Ca^{2+} influx. *Plant J* 2001,*27* (1), 1-12.

Sathyanarayanan, P. V.; Cremo, C. R.; Poovaiah, B. W. Plant chimeric Ca^{2+}/Calmodulin-dependent protein kinase. Role of the neural visinin-like domain in regulating autophosphorylation and calmodulin affinity.*J Biol Chem* 2000,*275* (39), 30417-30422.

Sato, Y.; Wada, M.; Kadota, A. External Ca^{2+} is essential for chloroplast movement induced by mechanical stimulation but not by light stimulation. *Plant Physiol* 2001,*127* (2), 497-504.

Schroeder, J. I.; Allen, G. J.; Hugouvieux, V.; Kwak, J. M.; Waner, D. Guard Cell Signal Transduction.*Annu Rev Plant Physiol Plant Mol Biol* 2001,*52*, 627-658.

Schroeder, J. I.; Kwak, J. M.; Allen, G. J. Guard cell abscisic acid signalling and engineering drought hardiness in plants. *Nature* 2001,*410* (6826), 327-330.

Schultz, J.; Copley, R. R.; Doerks, T.; Ponting, C. P.; Bork, P. SMART: a web-based tool for the study of genetically mobile domains. *Nucleic Acids Res* 2000,*28* (1), 231-234.

Schuurink, R. C.; Shartzer, S. F.; Fath, A.; Jones, R. L. Characterization of a calmodulin-binding transporter from the plasma membrane of barley aleurone. *Proc Natl Acad Sci U S A* 1998,*95* (4), 1944-1949.

Seigneurin-Berny, D.; Gravot, A.; Auroy, P.; Mazard, C.; Kraut, A.; Finazzi, G.; Grunwald, D.; Rappaport, F.; Vavasseur, A.; Joyard, J.; Richaud, P.; Rolland, N.HMA1, a new Cu-ATPase of the chloroplast envelope, is essential for growth under adverse light conditions. *J Biol Chem* 2006,*281* (5), 2882-2892.

Sharp, R. E.; Poroyko, V.; Hejlek, L. G.; Spollen, W. G.; Springer, G. K.; Bohnert, H. J.; Nguyen, H. T. Root growth maintenance during water deficits: physiology to functional genomics. *J Exp Bot* 2004,*55* (407), 2343-2351.

Shi, H.; Ishitani, M.; Kim, C.; Zhu, J. K. The *Arabidopsis thaliana* salt tolerance gene SOS1 encodes a putative Na$^+$/H$^+$ antiporter. *Proc Natl Acad Sci U S A* 2000,*97* (12), 6896-6901.

Shi, H.; Lee, B. H.; Wu, S. J.; Zhu, J. K. Overexpression of a plasma membrane Na$^+$/H$^+$ antiporter gene improves salt tolerance in *Arabidopsis thaliana*. *Nat Biotechnol* 2003,*21* (1), 81-85.

Shi, H.; Xiong, L.; Stevenson, B.; Lu, T.; Zhu, J. K. The *Arabidopsis* salt overly sensitive 4 mutants uncover a critical role for vitamin B6 in plant salt tolerance. *Plant Cell* 2002,*14* (3), 575-588.

Shi, H.; Zhu, J.K. Regulation of expression of the vacuolar Na$^+$/H$^+$ antiporter gene AtNHX1by salt stress and abscisic acid.*Plant Mol Biol* 2002, 50(3),543-50.

Shigaki, T.; Cheng, N. H.; Pittman, J. K.; Hirschi, K. Structural determinants of Ca^{2+} transport in the *Arabidopsis* H$^+$/Ca^{2+} antiporter CAX1.*J Biol Chem* 2001,*276* (46), 43152-43159.

Shigaki, T.; Hirschi, K. D. Diverse functions and molecular properties emerging for CAX cation/H$^+$ exchangers in plants.*Plant Biol (Stuttg)* 2006,*8* (4), 419-429.

Shinozaki, K.; Yamaguchi-Shinozaki, K. Molecular responses to dehydration and low temperature: differences and cross-talk between two stress signaling pathways. *Curr Opin Plant Biol* 2000,*3* (3), 217-223.

Stockinger, E. J.; Gilmour, S. J.; Thomashow, M. F.*Arabidopsis thaliana* CBF1 encodes an AP2 domain-containing transcriptional activator that binds to the C- repeat/DRE, a cis-acting DNA regulatory element that stimulates transcription in response to low temperature and water deficit. *Proc Natl Acad Sci U S A* 1997,*94* (3), 1035-1040.

Sugden, C.; Crawford, R. M.; Halford, N. G.; Hardie, D. G. Regulation of spinach SNF1-related (SnRK1) kinases by protein kinases and phosphatases is associated with phosphorylation of the T loop and is regulated by 5'-AMP. *Plant J* 1999,*19* (4), 433-439.

Sugden, C.; Donaghy, P. G.; Halford, N. G.; Hardie, D. G. Two SNF1-related protein kinases from spinach leaf phosphorylate and inactivate 3-hydroxy-3-methylglutaryl- coenzyme A reductase, nitrate reductase, and sucrose phosphate synthase in vitro. *Plant Physiol* 1999,*120* (1), 257-274.

Thion, L.; Mazars, C.; Nacry, P.; Bouchez, D.; Moreau, M.; Ranjeva, R.; Thuleau, P. Plasma membrane depolarization-activated calcium channels, stimulated by microtubule-depolymerizing drugs in wild-type *Arabidopsis thaliana* protoplasts, display constitutively large activities and a longer half-life in ton 2 mutant cells affected in the organization of cortical microtubules. *Plant J* 1998,*13* (5), 603-610.

Thomashow, M. F. PLANT COLD ACCLIMATION: Freezing Tolerance Genes and Regulatory Mechanisms. *Annu Rev Plant Physiol Plant Mol Biol* 1999,*50*, 571-599.

Tripathi, V.; Parasuraman, B.; Laxmi, A.; Chattopadhyay, D.CIPK6, a CBL- interacting protein kinase is required for development and salt tolerance in plants. *Plant J* 2009,*58* (5), 778-790.

Tripathi, V.; Syed, N.; Laxmi, A.; Chattopadhyay, D. Role of CIPK6 in root growth and auxin transport.*Plant Signal Behav* 2009,*4* (7), 663-665.

Tucker, E. B.; Lee, M.; Alli, S.; Sookhdeo, V.; Wada, M.; Imaizumi, T.; Kasahara, M.; Hepler, P. K. UV-A induces two calcium waves in *Physcomitrella patens. Plant Cell Physiol* 2005,*46* (8), 1226-1236.

Tuteja, N. Abscisic Acid and abiotic stress signaling.*Plant Signal Behav* 2007,*2* (3), 135-138.

Tuteja, N.; Mahajan, S. Calcium signaling network in plants: an overview. *Plant Signal Behav* 2007,*2* (2), 79-85.

Tuteja, N.; Sopory, S. K. Chemical signaling under abiotic stress environment in plants.*Plant Signal Behav* 2008,*3* (8), 525-536.

Urao, T.; Katagiri, T.; Mizoguchi, T.; Yamaguchi-Shinozaki, K.; Hayashida, N.; Shinozaki, K. Two genes that encode Ca^{2+}-dependent protein kinases are induced by drought and high-salt stresses in *Arabidopsis thaliana. Mol Gen Genet* 1994,*244* (4), 331- 340.

van den Wijngaard, P. W.; Bunney, T. D.; Roobeek, I.; Schonknecht, G.; de Boer, A. H. Slow vacuolar channels from barley mesophyll cells are regulated by 14-3-3 proteins. *FEBS Lett* 2001,*488* (1-2), 100-104.

Very, A. A.; Davies, J. M. Hyperpolarization-activated calcium channels at the tip of *Arabidopsis* root hairs.*Proc Natl Acad Sci U S A* 2000,*97* (17), 9801-9806.

Vidal, J.; Chollet, R. Regulatory phosphorylation of C4 PEP carboxylase.*Trends Plant Sci* 1997, 2, 230–237.

Volotovski, I. D.; Sokolovsky, S. G.; Molchan, O. V.; Knight, M. R. Second messengers mediate increases in cytosolic calcium in tobacco protoplasts. *Plant Physiol* 1998,*117* (3), 1023-1030.

Vos, J.W.; Safadi, F.; Reddy, A.S.; Hepler, P.K. The kinesin-like calmodulin binding protein is differentially involved in cell division. *Plant Cell*. 2000, 12(6), 979-990.

Wais, R.J.; Galera, C.; Oldroyd, G.; Catoira R.; Penmetsa, R.V.; Cook, D.; et al. Genetic analysis of calcium spiking responses in nodulation mutants of *Medicago truncatula. Proc Natl Acad Sci U S A*. 2000, 97(24), 13407-13412.

Walker, S. A.; Viprey, V.; Downie, J. A. Dissection of nodulation signaling using pea mutants defective for calcium spiking induced by nod factors and chitin oligomers. *Proc Natl Acad Sci U S A* 2000,*97* (24), 13413-13418.

Wan, B.; Lin, Y.; Mou, T. Expression of rice Ca^{2+}-dependent protein kinases (CDPKs) genes under different environmental stresses. *FEBS Lett* 2007,*581* (6), 1179-1189.

Wang, D.; Harper, J.F.; Gribskov, M. Systematic trans-genomic comparison of protein kinases between *Arabidopsis* and *Saccharomyces cerevisiae.Plant Physiol* 2003,,132(4), 2152-2165.

Wang, D.; Xu, Y.; Li, Q.; Hao, X.; Cui, K.; Sun, F.; Zhu, Y. Transgenic expression of a putative calcium transporter affects the time of *Arabidopsis* flowering. *Plant J* 2003,*33* (2), 285-292.

Watillon, B.; Kettmann, R.; Boxus, P.; Burny, A. Structure of a calmodulin-binding protein kinase gene from apple. *Plant Physiol* 1995,*108* (2), 847-858.

Weinl, S.; Held, K.; Schlucking, K.; Steinhorst, L.; Kuhlgert, S.; Hippler, M.; Kudla, J.A plastid protein crucial for Ca^{2+}-regulated stomatal responses.*New Phytol* 2008,*179* (3), 675-686.

White, P. J.; Bowen, H. C.; Demidchik, V.; Nichols, C.; Davies, J. M. Genes for calcium-permeable channels in the plasma membrane of plant root cells.*Biochim Biophys Acta* 2002,*1564* (2), 299-309.

White, P. J.; Broadley, M. R., Calcium in plants.*Ann Bot* 2003,*92* (4), 487-511.

Xiong, L.; Ishitani, M.; Lee, H.; Zhu, J. K.The *Arabidopsis* LOS5/ABA3 locus encodes a molybdenum cofactor sulfurase and modulates cold stress- and osmotic stress-responsive gene expression. *Plant Cell* 2001,*13* (9), 2063-2083.

Xiong, L.; Lee, B.; Ishitani, M.; Lee, H.; Zhang, C.; Zhu, J. K.FIERY1encoding an inositol polyphosphate 1-phosphatase is a negative regulator of abscisic acid and stress signaling in *Arabidopsis*. *Genes Dev* 2001,*15* (15), 1971-1984.

Xiong, L.; Schumaker, K. S.; Zhu, J. K. Cell signaling during cold, drought, and salt stress. *Plant Cell* 2002,*14 Suppl*, S165-183.

Xiong, L.; Zhu, J. K. Abiotic stress signal transduction in plants: Molecular and genetic perspectives. *Physiol Plant* 2001,*112* (2), 152-166.

Xiong, T. C.; Bourque, S.; Lecourieux, D.; Amelot, N.; Grat, S.; Briere, C.; Mazars, C.; Pugin, A.; Ranjeva, R. Calcium signaling in plant cell organelles delimited by a double membrane. *Biochim Biophys Acta* 2006,*1763* (11), 1209-1215.

Yang, G.; Shen, S.; Yang, S.; Komatsu, S.OsCDPK13, a calcium-dependent protein kinase gene from rice, is induced in response to cold and gibberellin. *Plant Physiol Biochem* 2003, 41, 369–374.

Yang, T.; Poovaiah, B. W. Calcium/calmodulin-mediated signal network in plants.*Trends Plant Sci* 2003,*8* (10), 505-512.

Yang, H.M.; Zhang, X.Y.; Tang, Q.L.; Wang, G.X. Extracellular calcium is involved in stomatal movement through the regulation of water channels in broad bean. *Plant Growth Regul* 2006, 50, 79–83.

Yoshida, R.; Hobo, T.; Ichimura, K.; Mizoguchi, T.; Takahashi, F.; Aronso, J.; Ecker, J. R.; Shinozaki, K. ABA-activated SnRK2protein kinase is required for dehydration stress signaling in *Arabidopsis*. *Plant Cell Physiol* 2002,*43* (12), 1473-1483.

Zhang, L.; Liu, B. F.; Liang, S.; Jones, R. L.; Lu, Y. T. Molecular and biochemical characterization of a calcium/calmodulin-binding protein kinase from rice. *Biochem J* 2002,*368* (Pt 1), 145-157.

Zhang, L.; Lu, Y. T. Calmodulin-binding protein kinases in plants.*Trends Plant Sci* 2003,*8* (3), 123-127.

Zhang, M.; Liang, S.; Lu, Y. T. Cloning and functional characterization of NtCPK4, a new tobacco calcium-dependent protein kinase.*Biochim Biophys Acta* 2005,*1729* (3), 174-185.

Zhou, D.Y.; Tian, Q.Y.; Li, L.H.; Zhang, W.H. Nitric oxide in involved in nitrate- induced inhibition of root elongation in Zea mays, *Ann Bot* 2007, (100), 497–503.

Zhu, J. K. Plant salt tolerance. *Trends Plant Sci* 2001,*6* (2), 66-71.

Zhu, J. K. Salt and drought stress signal transduction in plants. *Annu Rev Plant Biol* 2002,*53*, 247-273.

Zhu, S. Y.; Yu, X. C.; Wang, X. J.; Zhao, R.; Li, Y.; Fan, R. C.; Shang, Y.; Du, S. Y.; Wang, X. F.; Wu, F. Q.; Xu, Y. H.; Zhang, X. Y.; Zhang, D. P. Two calcium-dependent protein kinases, CPK4 and CPK11, regulate abscisic acid signal transduction in *Arabidopsis*. *Plant Cell* 2007,*19* (10), 3019-3036.

Zou, H.; Lifshitz, L. M.; Tuft, R. A.; Fogarty, K. E.; Singer, J. J. Visualization of Ca^{2+} entry through single stretch-activated cation channels. *Proc Natl Acad Sci U S A* 2002,*99* (9), 6404-6409.

In: Calcium Signaling
Editor: Masayoshi Yamaguchi

ISBN: 978-1-61324-313-8
©2012 Nova Science Publishers, Inc.

Chapter 4

CALMODULIN SIGNALING INSIDE-OUT: INTRACELLULAR AND EXTRACELLULAR CALMODULIN AND ITS INTERACTION WITH A MATRICELLULAR CYSTEINE-RICH CALMODULIN-BINDING PROTEIN

Danton H. O'Day[*,1,2], *Robert J. Huber*[2] *and Andres Suarez*[1]

[1]Department of Biology, University of Toronto at Mississauga,
Mississauga, Ontario, Canada
[2]Department of Cell and Systems Biology,
University of Toronto, Toronto, Ontario, Canada

ABSTRACT

The cysteine-rich extracellular calmodulin (CaM) binding protein (CaMBP) cyrA from *Dictyostelium* possesses 4 tandem EGF-like (EGFL) domains in its C-terminus. cyrA is secreted during development where it comprises one of many extracellular matrix proteins in the sheath of the multicellular pseudoplasmodium or slug. Proteolysis causes the release of cleavage products of cyrA of various sizes enriched in the EGFL domains and at least one of these domains (EGFL1) can feed back to inhibit cyrA proteolysis. cyrA shares sequence similarity to mammalian tenascin C (tenC) with EGFL1 showing high sequence identity to Ten14, an EGFL domain from tenC. Like Ten14, EGFL1 of cyrA enhances cell movement. In addition, EGFL1 has been shown to enhance both random cell motility and cAMP-mediated chemotaxis by activation of downstream calcium-dependent signalling via binding to the surface of *Dictyostelium* cells. Thus cyrA represents a true matricellular protein from lower eukaryotes and the only matricellular CaMBP identified to date for any organism. In keeping with the extracellular locale of cyrA, extracellular CaM has also been demonstrated in *Dictyostelium*. Extracellular CaM has been discovered in a small number of other organisms but its functions have not been well defined. Addition of CaM antagonists to *Dictyostelium* inhibits extracellular cyrA breakdown suggesting a role for extracellular CaM in protecting cyrA proteolysis. Using

---

* Correspondence to: Dr. Danton H. O'Day, Email: danton.oday@utoronto.ca

GFP-constructs as well as polyclonal antibodies against both the C- and N-termini of cyrA, the intracellular locales of this CaMBP and the ways it is processed to release fragments containing EGFL domains is coming to light. Here we review past work and present some novel data on this interesting novel matricellular CaMBP.

Keywords: Calmodulin, calmodulin-binding protein, cysteine-rich protein, matricellular, EGF-like repeat, cell motility, chemotaxis, development, Dictyostelium

INTRODUCTION

Dictyostelium discoideum is a primary model eukaryotic organism for the study of cell, developmental and molecular biology. Its genome has been sequenced and there is an online interactive database and stock centre (http://dictybase.org/). For these and a multitude of other reasons, *Dictyostelium* has become a classic and central model organism especially for the study of cell motility and chemotaxis (e.g., Bonner, 1944; Chisholm and Firtel, 2004; King and Insall, 2009; Raper, 1940). During growth in mixed culture, cells chemotactically respond to folic acid (FA) secreted by *E. coli*, allowing them to track down their food source. Removal of nutrients (bacteria or axenic medium) leads to starvation and a shift from FA chemotaxis to cAMP-mediated chemotaxis which underlies multicellular development. Chemotaxis generates large cell aggregates that transform into multicellular pseudoplasmodia or slugs. Each slug ultimately culminates into a mature fruiting body consisting of a stalk of dead stalk cells supporting a suspension of viable spores.

In spite of its central role as a model for the study of cell motility and chemotaxis, several issues remain unresolved or at least not widely accepted. While the primary focus on cell motility and chemotaxis has been on cAMP signalling, calcium signal transduction is also essential to random cell motility and chemotaxis in *Dictyostelium* just as it is in higher cells (Lombardi et al, 2008; Valeyev et al, 2009). Calcium ions only function by regulating calcium-binding proteins (CBPs) including calmodulin (CaM) but this area of research in cell motility and chemotaxis is less well documented (Gauthier and O'Day, 2001; Catalano and O'Day, 2008). On the other hand, the role of peptides (polypeptides? i.e. cleavage products of cyrA) as regulators of cell motility, as they are in mammalian cells, has only recently been revealed in *Dictyostelium* (Huber and O'Day, 2009). If a system is to serve as a model, it must reflect those organisms for which it is to serve as a model. Below we review the role of calcium/CaM signalling in cell motility and chemotaxis in *Dictyostelium* before focussing on Epidermal Growth Factor-like (EGFL) peptides and the cysteine-rich protein cyrA from which these bioactive peptides are derived. This work further supports that this lower eukaryote is a true model for the study of cellular motility and chemotaxis in higher organisms.

In *Dictyostelium*, research has shown that intracellular calcium fluxes mediate many processes including chemotaxis, cell-cycle, differentiation, fertilization, gametogenesis and germination (reviewed in Catalano and O'Day, 2008; Lombardi et al, 2008; O'Day, 2006; Valeyev et al, 2009). In all eukaryotes, changes in intracellular Ca^{2+} lead to effects mediated by CBPs. *Dictyostelium* possesses at least 13 CBPs containing the Ca^{2+}-binding EF-hand motif, such as CaM (Dharamsi et al, 2000). CaM (~17 kDa) is highly conserved in all eukaryotes (Klee and Vanaman, 1982). Encoded by the gene *calA*, *Dictyostelium* CaM

(DdCaM) is an essential protein that is highly identical and functionally equivalent to mammalian CaM (Clarke et al, 1990). Upon Ca^{2+} binding, CaM undergoes a large conformational change exposing two hydrophobic patches that allow for CaM-binding protein (CaMBP) interaction (James, *et al*, 1995). Ca^{2+}-dependent CaMBPs are loosely categorized primarily based on the position of conserved hydrophobic residues within a continuous CaM-binding domain (CaMBD; Persechini and Kretsinger, 1988; Ikura, 1992). CaM can also bind proteins in a Ca^{2+}-independent manner as "apocalmodulin" through IQ and IQ like motifs (Jurado et al, 1999). Other modes of CaM-binding exist, some of which involve discontinuous CaMBDs (Hoeflich and Ikura, 2003).

We have used the multifunctional CaM-binding overlay technique (CaMBOT), to profile essentially the full complement of Ca^{2+}-dependent and independent CaMBPs of *Dictyostelium* (Catalano and O'Day, 2008; O'Day, 2006). In *Dictyostelium* over four-dozen CaMBPs, many of which are differentially expressed during gametogenesis, fertilization and zygote differentiation as well as during chemotaxis and spore germination, have been revealed (O'Day et al, 2003; O'Day, 2006). Detailed profiling has demonstrated that certain Ca^{2+}-dependent and -independent CaMBPs are temporally associated with specific sub-cellular locales (i.e., nucleus, cytoplasm, cell membrane) and linked to specific CaM-dependent events such as cAMP-mediated chemotaxis (Gauthier and O'Day, 2001).

In addition to profiling, we have used CaMBOT to isolate cDNAs encoding CaMBPs from an expression library from multicellular development of *Dictyostelium* (O'Day, 2003, 2006). Recently, we critically reviewed the complete literature on all known CaMBPs in *Dictyostelium* (Catalano and O'Day, 2008). Using CaMBOT, we have isolated genes encoding several novel CaMBPs including nucleolar nucleomorphin, cysteine-rich cyrA, and cmbB. Genes encoding several known CaMBPs were isolated (e.g., calcineurin) plus some that were not previously known to be CaMBPs (i.e., phosphoglycerate kinase, histone H1 and thymidine kinase). In total, CaMBPs have been identified in *Dictyostelium* that show association with the nucleus (e.g., nucleomorphin, H1), nucleolus (e.g., nucleomorphin, eukaryotic translation initiation factor 6; hsp32, heat shock protein 32, tumor necrosis factor receptor-associated protein 1), cytoplasm (e.g., cyrA, cmbB, calcineurin), ribosomes (e.g., ribosomal subunit protein L19), cytoskeleton (e.g., MHCKA/B, WW domain-containing protein, various myosins, spectrin), contractile vacuole (e.g., vwkA) and the cell cortex (e.g., DdGAP1; Catalano and O'Day, 2008, 2011). More recently, cyrA was found to be secreted as an extracellular CaMBP (Suarez et al, 2011a). In short there is a wide distribution of CaMBPs inside and outside of *Dictyostelium* cells.

Of these CaMBPs, cyrA is of special interest because it is the first extracellular CaMBP to be discovered in *Dictyostelium* and it fits the requirements to be designated as a true matricellular protein, making it the first matricellular protein identified in a lower eukaryote (Suarez et al, 2011a). These results raise questions not only as to why a CaMBP would be secreted to work extracellularly but also what role CaM could be playing in the regulation of a protein that localizes both intracellularly and extracellularly. Here we will attempt to answer these questions.

CURRENT RESEARCH

The ECM of *Dictyostelium* and cyrA

The multicellular slug is surrounded by a slime sheath that is comprised of proteins and carbohydrates equivalent to the extracellular matrix (ECM) of higher organisms (Wilkins and Williams, 1995). Several proteins have been identified in this ECM: cysteine-rich ecmA and ecmB, the glycoprotein sheathins (ecmC-E) and the small ecmF proteins that comprise a multigene family (Jermyn and Williams, 1991; Shimada et al, 2004; Wilkins and Williams, 1995). More recently, we have identified the cysteine-rich CaMBP cyrA as a component of the ECM of *Dictyostelium* (Suarez et al, 2011a).

A. The Primary Domains of cyrA

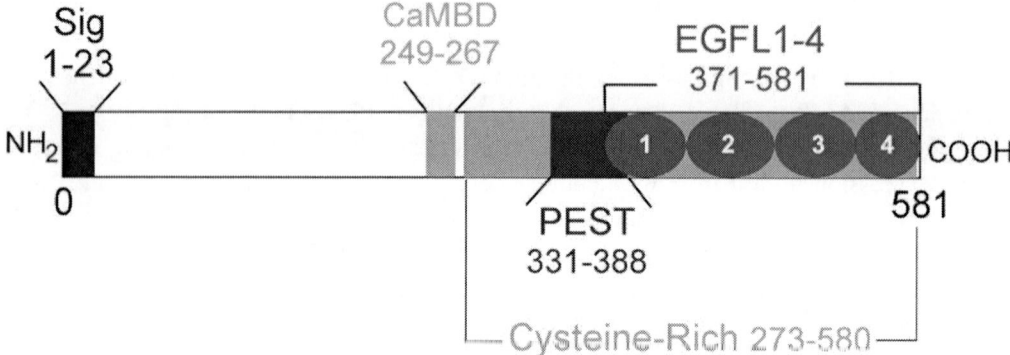

B. EGF vs cyrA EGFL Domains

```
EGF     -NSDSECPL---SHDGYCLHDGVCMYIEALDKYACNCVVGYIGERCQYRDLKWWEL-R
EGFL1   CDDNDECTVDECSITSGCKHT-VCPNLDKSGKTEIKCVNG----KCQTRVKSPCELMT
        :.:.**.:    *   . * *   **   ::   .*   :** *    :** *   .  **


EGF     NSDSECPLSHDGYCLHD-------GVCMYIEALDKYACNCVVGYIGERCQYR--------DLKWWELR
EGFL2   -----CP--NDTICIENYNNTKNLTICLPIEC-SINNCDDGNGCTIDSCNFQTGFCDHLLCPNKYLDPI
             **   :*   *:.:         :*: **. .   *:   *    : *:::       *: :


EGF     NSDSECPLSHDGYCLHDGV--CMYIEALDKYACNCVVGY--IGERCQYRDLKWWELR
EGFL3   TNHTTIPICENNKCVNKTLSNCEYFKC-DHPNEICIEEKITLAPKCVHFD--SGCL-
        ...:  *:..:. *::. :  * *::. *:    *:     :. :* : *      *


EGF     --NSDSECPLSHDGYCLH----DGVCMYIEALDKYACNCVVGYIGERCQYRDLKWWELR
EGFL4   CVHFDSGCLSCSDLNCQSLTSPNSRCKYIE-MDNQKLRCK-GSVGSCCPYLPTCY----
          : ** *  . *  *     :. * *** :*:   .*  *  :*. * *    :
```

Figure 1. Domains of cyrA. (A) Major Domains: A signal sequence (Sig), a calmodulin binding domain (CaMBD), an extensive cysteine-rich region, a PEST region and four tandem EGF-like repeats (EGFL1-4) are shown. (B) Alignment of EGFL1-4 repeats with human EGF. Fully conserved (*), strong group conserved (:), and weak group conserved (.).

Using CaMBOT, a cDNA (p64) was isolated from developing cells of *Dictyostelium*. DNA sequencing of p64 revealed an insert of 1841 bp with the open reading frame encoding a novel 581 amino acid polypeptides with a predicted molecular weight of 64.8 kDa. Because we used a recombinant CaM probe to the isolate the cDNA, the protein was originally designated CaMBP64 (Genbank accession number AAG34703.1). However, following the nomenclature system for *Dictyostelium* proteins, since CaMBP64 is a cysteine-rich protein it was renamed cyrA (DDB0231648; www.dictybase.org; Suarez et al, 2011a). Sequence analysis predicted an N-terminal signal sequence (Figure 1, A).

A PEST sequence (rich in proline, glutamic acid, serine and threonine) exists within the protein suggesting it is a site for proteolytic processing (Figure 1, A; Rogers et al., 1986). Cysteine-rich proteins are often found as components of the ECM, many of which possess EGFL repeats (Figure 1, A; Rao et al., 1995). The extensive cysteine-rich region of cyrA houses four EGFL domains (EGFL1-4) which have special relevance to the function of cyrA as discussed below. A single CaM-binding domain (^{249}DIFVIMRTGFKGVLQINFR^{267}) was also identified (Figure 1,A). Calcium-dependent and -independent CaM binding was experimentally verified (Suarez et al, 2011a). Within the CaMBD are a number of calcium-dependent binding motifs while calcium-independent binding would be mediated by the IQ motif (^{262}LQINFR^{267}) that is located within the full CaMBD.

The EGFL Repeats of *Dictyostelium*

EGFL repeats share sequence similarity with EGF. Their conformational structure, like that of EGF, is determined by the disulphide bridges between conserved cysteine residues (Taylor et al, 1972; Zanuttin et al., 2004). It has been suggested that *Dictyostelium* possesses more EGFL repeat regions than any other sequenced eukaryote (Glöckner et al., 2002). For example, the ECM proteins ecmA and B mentioned above both contain EGFL repeats. The novel adhesion receptor SadA contains three conserved EGFL repeats with strong sequence similarity to those present in integrins and tenascins (Fey et al., 2002). Tenascins comprise a large family of ECM glycoproteins. One member of this protein family, tenascin C, contains EGFL repeats numbered from Ten1 to Ten14 (Ikuta et al., 1998; Iyer et al., 2007) which have been shown to be involved in the regulation of cell motility via binding to the EGF receptor (EGFR; Iyer et al., 2007, 2008; Prieto et al., 1992; Swindle et al., 2001). ECM proteins, such as tenascin C, that mediate specific cellular processes are referred to as matricellular proteins as mentioned above.

EGFL1 of cyrA Enhances Random Cell Motility and Chemotaxis

In order to study the function of the EGFL repeats in *Dictyostelium*, Huber and O'Day (2009) prepared un-labelled and FITC-labelled peptides (DdEGFL1) equivalent to the first 18 amino acids of EGFL1 of cyrA. The EGFL1 repeat of cyrA is very similar to Ten14, which enhances cell motility in mammalian cells. Cell assays carried out in the presence DdEGFL1 showed that the rates of both random cell motility and cAMP-mediated chemotaxis were increased over controls lacking the peptide. Since FITC-DdEGFL1 does not enter cells it suggests the peptide enhances cell movement via binding to a receptor on the cell membrane. That work also showed that DdEGFL1 treatment led to the sustained threonine phosphorylation of a 200kDa protein. Subsequently, it was revealed that with starvation

Dictyostelium cells show increased responsiveness to DdEGFL1 treatment (Suarez et al, 2011a).

Calcium Signalling is Required for DdEGFL1 Function

Exposing DdEGFL1 treated cells to agents that interfere with calcium function indicate that the enhancement of movement requires intracellular calcium (Figure 2; modified from Huber and O'Day, 2011).

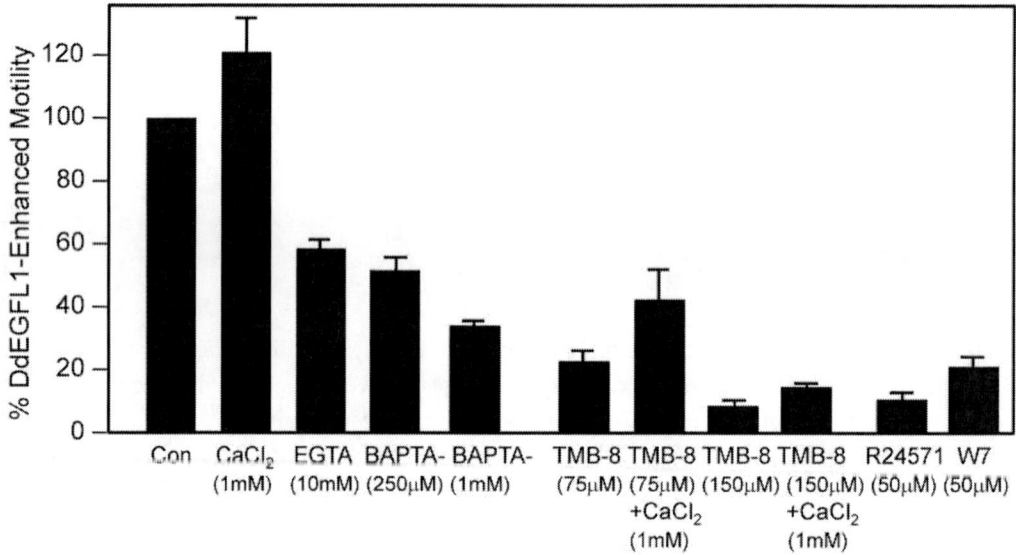

Figure 2. Calcium involvement in the DdEGFL1 enhancement of cell motility in *Dictyostelium*. Random cell motility was assessed using the radial bioassay (Huber and O'Day, 2009, 2011). Agents were added to the agar along with DdEGFL1 prior to plating cells for the bioassay. The rate of movement was assessed after 4 hours and compared to the movement of cells on plates containing DdEGFL1 only. EGTA and BAPTA are calcium chelating agents. TMB-8 inhibits calcium release from intracellular stores. R24571 (calmidazolium) and W7 are CaM antagonists with different modes of action. All shown data for differences between control and treated cells are statistically significant (p-value < 0.05).

Continued work further revealed that DdEGFL1 binding to the cell membrane affects downstream signalling pathways leading to calcium mobilization (Huber and O'Day, 2011). Pharmacological studies have revealed that both PI3K and PLA2 –mediated signalling is important for DdEGFL1 enhanced motility (Huber and O'Day, 2009). These two proteins have been shown to mediate cAMP-mediated chemotaxis in parallel compensatory pathways (Chen et al., 2007; van Haastert et al., 2007). Both experimental studies and computer modelling have shown that intracellular levels of cAMP and Ca^{2+} are tightly inter-connected during chemotactic aggregation in *Dictyostelium* (e.g., Valeyev et al, 2009). Our research further supports this notion in terms of DdEGFL1 function. DdEGFL1 stimulates a positive regulatory mechanism involving calcium signalling (Figure 3). This calcium signalling leads to an increase in the amount of cytoskeletal myosin II heavy chain (MHC) and actin (Huber and O'Day, 2011). In contrast, cAMP signalling appears to act as a feed-back, brake mechanism.

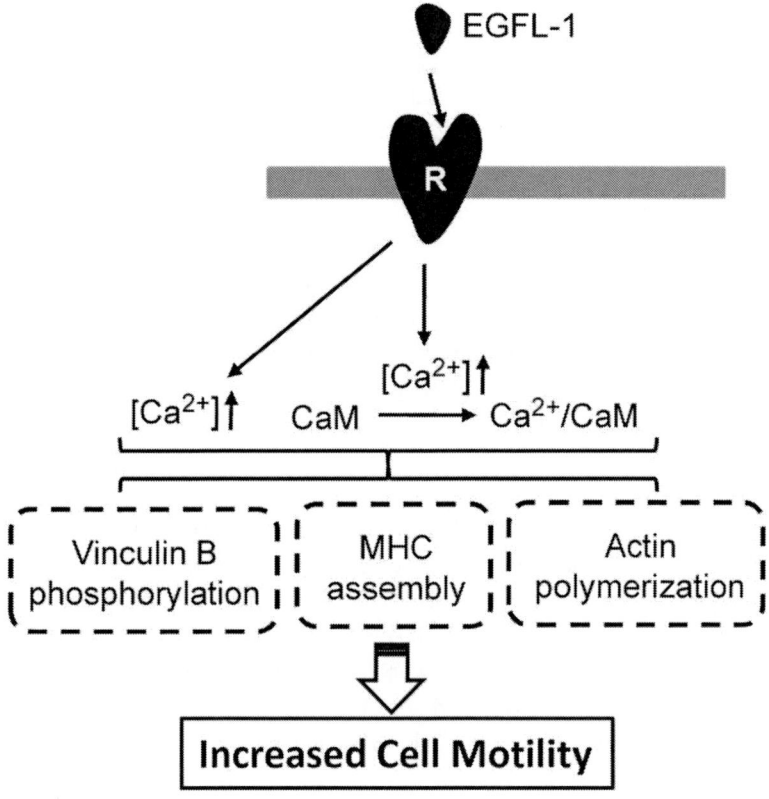

Figure 3. DdEGFL1 enhances cell motility via calcium signalling. The activation of calcium signalling leads to the activation of CaM and other as yet unidentified processes. These in turn lead to the assembly of MHC and the polymerization of actin, while the phosphorylation of vinculin B allows for their attachment to the cell membrane.

DdEGFL1 Treatment Regulates the Cytoskeleton

DdEGFL1 signalling also leads to the increased threonine phosphorylation of a ~200kDa protein. Studies with mutants revealed this protein was not MHC. To determine which protein was phosphorylated by DdEGFL1 treatment, immunoprecipitation with anti-phosphothreonine was carried out. The immunoprecipitate was subjected to SDS-PAGE and immunoblotting followed by sequencing of target bands by mass spectroscopy. This led to the identification of the band as vinculin B, a protein that links cell adhesion molecules to the actin cytoskeleton and functions in the motility of a diversity of normal and cancer cells (Huber and O'Day, manuscript in preparation). Thus DdEGFL1 enhances random cell motility and chemotaxis via calcium signalling that leads to stimulating significant increases in the expression of both polymeric MHC and actin in the cytoskeleton plus the phosphorylation of the linker protein vinculin B. Together these events lead to an increase in the rate of cell motility.

CyrA is Cleaved to Release EGFL-rich C-terminal Fragments

The questions that now arise are: do the EGFL repeats within cyrA work in the context of the full length protein or as smaller fragments, and what is the role of extracellular CaM in regulating this CaMBP? The proteolytic cleavage of matricellular proteins to release smaller signalling polypeptide fragments is of biomedical importance and has been actively studied in mammalian cells. For example, the EGFL repeats in the ectodomain of laminin-5 are shed by proteolytic activity to increase cell motility (Carpenter and Cohen, 1990; Giannelli et al., 1997; Schenk et al., 1993). Only recently has it become clear that proteolytic processing to release bioactive peptides also occurs in *Dictyostelium*. The secreted AcbA protein is cleaved by a membrane-bound serine protease (TagC) releasing spore differentiation factor-2 (SDF-2; Anjard and Loomis, 2005). SDF-2 binds to a receptor on prespore cells to induce spore cell differentiation. However, the processing of matricellular cyrA to yield biologically active EGFL-containing peptides is only now coming to light.

To understand the intracellular localization of cyrA, antibodies were generated against peptide sequences within the protein (Suarez et al, 2011a,b). Western blotting of cell extracts and the extracellular medium during development revealed that cyrA is processed both intracellularly and extracellularly to release two major C-term fragments containing the EGFL sequences as well as the CaMBD: a 45kDa and 40kDa fragment (Figure 4). Treatment of cells with DdEGFL1 inhibited the cleavage of cyrA suggesting an enzymatic feedback mechanism. Since both intracellular and extracellular cyrA fragments bind to CaM, CaM antagonists were also tested to determine what effect they would have on cyrA processing (Suarez et al, 2011a,b). Inhibition of CaM by these antagonists led to increased processing suggesting that CaM serves to protect cyrA from proteolytic digestion.

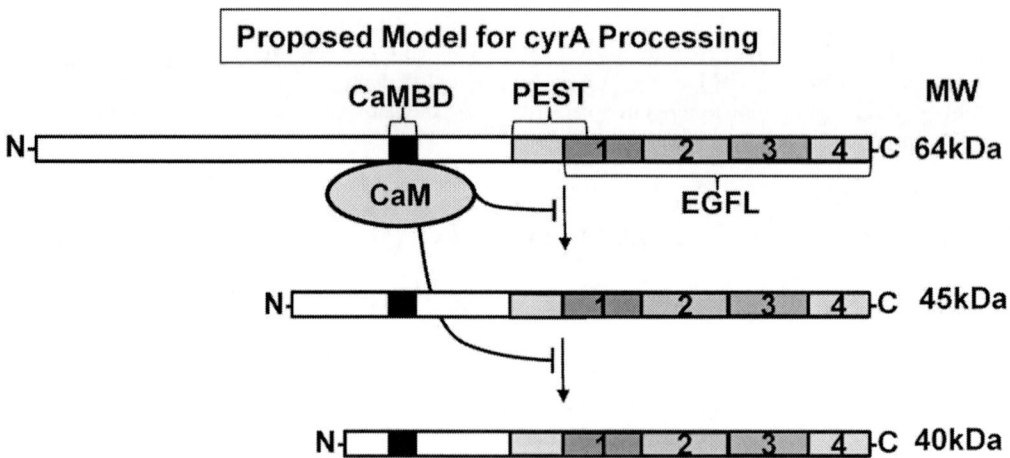

Figure 4. A model for the processing of cyrA and the protective function of calmodulin. Processing of cyrA leads to two major CaM binding fragments of 40 and 45kDa. Each of these contains all of the EGFL domains suggesting they are both biologically active and may be the primary mode of action for EGFL1.

The release of large polypeptide fragments containing bioactive domains is not uncommon. For example, incomplete proteolysis of pre-proEGF produces larger pro-EGF peptide fragments that still bind to the EGFR to activate it (Dempsey et al., 1997; Mroczkowski et al., 1989). The full processing of cyrA and the biological activity of the various proteolytic fragments is under investigation.

CaM is Present both Intracellularly and Extracellularly During Development

Western blotting of cell extracts and extracellular medium showed that CaM was present both inside and outside of *Dictyostelium* cells during development (Suarez et al, 2011c). While this research was ongoing, a proteomics study revealed the presence of extracellular DdCaM in *Dictyostelium* but no other insight into this localization was provided (Bakthavatsalam and Gomer, 2010). The literature on extracellular CaM is scarce but has been shown in several different organisms. In plants extracellular CaM regulates pollen germination and germ tube formation (Ma et al., 1999). Ikezaki et al. (1999) not only gave an overview of the literature on extracellular CaM as a ubiquitous extracellular regulatory protein in mammals, but they also showed that it enhances vasodilation in mammals. In frogs, extracellular CaM was shown to inhibit sensory axon outgrowth and injury-induced proliferation of non-neuronal cells (Remgard et al., 1995). To add to this knowledge, we have revealed an extracellular function for *Dictyostelium* CaM: it binds to matricellular cyrA and in doing so protects it from proteolytic cleavage. Whether this is the primary function of extracellular CaM or a result of CaM-binding to cyrA for other biological reasons remains to be determined.

Based on the extensive research of a multitude of others and our previous research on the intracellular and extracellular localization of CaMBPs and CaM, it is clear that CaM not only undergoes dramatic changes based on calcium-binding, but it also undergoes a diversity of translocations in order to regulate its target proteins (Fig. 5).

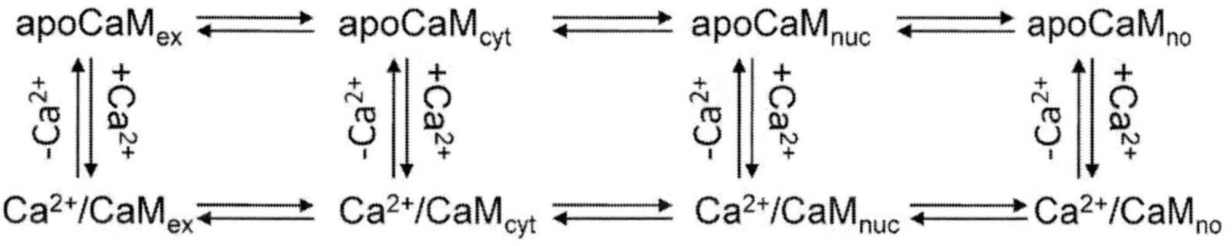

$$apoCaM_{ex} \rightleftarrows apoCaM_{cyt} \rightleftarrows apoCaM_{nuc} \rightleftarrows apoCaM_{no}$$

$$Ca^{2+}/CaM_{ex} \rightleftarrows Ca^{2+}/CaM_{cyt} \rightleftarrows Ca^{2+}/CaM_{nuc} \rightleftarrows Ca^{2+}/CaM_{no}$$

Figure 5. A summary of calmodulin (CaM) translocations. CaM exists in two primary states calcium-free or apo-CaM and Ca^{2+}/CaM. A diversity of work by many researchers has shown that CaM and/or its target proteins (CaMBPs) are localization in various intracellular compartments (nucleolus, no; nucleus, nu; cytoplasm, cyt) as well as extracellularly (ex).

How these translocations occur and are regulated remains to be elucidated. As an example, while CaM is known to exist in both the nucleoplasm and cytoplasm of eukaryotes it does not possess a nuclear localization signal (NLS) so must be transported by other mechanisms. Clearly this is a field that needs more attention.

Conclusion

CyrA is a True Matricellular Calmodulin-Binding Protein

The cysteine-rich protein cyrA of *Dictyostelium* fulfills the criteria to establish it as a true matricellular protein. It is secreted from cells where it becomes a component of the extracellular matrix (slime sheath) and it possesses EGFL domains that regulate random cell motility and chemotaxis. Previously, matricellular proteins were solely a mammalian entity. Thus cyrA shows that true matricellular proteins exist in lower eukaryotes as well. Data on cyrA processing suggests it behaves like other matricellular proteins by releasing polypeptide fragments containing functional EGFL repeats. Furthermore, cyrA proteolysis is regulated through feed-back by the EGFL peptide DdEGFL1. The role of CaM is of further interest since no matricellular protein has been shown previously to be a CaM-binding protein. Finally, the demonstration that CaM is secreted by *Dictyostelium* cells where it can protect cyrA against proteolysis opens new avenues for research on the extracellular functions of this fundamental signalling protein. The key now is to find the receptor for this matricellular protein.

A Developmental Model for cyrA Function

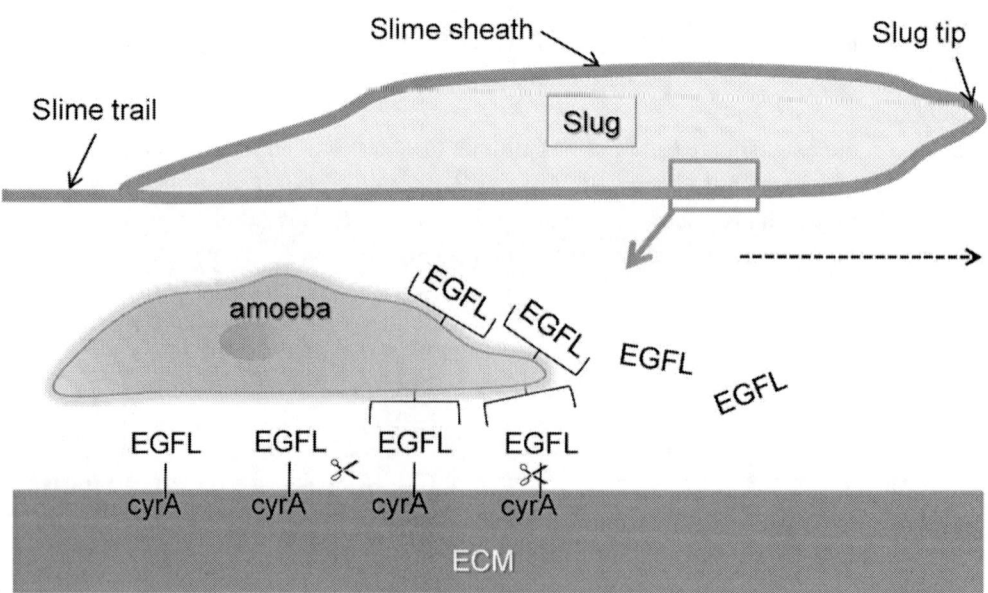

Figure 6. A proposed model for cyrA generated EGFL function. The multicellular slug of *Dictyostelium* has a tip that regulates morphogenesis through various events including the secretion of cAMP and of slime sheath components including cyrA. Sheath is also left behind as the slug migrates. Proteolytic cleavage (scissors) of cyrA (CyrA-EGFL) releases C-terminal products containing EGFL domains. The released C-terminal EGFL sequences bind, as part of larger cleavage products or after further cleavage to smaller EGFL units, to the cell surface to stimulate the movement of cells within the slug. The arrow indicates the direction of movement of both the slug and EGFLrepeat-responsive amoebae.

The results of this chapter can be put into a developmental context. The *Dictyostelium* slug is a model of tissue organization consisting of different cell types surrounded by a slime sheath (Wilkins and Williams, 1995). The slug tip controls slug movement and morphogenesis at least in part through the secretion of cAMP and new sheath synthesis (Matsukuma and Durston, 1979; Raper, 1940; Sternfeld and O'Mara, 2005; Wilkins and Williams, 1995). As part of the many functions of the ECM and its constituent proteins in slug movement and morphogenesis, we propose that the release of EGFL repeat-containing cyrA cleavage products serves to enhance cell motility working with cAMP to drive the forward movement of the slug (Fig. 6).

ACKNOWLEDGMENT

This chapter was supported by a grant from the Natural Sciences and Engineering Council of Canada (DHO'D; A6807).

REFERENCES

Anjard, C. and W. F. Loomis, 2005. Peptide signaling during terminal differentiation of *Dictyostelium*. *Proc. Natl. Acad. Sci.* (USA) 102: 7607-7611.

Bakthavatsalam, D. and R.H. Gomer, 2010. The secreted proteome profile of developing *Dictyostelium discoideum* cells. *Proteomics* 10: 2556-2559.

Bonner, J. T., 1994. The migration stage of *Dictyostelium*: Behavior without muscles or nerves. *FEMS Microbiol. Lett.* 120: 1-8.

Carpenter, G. and S. Cohen, 1990. Epidermal growth factor. J. Biol. Chem. 265: 7709-7712.

Catalano, A. and D. H. O'Day, 2008 Calmodulin-binding proteins in the model organism *Dictyostelium*: A complete and critical review. *Cell. Signal.* 20: 277-291.

Catalano, A. and D. H. O'Day, 2011. Nucleolar localization and identification of nuclear/nucleolar localization signals of the calmodulin-binding protein nucleomorphin during growth and mitosis in *Dictyostelium*. *Histochem. Cell. Biol.* 135: 239-249.

Chen, L.F., M. Iijima, M. Tang, M.A. Landree, Y.E. Huang, Y. Xiong, P.A. Iglesias and P.N. Devreotes, 2007. PLA2 and PI3K/PTEN pathways act in parallel to mediate chemotaxis. *Dev. Cell* 12: 603-614.

Chisholm, R. L. and R.A. Firtel, 2004. Insights into morphogenesis from a simple developmental system. *Nature Rev. Molec. Cell Biol.* 5: 531-541.

Clarke, M., 1990. Calmodulin structure, localization and expression in *Dictyostelium discoideum* In: O'Day (ed.), Ca^{2+} *as an Intracellular Messenger in Eukaryotic Microbes*, ASM Press, Washington, D.C., pp. 375-391.

Dempsey, P. J., K.S. Meise, Y. Yoshitake, K. Nishikawa and R.J. Coffey, 1997. Apical enrichment of human EGF precursor in Madin-Darby canine kidney cells involves preferential ectodomain cleavage sensitive to a metalloprotease inhibitor. *J. Cell Biol.* 138: 747-758.

Dharamsi, A., D. Tessarolo, B. Coukell and J. Pun, 2000. CBP1 associates with the *Dictyostelium* cytoskeleton and is important for normal cell aggregation under certain developmental conditions, *Exp. Cell Res.* 258: 298-309.

Fey, P., S. Stephens, M.A. Titus and R.L. Chilsolm, 2002. SadA, a novel adhesion receptor in *Dictyostelium*. *J. Cell Biol.* 159: 1109-1119.

Giannelli, G., J. Falk-Marzillier, O. Schiraldi, W.G. Stetler-Stevenson and V. Quaranta, 1997. Induction of cell migration by matrix metalloprotease-2 cleavage of laminin 5. *Science* 277: 225-228.

Gauthier, M. L. and D. H. O'Day, 2001. Detection of calmodulin-binding proteins and calmodulin-dependent phosphorylation linked to calmodulin-dependent chemotaxis to folic and cAMP in *Dictyostelium*. *Cell. Signal.* 13: 575-584.

Glockner, G., L. Eichinger, K. Szafranski, J.A. Pachebat, A.T. Bankier, P.H.L. Dear, R. Lehmann, C, Baumgart, G. Parra, J.F. Abril, R. Giogo, K. Kumpf, B. Tunggal, E. Cox, M.A. Quail, M. Platzer, A. Rosenthal and A.A. Noegel, 2002. Sequence and analysis of chromosome 2 of *Dictyostelium discoideum*. *Nature* 418: 79-85.

Huber, R. and D. H. O'Day, 2009. An EGF-like peptide sequence from *Dictyostelium* enhances cell motility and chemotaxis. Biochem. Biophys. Res. Commun. 379: 470-475.

Huber, R. and D. H. O'Day, 2011. EGF-like peptide-enhanced cell motility in *Dictyostelium* functions independently of the cAMP-mediated pathway and requires active Ca^{2+}/calmodulin signalling. *Cell. Signal.* 23: 731-738.

Hoeflich K.P. and M. Ikura, 2002. Calmodulin in action: diversity in target recognition and activation mechanisms. *Cell* 108: 739-742.

Ikezaki, H., Patel, M., Onyuksel, H., Akhter, S. R., Gao, X. P. and Rubenstein, I. (1999) Exogenous calmodulin potentiates vasodilation elicited by phospholipid-associated VIP in vivo. *Am. J. Physiol.: Regulat., Integrat. Comp. Physiol.* 276: 1359-1365.

Ikuta, T., N. Sogawa, H. Ariga, T. Ikemura and K. I. Matsumoto, 1998. Structural analysis of mouse tenascin-X: evolutionary aspects of reduplication of FNIII repeats in the tenascin gene family. *Gene* 217: 1-13.

Ikura, M., G. M. Clore, A. M. Gronenborn, G. Zhu, C. B. Klee, and A. Bax, 1992. Solution structure of a calmodulin-target peptide complex by multidimensional NMR, *Science* 256: 632-638.

Iyer, A. K. V., T. T. Kien, C. W. Borysenko, M. Cascio, C. J. Camacho, H. C. Balir, L. Bahar and A. Wells, 2007. Tenascin cytotactin epidermal growth factor-like repeat binds epidermal growth factor receptor with low affinity. *J. Cell Physiol.* 211: 748-758.

Iyer, A. K. V., K. T. Tran, L. Griffith and A. Wells, 2008. Cell surface restriction of EGFR by a tenascin cytotactin-endoded EGF-like repeat is preferential for motility-regulated signaling. *J. Cell Physiol.* 214: 504-512.

James, P., T. Vorherr and E. Carafoli, 1995.Calmodulin-binding domains: just two faced or multi-faceted? *Trends Biochem. Sci.* 20: 38-42.

Jermyn, K. A. and Williams, J. G. (1991) An analysis of culmination in *Dictyostelium* using prestalk and stalk-specific cell autonomous markers. *Development* 111: 779-787.

Jurado, L. A., S. P. Chockalingham and H. W. Jarrett, 1999. *Apocalmodulin, Physiol. Rev.* 79: 661-682.

King, J.S. and R.H. Insall, 2009. Chemotaxis: finding the way forward with *Dictyostelium*. *Trends Cell Biol.* 19: 523-530.

Klee, C. and T. Vanaman, 1982. Calmodulin, *Adv. Protein Chem.* 35: 213-321.

Lombardi, M.L., D.A. Knecht and J. Lee, 2008. Mechano-chemical signalling maintains the rapid movement of *Dictyostelium* cells. *Exp. Cell Res.* 314: 1850-1859.

Ma, L., X. Xu and D. Sun, 1999. The presence of a heterotrimeric G protein and its role in signal transduction of extracellular calmodulin in pollen germination and tube growth. *Plant Cell* 11: 1351-1362.

Matsukuma, S. and A.J. Durston, 1979. Chemotactic cell sorting in *Dictyostelium discoideum*. *J. Embryol. Exp. Morphol.* 50: 243-251.

Mroczkowski, B., M. Reich, K. Chen, G. I. Bell, and S. Cohen, 1989. Recombinant human epidermal growth factor precursor is a glycosylated membrane protein with biological activity. *Mol. Cell Biol.* 9: 2771-2778.

O'Day, D. H., 2003. CaMBOT: profiling and characterizing calmodulin-binding proteins. *Cell. Signal.* 15: 347-354.

O'Day, D.H., 2006. Calmodulin-Mediated Signaling in *Dictyostelium discoideum*: CaMBOT Isolation and Characterization of the Novel Poly-Domain Protein Nucleomorphin and Other Calmodulin Binding Proteins. Chapter 6, In: *Focus on Cellular Signalling Research*, Dorothy T.

Persechini, A. and R. H. Kretsinger, 1988. The central helix of calmodulin functions as a flexible tether, *J. Biol. Chem.* 263: 12175-12178.

Prieto, A.L., C. Andersson-Fisone and K.L. Crossin, 1992 Characterization of multiple adhesive and counteradhesive domains in the extracellular matrix protein cytotactin, *J. Cell Biol.* 119: 663-678.

Rao, Z., P.Handford, M. Mayhew, V. Knott, G. C. Browniee, and D. Stuart, 1995. The structure of a Ca^{2+}-binding epidermal growth factor-like domain: Its role in protein-protein interactions. *Cell* 82: 131-141.

Raper, K.B, 1940. Pseudoplasmodium formation and organization in *Dictyostelium discoideum*. *J. Elisha Mitchell Sci. Soc.* 56: 241-282.

Remgard, P., P. A. R. Ekstrom, P. Wiklund, and A. Edstrom, 1995. Calmodulin and in vitro regenerating frog sciatic nerves: Release and extracellular effects. *Europ. J. Neurosci.* 7: 1386-1392.

Rogers S. W., R. Wells and M. Rechsteiner, 1986. Amino acid sequences common to rapidly degraded proteins: The PEST hypothesis. *Science* 234: 364-368.

Schenk, S., E.Hintermann, M. Bilban, N. Koshikawa, C. Hojilla, R. Khokha and V. Quaranta, 1993. Binding to EGF receptor of a laminin-5 EGF-like fragment liberated during MMP-dependent mammary gland involution. *J. Cell Biol.* 161: 197-209.

Shimada, N., K. Nishio, M. Maeda, H. Urushiara and T. Kawata, 2004. Extracellular matrix family proteins that are potential targets of Dd-STATa in *Dictyostelium discoideum*. *J. Plant Res.* 117: 345-353.

Sternfeld, J. and R. O'Mara, 2005. Aerial migration of the *Dictyostelium* slug. *Dev. Growth Differen.* 47: 49-58.

Suarez, Andres, Robert J. Huber, Michael Myre, Danton H. O'Day, 2011a. An extracellular matrix, calmodulin-binding protein from *Dictyostelium* with EGF-Like repeats that enhance cell motility. *Cell. Signal* 23: xxx-xxx.

Suarez, Andres, Robert J. Huber, Danton H. O'Day, 2011b. Intracellular and extracellular processing of Dictyostelium cyrA, a novel matricellular, calmodulin-binding protein with EGF-like repeats that enhance cell motility. (manuscript submitted).

Suarez, Andres, Robert J. Huber, Danton H. O'Day, 2011c. Intra- and extra-cellular calmodulin from Dictyostelium during growth and development. (unpublished results).

Swindle, S. C., K. T..Tran, T. D. Johnson, P. Banerjee, A. M. Mayes, L. Griffith and A. Wells, 2001. Epidermal growth factor (EGF)-like repeats of human tenascin-C as ligands for EGF receptor. *J. Cell Biol.* 154: 459-468.

Taylor, J.M., W.M. Mitchell and S. Cohen, 1972. Epidermal growth factor: physical and chemical properties. *J. Biol. Chem.* 247: 5928-5934.

Valeyev, N.V., J.-S. Kim, J.S. Heslop-Harrison, I. Postlewaite, N.V. Kotov and D.G. Bates, 2009. Computational modelling suggests dynamic interactions between Ca^{2+}, IP3 and G protein-coupled modules are key to robust *Dictyostelium* aggregation. *Mol. Biosyst.* 5: 612-628.

Van Haastert, P.J.M., I. Keizer-Gunnink and A. Kortholt, 2007. Essential role of PI3-kinase and phospholipase A2 in *Dictyostelium discoideum* chemotaxis. *J. Cell. Biol.* 177: 809-816.

Wilkins, M.R. and K.L. Williams, 1995. The extracellular matrix of the *Dictyostelium discoideum* slug. *Experientia* 51: 1189-1196.

Zanuttin, F., C. Guarnaccia, A. Pintar and S. Pongor, 2004. Folding of epidermal growth factor-like repeats from human tenascin studied through a sequence frame-shift approach. *European J. Biochem.* 271: 4229–4240.

In: Calcium Signaling
Editor: Masayoshi Yamaguchi

ISBN: 978-1-61324-313-8
©2012 Nova Science Publishers, Inc.

Chapter 5

A TALE OF THREE PROTEINS: THE NUCLEOLAR CALMODULIN-BINDING PROTEIN NUCLEOMORPHIN AND ITS BINDING PARTNERS CALCIUM-BINDING PROTEIN 4A AND PUROMYCIN-SENSITIVE AMINOPEPTIDASE

*Danton H. O'Day[*1,2], Andrew Catalano[1] and Yekaterina Poloz[1]*

[1]Department of Cell and Systems Biology,
University of TorontoToronto, Ontario, Canada
[2]Department of Biology, University of Toronto at Mississauga,
Mississauga, Ontario, Canada

ABSTRACT

Nucleomorphin (NumA1) was isolated and verified as a calmodulin-binding protein from multicellular development of *Dictyostelium*. NumA1 shares many attributes with mammalian nucleolar proteins including multiple nuclear and nucleolar localization signals (NLSs and NoLSs) and an extensive acidic glu/asp repeat. Initial studies showed that deletion of the glu/asp repeat resulted in the formation of multiple nuclei suggesting a role for NumA1 in regulating nuclear number and continued work suggests a cell cycle role for this protein. Yeast two hybrid (Y2H) and co-immunoprecipitation studies revealed that calcium-binding protein 4a (CBP4a /cbpD1) binds to the acidic glu/asp domain of NumA1 in a calcium-dependent manner. NumA1 was subsequently verified as a nucleolar protein when treatments with actinomycin-D (AM-D) caused the loss of NumA1 patches adjacent to the nuclear envelope. In keeping with its nucleolar localization, immunolocalization studies revealed that NumA1 disappears from the nucleolar patches during prophase when the nucleolus dissociates and reappears during telophase with nucleolar reformation. As a NumA1-binding partner, CBP4a also localizes to the nucleolus and, similarly, AM-D treatment causes its loss from the nucleolus. CBP4a also disappears and reappears during prophase and telophase respectively, as befits a nucleolar protein. During mitosis, CBP4a exists as multiple islands throughout

---

* Correspondence to Dr. Danton H. O'Day: Email: danton.oday@utoronto.ca

the nucleoplasm. Treatment of cells with BAPTA, a chelator of calcium, causes CBP4a to be dislodged from the nucleolus which fits with its role as a calcium-dependent NumA1-binding protein. A second NumA1-binding protein was also revealed by Y2H which does not require calcium: *Dictyostelium* puromycin-sensitive aminopeptidase (DdPsa). DdPsa localizes predominantly to the nucleus. The relationship of NumA1 to mammalian nucleolar proteins and further insight into the role of calcium signalling is discussed in the light of additional data.

Keywords: Nucleomorphin; calcium-binding protein 4a; puromycin-sensitive amino-peptidase; nucleolus; nucleoplasm; NLS; NoLS; mitosis; actinomycin-D; Dictyostelium.

INTRODUCTION

The Nucleolus and Nucleolar Proteins

The nucleolus is primarily an interphase component of the nucleus that dissociates during prophase reforming during telophase in mammalian cells (McClintock, 1934). Innumerable studies have verified the functional link between nucleoli and rRNA synthesis (Bartova et al, 2010). Actinomycin-D (AM-D) treatment, which inhibits RNA polymerase, leads to nucleolar dissolution while removal of the drug results in nucleolar reformation. The nucleolus is more than a ribosome factory: it is a multifunctional intranuclear domain involved in the cell cycle, viral replication, molecular sequestering, stress responses, chaperone activity and centrosome function (e.g., Bartova et al, 2010; Pederson and Tsai, 2009; Sirri et al, 2008). Proteomic studies have revealed the presence of about 700 nucleolar proteins, many uncharacterized (Anderson et al, 2005). The sequence of events of the nucleolar cycle remains to be clarified. While nucleolar reformation during telophase occurs around the nucleolar organizer region (rDNA), it is not regulated solely by the re-starting of rRNA transcription (Prieto and McStay, 2008; Sirri, et al, 2002; Sirri et al, 2008).

Many major, multifunctional nucleolar proteins, including B23 (nucleophosmin/NPM1 /NO38/ numatrin), fibrillarin, nucleostemin, and nucleolin, share two primary elements: one or more extensive acidic glu/asp tracts plus one or more nuclear and nucleolar localization signals (NLSs, NoLSs; Mongelard and Bouvet, 2007; Pederson and Tsai, 2009; Sirri et al, 2008). While it is not yet understood how NoLSs, especially those that co-function as NLSs, direct specific proteins to the nucleolus it is clear that they are effective and critical. A single amino acid change in the NoLS of ribosomal protein S19 stops its nucleolar localization leading to Diamond-Black-fan anemia (DBA; Kundu-Michalik, 2008). The acidic glu/asp tract binds to NLS/NoLS sequences potentially serving to recruit other nucleolar proteins (Okuwaki, 2008; Xue and Melese, 1994). In spite of our widespread and growing knowledge of the non-ribosomal functions of nucleoli in mammalian cells, comparatively little is known about nucleolar proteins and their alternative functions in lower eukaryotes.

The Nucleolus of *Dictyostelium*

Dictyostelium discoideum is a central model organism for studying the fundamental events of cell and developmental biology and for understanding many major diseases. Its genome has been sequenced plus there is an interactive database and extensive stock centre of strains (wild type, deletion mutants, knock-outs, etc.) and plasmids (http://dictybase.org/). The organism grows and multiplies as a unicellular, haploid amoeba with starvation leading to multicellular development making it especially amenable to experimental intervention and mutant generation. Most eukaryotes have a nucleolus with distinct regions (fibrillar component or FC, dense fibrillar component or DFC, and granular component or GC) as seen in transmission electron microscope images, however *Dictyostelium* has a bipartite nucleolus, with overlapping DFC and GC regions, similar to that observed in dinoflagellates, ascomycetes, yeast and *Xenopus* (Benichou et al, 1983; Louvet et al, 2005; Maeda and Takeuchi, 1969). As in other organisms, the *Dictyostelium* nucleolus is organized around rDNA genes functioning as a ribosomal rRNA synthesizing/processing centre coupled with pre-ribosome biogenesis (Benichou et al, 1983; Moerman and Klein, 1995). However, it is organized around extrachromosomal, linear rDNA strands as opposed to chromosomal rDNA cistrons (Cockburn et al, 1978).

The organization of nucleoli around extrachromosomal linear or circular rDNA genes is not uncommon in lower eukaryotes. In the parasitic *Entamoeba histolytica* and *E. invadens*, the extrachromosomal rDNA organizes nucleolar patches adjacent to the inner nuclear membrane that are essentially identical to *Dictyostelium* nucleoli (Jhingan et al, 2009). Many other organisms show a *Dictyostelium*-like nucleolar structure including fungi (e.g., *Physarum*) and protozoa (e.g., *Paramecium, Euglena, Entamoeba histolytica*; Jhingan et al, 2009). In fact the evolution of nucleoli remains enigmatic as does the primary role of rDNA genes in organizing the non-ribosomal functions of the nucleolus in all species (McKeown and Shaw, 2009). Thus the study of nucleolar structure and function in a diversity of organisms is essential. While it is firmly established that the rDNA sequences organize nucleoli within the nucleoplasm in other organisms, it is not clear how non-chromosomal rDNA can establish nucleolar patches adjacent to the inner nuclear membrane in *Dictyostelium*. Do specific nucleolar proteins associate first with the inner nuclear membrane serving as a primary anchor for these untethered rDNA sequences? Understanding how proteins with different functions enter and exit nucleoli under various conditions can reveal candidates for this potential role.

There are a multitude of reasons for studying the structure and function of nucleoli. For example, in a diversity of organisms including humans, a large number of viral proteins translocate between the nucleoplasm and nucleolus as part of the viral replication cycle diverting nucleolar proteins from their normal cellular functions (Tarapore et al, 2006). In humans, these include proteins from HIV, HPV, HSV-1/2, influenza virus and more (e.g., Szebeni et al, 1997). Insight into the diverse roles of nucleolar proteins and their translocation to different intracellular locales will benefit from studies of nucleologenesis in different organisms and model systems (Pederson and Tsai, 2009; Prieto and McStay, 2008; Sirri et al, 2008).

While novel approaches (e.g., pseudo-NORs; Prieto and McStay, 2008) can help us understand some of the elements involved in nucleolar formation and function, we believe that using a model biological system (i.e., *Dictyostelium*) more likely will elucidate the normal nucleolar cycle events as well as the biological function of specific nucleolar proteins. There were a number of classic studies on nucleolar structure in *Dictyostelium* published decades ago but only a small number of nucleolar proteins have been characterized to any significant degree including: Eif6 (eukaryotic translation initiation factor 6), Hsp32 (heat shock protein 32), TRAP1 (tumour necrosis factor receptor-associated protein 1) and nucleomorphin (Balbo and Bozzaro, 2006; Catalano and O'Day, 2011b; Moerman and Klein, 1998; Myre and O'Day, 2002; Yamaguchi et al, 2005).

CURRENT RESEARCH

Nucleomorphin: A Nuclear and Nucleolar Calmodulin-Binding Protein

Nucleomorphin is a BRCT-domain family member, calmodulin-binding protein that regulates nuclear number and is thought to be involved in the cell cycle in *Dictyostelium discoideum* (Catalano and O'Day, 2008; Myre and O'Day, 2002; 2004b; O'Day, 2006). As shown in Figure 1, immunolocalization studies have shown that NumA1 resides in the nucleolus and to a lesser extent the nucleoplasm (Myre and O'Day, 2002; Catalano and O'Day, 2011b). This nucleolar localization was verified when NumA1 could no longer be demonstrated in nuclear patches (nucleoli) after treatment with actinomycin D (AM-D) which inhibits rRNA synthesis and results in the loss of nucleoli (Figure 1).

The experimentally verified functional domains of NumA1 are shown in Figure 2. Several attributes link NumA1 to mammalian nucleolar proteins: an extensive acidic glu/asp tract, one or more NLS/NoLSs, and many phosphorylation sites, plus translocation between the nucleolus and nucleus. Attempts to knock-out NumA1 have been unsuccessful suggesting that it is an essential protein. Overexpression of GFP-NumA1Δglu/asp constructs, lacking the extensive acidic tract, results in highly multinucleate cells without affecting NumA1 localization (Myre and O'Day, 2002). Similarly, the calmodulin-binding domain (^{182}LQKQQKIYKDLERF^{195}) is not involved in nucleolar localization (Myre and O'Day, 2005).

Expression of other deletion constructs coupled with peptide studies showed that functional NLSs (classical: NLS-1, ^{31}PKSKKKF^{37}, NLS-2, ^{61}RPRK^{64}; bipartite: NLS-4, ^{48}KKSYQDPEIIAHSRPRK^{64}) that are only present in the N-term also act as NoLSs. (Figure 2; Myre and O'Day, 2005; Catalano and O'Day, 2011b).

C-Term NLS-3 (^{246}PTKKRSL^{252}) peptides can also translocate to the nucleolus but deletion constructs revealed it is not a functional NLS in NumA1 (Myre and O'Day, 2005). Work by others showed that certain mammalian nucleolar protein localization involves NLSs binding to acidic glu/asp sequences (Okuwaki, 2008; Xue and Melese, 1994). Mixtures of FITC-NLS/NoLS peptides with FITC-glu/asp peptides also suggest a direct interaction between the two (Catalano and O'Day, 2011b).

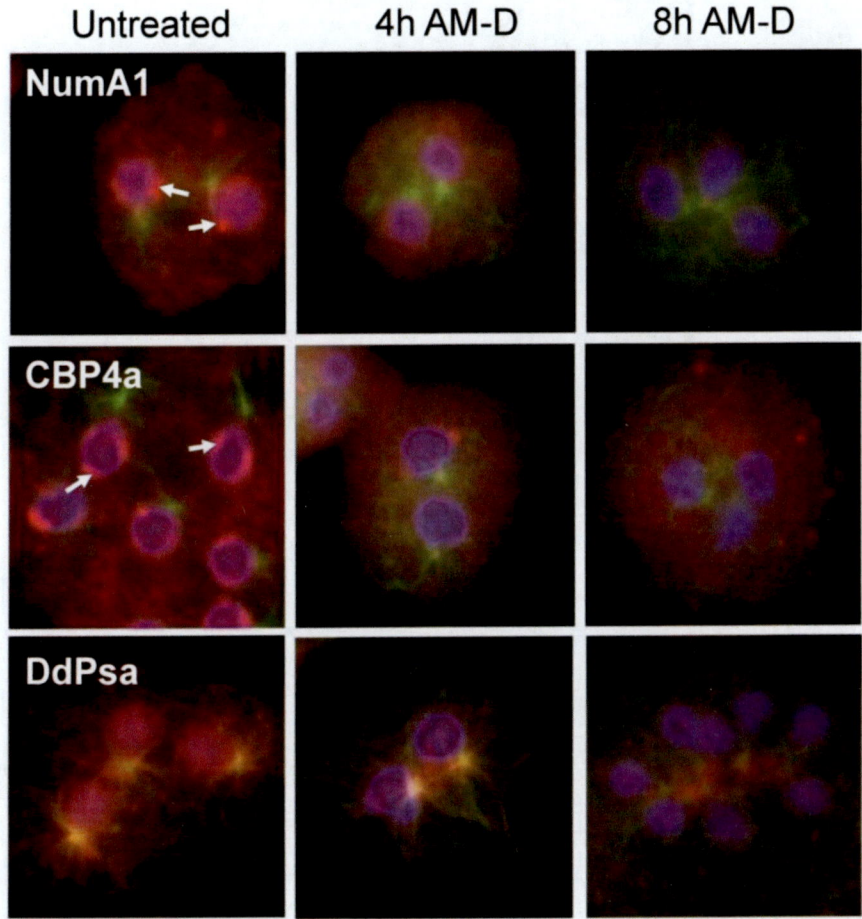

Figure 1. Effect of actinomycin D (AM-D) on localization of nucleomorphin (NumA1), calcium-binding protein 4a (CBP4a) and puromycin-sensitive aminopeptidase (DdPsa). Cells were untreated (first column) or treated with 0.05 mg/mL AM-D for either 4 hours or 8 hours after which they were fixed in ultracold methanol. Cells on coverslips were probed with anti-NumA1, -CBP4a or -DdPsa followed by Alexa Fluor 555® goat anti-rabbit (red), then with anti-α-tubulin followed by Alexa Fluor 488® goat anti-mouse (green). Coverslips were slide-mounted with Prolong Antifade containing DAPI to stain nuclear DNA (blue). A Nikon 100x oil-immersion objective lens with an aperture of 1.30 was used to capture images.

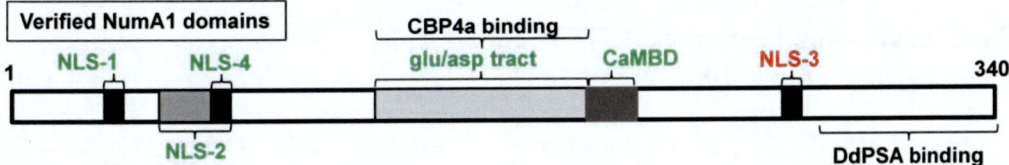

Figure 2. Domain structure diagram of nucleomorphin (NumA1). The verified functional domains are shown in green. The nuclear localization signals 1, 2 and 4 (NLS-1, -2, -4) are functional in NumA1 while NLS-3 is not. The calmodulin-binding domain (CaMBD) possesses Ca^{2+}-dependent and -independent binding motifs for calmodulin. NumA1 interacts with in a calcium-dependent manner with calcium-binding protein 4a (CBP4a) via its extensive, acidic glu/asp tract. It also interacts with puromycin-sensitive aminopeptidase (DdPsa) via its C-term.

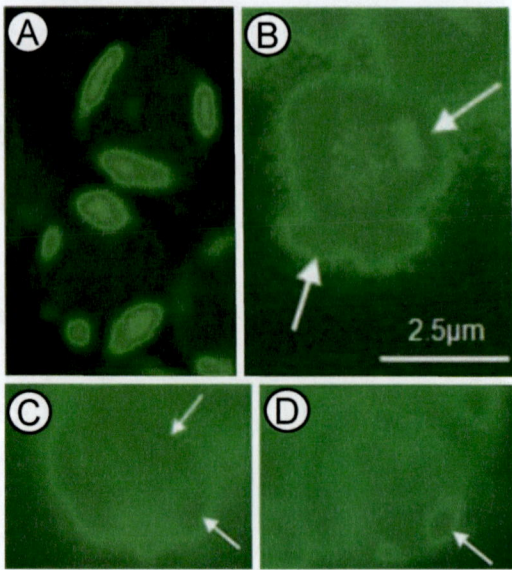

Figure 3. Localization of FITC-nuclear/nucleolar localization signal (NLS/NoLS) peptides. A. Cells incorporate the FITC- NLS/NoLS peptides into the nucleolus. B. A similar localization occurs when isolated nuclei were treated with these peptides. The full nucleolar patch localization of NLS/NoLS peptides (C) is restricted to small intranuclear patches, if cells are treated with actinomycin D for 4 hours prior to peptide treatment (D).

This bi-functional role of NLSs as NoLSs is common in mammalian nucleolar proteins. Of particular note is that the NoLS peptides typically localize as patches equivalent to natural nucleoli in both intact cells and isolated nuclei suggesting that while they target NumA1 to the nucleolus they likely are doing so via association with previously existing nucleolar proteins (Figure 3). More interesting is evidence that prior treatment with AM-D doesn't completely eradicate the nucleolar localization of these peptides but instead leads to the localization within smaller patches adjacent to the inner nuclear envelope (Figure 3, C,D). This further indicates that in *Dictyostelium* nucleolar domains adjacent to the inner nuclear envelope are present that serve to organize the nucleolus. Whether this is due to the localization of rDNA sequences, as occurs in mammalian cells, remains to be elucidated. Additional insight into the behaviour of NumA1 was gained from immunolocalization studies during mitosis as discussed below.

Calcium-Binding Protein 4a (CBP4a/cpdD1): A Nucleolar NumA1-Binding Partner

Yeast 2 hybrid (Y2H) and co-immunoprecipitation showed Ca^{2+}-binding protein CBP4a (cbpD1) binds to the glu/asp tract in NumA1 in a Ca^{2+}-dependent manner (Myre and O'Day, 2004a). CBP4a is primarily a prestalk (PstO) cell protein (Dorywalska et al, 2000; Maeda et al, 2003; Sakamoto et al, 2003). CBP4a has two identified but not verified functional domains: an NLS and an FHA-binding motif (Catalano and O'Day, 2011a). FHA domains are DNA-binding sequences found in nuclear and nucleolar protein kinases and transcription factors. For example, JUMU is an FHA domain transcription factor involved in the nucleolar structure and function in *Drosophila* (Hofmann et al, 2010). CBP4a and NumA1 co-localize and AM-D treatment causes the loss of their nucleolar localization (Figure 1; Catalano and O'Day, 2011a). In contrast, treatment with BAPTA, a calcium chelator, causes the loss of nucleolar CBP4a but not NumA1 showing that while they interact, their nucleolar localization is differentially regulated (Figure 4).

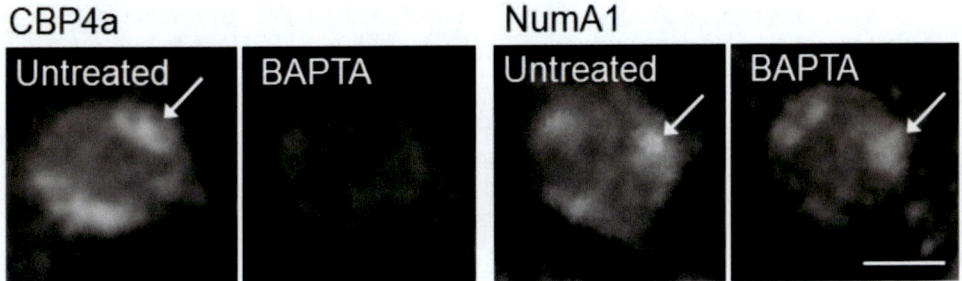

Figure 4. Effect of calcium chelation on immunolocalization of calcium-binding protein 4a (CBP4a) and nucleomorphin (NumA1). High magnification images of nucleus demonstrates CBP4a and NumA1 in nucleolar patches (arrows) in untreated cells. After treatment, CBP4a is no longer detectable in the nucleolus while the nucleolar localization of NumA1 remains unchanged. Scale bar represents 2 μM.

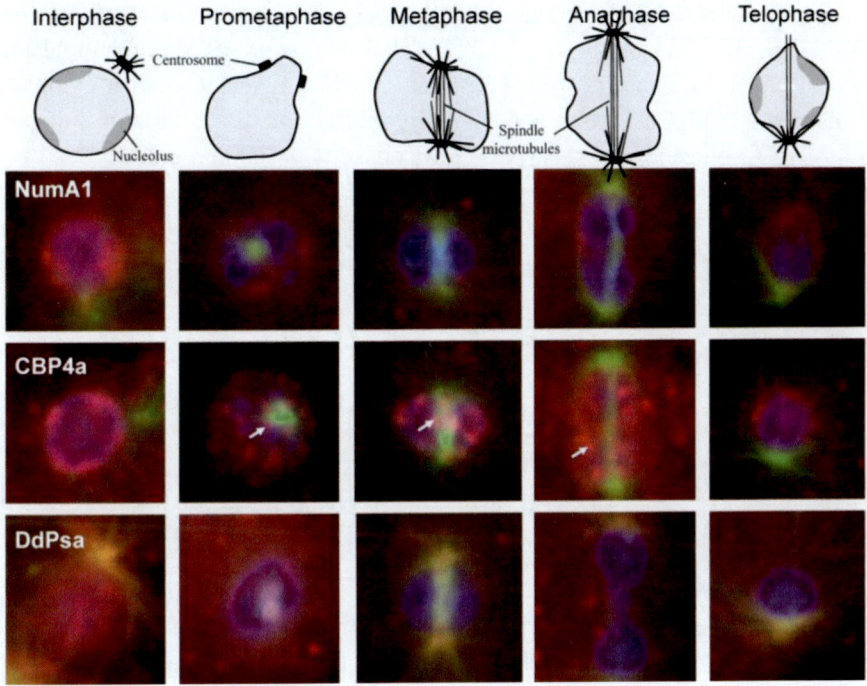

Figure 5. The immunolocalization of nucleomorphin (NumA1), Calcium-binding protein 4a (CBP4a) and *Dictyostelium* puromycin-sensitive aminopeptidase (DdPsa) during mitosis. *Dictyostelium* undergoes a closed mitosis in which the nuclear envelope remains intact (top images). Immunolocalization using polyclonal antibodies was used to follow the localization of NumA1, CBP4a and DdPsa (red) throughout mitosis. The nucleus (DNA) was stained with DAPI (blue) and microtubules were localized with anti-α-tubulin (green). The arrows indicate the CBP4a islands that form during mitosis. Images were captures with a Nikon 100x oil-immersion objective lens with an aperture of 1.30.

Nucleolar Localization of NumA1 and CBP4a during Mitosis

The NumA1/CBP4a relationship changes during mitosis (Figure 5). With nucleolar dissolution, NumA1 moves into the nucleoplasm until nucleolar reformation (Catalano and O'Day, 2011b). In contrast, CBP4a disassembles into many large, diffuse nucleoplasmic "islands" which later align at the nuclear periphery for nucleolar reformation during telophase (Figure 5). The developmental regulation of NumA1 and CBP4a further emphasizes their inter-relationship. *Dictyostelium* development is controlled by various differentiation factors including cAMP, ammonia and differentiation inducing factor 1 (DIF-1; reviewed in O'Day et al, 2009). DIF-1 mediates stalk cell differentiation, ammonia is involved in spore formation and cAMP functions differentially in both. During development, expression of both NumA1 and CBP4a is stimulated by DIF-1 and repressed by cAMP/ammonia showing a co-regulation related to their link to stalk cell differentiation (Maeda et al, 2003; O'Day et al, 2009).

Puromycin-Sensitive Aminopeptidase (PSA): A Second NumA1-Binding Partner

Aminopeptidases have been identified in bacteria, fungi, plants and animal tissues, which suggest they are involved in a number of biological processes such as hormone production and peptide digestion (Takahashi, *et al*, 1987; Takahashi, *et al*, 1989). Immunohistochemical analysis of PSA from a murine neuroblastoma cell line revealed both a cytoplasmic and nuclear distribution in COS cells and 3T3 fibroblasts and both bestatin and puromycin, inhibitors of PSA, blocked mitosis at the G_2/M phase (Constam, *et al*, 1995). This suggests that PSA is involved in proteolytic regulation of the cell cycle (Constam, *et al*, 1995). In keeping with this, PSA has also been implicated in a similar role in rat primary hepatocytes (Takahashi, *et al*, 1989).

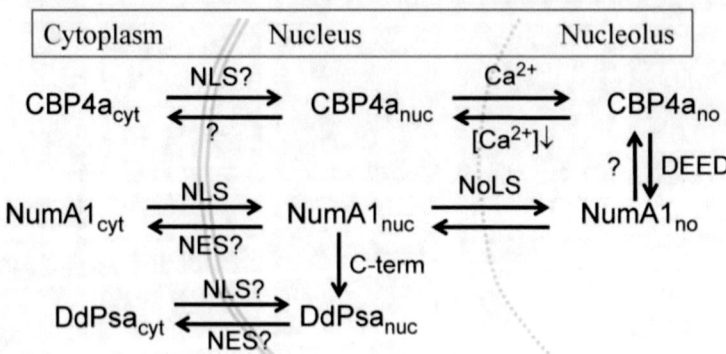

Figure 6. A working model for the interaction between nucleomorphin (NumA1) and its binding proteins calcium-binding protein 4a (CBP4a) and *Dictyostelium* puromycin-sensitive aminopeptidase (DdPsa). NumA1 binds to CBP4a in a calcium-dependent manner via an acidic glu/asp domain. Their co-localization in the nucleolus is calcium-dependent since chelation of calcium leads to the loss of nucleolar CBP4a but not NumA1. The nuclear relationship between NumA1 and CBP4a remains to be elucidated but the two proteins do show differing patterns of nucleoplasmic localization during mitosis. Nuclear DdPsa binds to NumA1 via its C-term. While the NLS/NoLSs of NumA1 have been defined, none have been experimentally demonstrated for DdPsa or CBP4a.

There is accumulating evidence suggesting a cell cycle role for *Dictyostelium* NumA1. The identification of a BRCT-domain containing isoform, the finding that overexpression of dominant negative mutants lacking the glu/asp tract induces highly multinucleate cells, and the identification of a specific interaction with *Dictyostelium* PSA (DdPsa), suggest that proteolytic regulation of NumA1 may be required for proper progression through the cell cycle.

Y2H screening revealed DdPsa as a second binding partner of NumA1. Binding is associated with the C-term (216-340) region of NumA1. While NumA1 and another binding protein, CBP4a, were found by immunolocalization to reside predominantly in the nucleolus, DdPsa localizes mainly to the nucleoplasm but not the nucleolus (Figures 1, 5). GFP-DdPsa also localizes to the nucleoplasm but not the nucleolus (Catalano, *et al*, 2011). During mitosis GFP-DdPsa leaves the nucleoplasm and redistributes throughout the cytoplasm. During development, DdPsa co-localizes with NumA1 in the nucleoplasm of both prestalk and prespore cells (Poloz et al, 2011).

The Translocation of NumA1, CBP4a and DdPsa

The results to date suggest that NumA1 interacts primarily with CBP4a in the nucleolus and with DdPsa in the nucleoplasm (Figure 6). NumA1 and CBP4a localize primarily to the nucleolus but during mitosis show different patterns of localization in the nucleoplasm. At some time (e.g., at least when they are synthesized), the proteins must exist in the cytoplasm using verified NLSs (NumA1) or identified but not verified NLSs (CBP4a, DdPsa) to enter the nucleus. Calcium signalling plays at least a part in the nucleolar localization of CBP4a but what role it and calmodulin have in the function or functional relationship of NumA1, CBP4a and DdPsa remain to be elucidated. Similarly in addition to verifying the NLSs of CBP4a and DdPsa the nuclear export signals (NES) remain to be determined. In spite of this CBP4a and NumA1 represent the first interacting nucleolar proteins in *Dictyostelium*.

ACKNOWLEDGMENT

This chapter was supported by a grant from the Natural Sciences and Engineering Council of Canada (DHO'D; A6807).

REFERENCES

Anderson, J.S., Y.W. Lam, A.K. Leung, S.E. Ong, C.E. Lyon, A.I. Lamondet and M. Mann, 2005. Nucleolar proteome dynamics. *Nature* 433: 77-83.

Balbo, A. and S. Bozzaro, 2006. Cloning of *Dictyostelium* eIF6 (p27[BBP]) and mapping its nucle(ol)ar localization subdomains. *Eur. J. Cell Biol.* 85: 1069-1078.

Bartova, E., A. H. Horakova, R. Uhlirova, I. Raska, G. Gliova, D. Orlova and S. Kozubek, 2010. Structure and epigenetics of nucleoli in comparison with non-nucleolar compartments. *J. Histochem. Cytochem.* 58: 391-403.

Benichou J.-C., B. Quiviger and A. Ryter, 1983. Cytochemical study of the nucleolus of the slime mold *Dictyostelium discoideum*. *J. Ultrastruct. Res*. 84: 60-66.

Catalano, A. and D. H. O'Day, 2008. Calmodulin-binding proteins in the model organism *Dictyostelium*. A complete and critical review. *Cell Signal*. 20: 277-291.

Catalano, A. and D. H. O'Day, 2011a. Calcium-dependent nucleolar localization of *Dictyostelium* calcium-binding protein 4a (CBP4a/cbpD1) during interphase and mitosis Manuscript in preparation.

Catalano, A. and D. H. O'Day, 2011b. Nucleolar localization and identification of nuclear/nucleolar localization signals of the calmodulin-binding protein nucleomorphin during growth and mitosis in *Dictyostelium*. *Histochem. Cell Biol*. 135: 239-249.

Catalano, A., Poloz, Y. and O'Day, D.H. (2011). Puromycin sensitive-aminopeptidase from *Dictyostelium*: A nuclear-binding partner of the calmodulin-binding protein nucleomorphin. Manuscript in preparation.

Cockburn, A. F., W. C. Taylor and R. A. Firtel, 1978. Dictyostelium rDNA of non-chromosomal palindromic dimers containing 5S and 36S coding regions. *Chromosoma* 70: 19-29.

Constam, D. B., A. R. Tobler, A. Rensing-Ehl, I. Kemler, L. B. Hersh, and A. Fontana, 1995. Puromycin-sensitive aminopeptidase: Sequence analysis, expression, and functional characterization, *J. Biol. Chem*. 270: 26931-26939.

Dorywalska M, B. Coukell and A. Dharamsi, 2000. Characterization and hetereologous expression of cDNAs encoding two novel closely related Ca^{2+}-binding proteins in *Dictyostelium discoideum*. *Biochim. Biophys. Acta. Mol. Cell. Res*. 1496: 356-361.

Hofmann, A., M. Brunner, A. Schwendemann, M. Strodicke, S. Karberg, A. Klebes, H. Saumweber and G. Korge, 2010. The winged-helix transcription factor JUMU regulates development, nucleolus morphology and function, and chromatin organization in *Drosophila melanogaster*. *Chromosome Res*. 18: 307-324.

Jhingan, G. D., S. K. Panigrahi, A. Bhattacharya and S. Bhattacharya, 2009. The nucleolus in *Entamoeba histolytica* and *Entamoeba invadens* is located at the nuclear periphery. *Mol. Biochem. Parasitol*. 167: 72-80.

Kundu-Michalik, S., M.-A. Bisotti, E. Lipsius, A. Bauche, A. Kruppa, T. Klokow, G. Kammler and J. Kruppa, 2008. Nucleolar binding sequences of the ribosomal protein S6e family reside in evolutionary highly conserved peptide clusters. *Mol. Biol. Evol*. 25: 580-590.

Louvet, E., H. R. Junera, S. Le Panse, and D. Hernandez-Verdun, 2005. Compartmentation of the nucleolar processing proteins in the granular component is a CK2-driven process. *Mol. Biol. Cell* 17: 2537-2546.

Maeda, M., H. Sakamoto, N. Iranfar, D. Fuller, T. Maruo, S. Ogihara, T. Morio, H. Urushihara, Y. Tanaka and W. F. Loomis, 2003. Changing patterns of gene expression in *Dictyostelium* prestalk cell subtypes recognized by in situ hybridization with genes from microarray analyses. *Eukaryotic Cell*. 2: 627-637.

Maeda, Y. and Takeuchi, I, 1969. Cell differentiation and fine structures in development of cellular slime molds. *Dev. Growth Diff*. 11: 232-245.

McClintock, B., 1934. The relation of a particular chromosomal element of the development of the nucleoli in *Zea mays*. Z. Zellforsch. *Mikrosk Anat*. 21: 294-328.

McKeown, P.C. and P.J. Shaw, 2009. Chromatin: linking structure and function in the nucleolus. *Chromosoma* 118: 11-23.

Moerman, A. M. and C. Klein, 1998. *Dictyostelium discoideum* Hsp32 is a resident nucleolar heat-shock protein. *Chromosoma* 107: 145-154.

Mongelard, F. and P. Bouvet, 2007. Nucleolin: a multifaceted protein. *Trends Cell Biol* 17: 80-85.

Myre, M. A., and D. H. O'Day, 2002. Nucleomorphin. A novel, acidic, nuclear calmodulin-binding protein from *Dictyostelium* that regulates nuclear number. *J. Biol. Chem.* 277: 19735-19744.

Myre, M. A., and D. H. O'Day, 2004a. *Dictyostelium* calcium-binding protein 4a interacts with nucleomorphin, a BRCT-domain protein that regulates nuclear number. *Biochem. Biophys. Res. Commun.* 322: 665-671.

Myre, M. A., and D. H. O'Day, 2004b. *Dictyostelium* nucleomorphin is a member of the RCT-domain family of cell cycle checkpoint proteins. *Biochim. Biophys. Acta.* 1675: 192-197.

Myre, M. A., and D. H. O'Day, 2005. An N-terminal nuclear localization sequence but not the calmodulin-binding domain mediates nuclear localization of nucleomorphin, a protein that regulates nuclear number in *Dictyostelium*. *Biochem. Biophys. Res. Commun.* 332: 157-166.

O'Day, D. H. (2006). Calmodulin-mediated signaling in *Dictyostelium discoideum*: CaMBOT isolation and characterization of the novel poly-domain protein nucleomorphin and other calmodulin-binding proteins. In *Focus on Cellular Signalling Research, Chapter 6.* (ed. D. T. Leeds). Hauppage, New York: Nova Science Publishers Inc.

O'Day, D. H., Y. Poloz and M. A. Myre, 2009. Differentiation inducing factor-1 (DIF-1) induces gene and protein expression of the *Dictyostelium* nuclear calmodulin-binding protein nucleomorphin. *Cell. Signal.* 21: 317-323.

Okuwaki, M. 2008. The structure and function of NPM1/nucleophosmin/B23, a multifunctional nucleolar acidic protein. *J. Biochem.* 143: 441-448.

Pederson, T. and R. Y. Tsai 2009. In search of nonribosomal nucleolar protein function and regulation. *J. Cell Biol.* 184: 771-776.

Poloz, Y., A. Catalano, and D.H.O'Day, 2011. *Enzymatic activity and developmental expression and localization of Dictyostelium puromycin-sensitive aminopeptidase.* Manuscript in preparation.

Prieto, J.-L. and B. McStay, 2008. Pseudo-NORs: a novel model for studying nucleoli. *Biochim. Biophys. Acta* 1783: 2116-2123.

Sakamoto, H., K. Nishio, M. Tomisako, H. Kuwayama, T. Yoshimasa, I. Suetake, S. Tajima, S. Ogihara, B. Coukell, and M. Maeda, 2003. Identification and characterization of novel calcium-binding proteins of *Dictyostelium* and their spatial expression patterns during development. *Develop Growth Differ.* 45: 507-514.

Sirri, V., D. Hernandez-Verdun and P. Roussel, 2002. Cyclin-dependent kinases govern formation and maintenance of the nucleolus. *J. Cell Biol.* 156: 969-981.

Sirri, V., S. Urcuqui-Inchima, P. Roussel, and D. Hernandez-Verdun, 2008. Nucleolus: the fascinating nuclear body. Histochem. *Cell Biol.* 129: 13-31.

Szebeni, A., B. Mehrota, A. Baumann, S. A. Adam, P. T. Wingfield and M. O. J. Olson, 1997. Nucleolar protein B23 stimulates nuclear import of the HIV-1 rev protein and NLS-conjugated albumen. *Biochemistry* 36: 3941-3949.

Takahashi, S. L., H. Kato, A. Takahashi, T. Noguchi, and H. Naito, 1987, The important role of bestatin- and leupeptin-sensitive proteases in the protein degradation pathway *in vivo*, *Int. J. Biochem.* 19: 401-412.

Takahashi, S., Y. Ohishi, H. Kato, T. Noguchi, H. Naito, T. Aoyagi, and H. Umezawa, 1989. The effects of bestatin, a microbial aminopeptidase inhibitor, on epidermal growth factor-induced DNA synthesis and cell division in primary cultured hepatocytes of rats. *Exp. Cell Res.* 183: 399–412.

Tarapore, P., K. Shimura, H. Suzuki, Y. Tokuyama, S.-H. Kim, A. Mayeda and K. Fukasawa, 2006. Thr199 phosphorylation targets nucleophosmin to nuclear speckles and represses pre-mRNA processing. *FEBS Lett.* 580: 399-409.

Xue, Z. and T. Melese, 1994. Nucleolar proteins that bind NLSs: a role in nuclear import or ribosome biogenesis? *Trends Cell Biol* 4: 414-417.

Yamaguchi, H., T. Morita, A. Amagai and Y. Maeda, 2005. Changes in spatial and temporal localization of *Dictyostelium* homologues of TRAP 1 and GRP94 revealed by immunoelectron microscopy. *Exp. Cell Res.* 303: 415-424.

ISBN: 978-1-61324-313-8
©2012 Nova Science Publishers, Inc.

Chapter 6

INSIGHTS INTO THE ROLE OF INTRACELLULAR CA^{2+} CONCENTRATION CHANGES DURING IN VITRO CHONDROGENESIS

Csaba Matta[1], János Fodor[2], Tamás Juhász[1] and Róza Zákány[1]

[1]Department of Anatomy, Histology and Embryology; University of Debrecen Medical and Health Science Centre, Debrecen, Hungary
[2]Department of Physiology; University of Debrecen Medical and Health Science Centre, Debrecen, Hungary

ABSTRACT

It is well established that Ca^{2+} signalling mediates the effects of mechano-transduction in chondrocytes of mature articular cartilage. However, little is known about the precise regulation of Ca^{2+} homeostasis in differentiating cells of developing hyaline cartilage. Therefore, our research group is committed to characterise the Ca^{2+} homeostasis and to map the 'Ca^{2+}toolkit' of differentiating chondrogenic mesenchymal cells. High density cell culture system (HDC) established from chondrogenic mesenchymal cells isolated from limb buds of 4-day-old chicken embryos is a well-known model of *in vitro* cartilage differentiation, in which a spontaneous cartilage formation occurs in 6 days. We measured cytosolic free Ca^{2+} concentration ([Ca^{2+}]$_i$) in cells of HDC on different days of culturing. After an initial value of 80 nM, a significant transient elevation was detected in Fura-2-loaded cells on day 3 of culturing, when the majority of cells differentiate into chondroblasts and chondrocytes. This 140 nM peak of cytosolic Ca^{2+} concentration is a result of increased Ca^{2+} influx and is found to be indispensable to proper chondrogenesis, because elimination of extracellular Ca^{2+} abolished the Ca^{2+} peak of day 3 and inhibited cartilage formation. Uncontrolled Ca^{2+} influx evoked by a Ca^{2+} ionophore (A23187) exerted dual effects on chondrogenesis in a concentration-dependent manner; low concentration of the ionophore increased [Ca^{2+}]$_i$ up to 150 nM and facilitated cartilage formation, whereas high concentration of this compound elevated it over 250 nM and almost totally blocked cartilage formation. Proliferation of chondrogenic cells was more sensitive to modulation of [Ca^{2+}]$_i$ then the viability of cells. Although chondrogenic cells express both IP$_3$ and ryanodine receptors and can release Ca^{2+} from intracellular stores, these stores proved to play a minor role in the Ca^{2+} homeostasis of these cells. As the inhibition of the Ca^{2+}-calmodulin sensitive protein phosphatase calcineurin impeded the $[Ca^{2+}]_i$-peak in chondrogenic cells and reduced cartilage formation, we propose its contribution in the regulation of $[Ca^{2+}]_i$ in these cells. We also found that chondrogenic cells secreted ATP and administration of ATP to the culture medium evoked Ca^{2+} transients exclusively in the presence of extracellular Ca^{2+} and on day 3 of culturing. Moreover, ATP caused elevated protein expression of the chondrogenic transcription factor Sox9 and also stimulated cartilage matrix production. ATP may exert these functions via acting through purinergic receptors; and indeed, expression of both ionotropic (P2X) and metabotropic (P2Y) purinergic receptors were detected. Metabotropic purinergic receptor agonist UTP caused a low level (60 nM) transient elevation of $[Ca^{2+}]_i$ in 3-day-old HDC, without having an influence on cartilage matrix production. Application of suramin, which blocks all P2X receptors but not $P2X_4$, did not impede the effects of ATP; furthermore, $P2X_4$ appeared in the plasma membrane of differentiating cells only from day 3. In summary, chondrogenic cells possess a set of different molecules which enable them to modulate their Ca^{2+} homeostasis and $[Ca^{2+}]_i$ was found to be kept in a narrow range during chondrogenesis. We present evidence on a significant new regulatory mechanism of chondrogenesis with revealing the role of Ca^{2+} influx of chondrogenic cells via $P2X_4$ purinergic receptors.

Keywords: Calcium signalling, chondrogenesis, $P2X_4$ purinergic receptors

INTRODUCTION

Hyaline cartilage is one of the basic supporting tissues of the vertebrate skeleton; it provides the source of bones formed by endochondral ossification; enables longitudinal growth of bones at the epiphyseal growth plates; supports the wall of various internal organs; and covers the articular surfaces of joints [1]. Cartilage is derived from the embryonic connective tissue during a process which starts with the condensation and nodule formation of chondroprogenitor mesenchymal cells. Besides transient appearance of Ca^{2+} dependent intercellular junctions (N-CAM and N-cadherin) [2], the differentiation of chondroprogenitor cells into chondroblasts is controlled by various growth factors and other signal molecules (*e.g.* FGF, BMPs, Wnt, and members of Sox transcription factor family) [3]. The extracellular matrix (ECM) surrounding the chondrogenic mesenchymal cells is also subject to profound changes: cells secrete cartilage-specific matrix components, such as collagen type II and aggrecan soon after their differentiation [4]. The characteristic composition and organization of ECM is pivotal for maintaining the appropriate morphology and function of chondrocytes [5]. The expression of the cartilage-specific matrix components is regulated by Sox9 transcription factor [6].

High density cell culture system (HDC) established from chondrogenic mesenchymal cells isolated from limb buds of 4-day-old chicken embryos is a well-known model of *in vitro* cartilage differentiation [7]. In HDC, spontaneous differentiation of chondroprogenitor mesenchymal cells into mature chondrocytes and formation of cartilage matrix occurs by mimicking all steps of embryonic cartilage formation within a 6-day-long culturing period; differentiation into chondroblasts and chondrocytes occurs predominantly on day 3. This simple model provides a means of investigating molecular steps leading to *in vitro* chondrogenesis [8].

Calcium ion is probably the most ubiquitous cellular signal. The intracellular free Ca^{2+} concentration ($\sim 10^{-7}$ M) is 10^4 times lower than that of the extracellular fluid. This considerable gradient provides the potential for the influx of Ca^{2+} into cells, where it can act as a second messenger, mostly via various modulatory Ca^{2+}-binding proteins, *e.g.* calmodulin. A wide range of stimuli can lead to the increase of cytosolic Ca^{2+} concentration either from the extracellular space or from intracellular stores. The elevated cytosolic free [Ca^{2+}] exerts numerous specific changes in various cellular functions, such as activation of calcium sensitive protein kinases and protein phosphatases, which, in turn, regulate other processes, such as proliferation or differentiation [9]. Signalling pathways governing cartilage differentiation are partially regulated by Ca^{2+} sensitive enzymes, for example PKCalpha, a Ser/Thr specific protein kinase [10], or calcineurin, a Ser/Thr-specific phosphoprotein phosphatase [11,12]. PKCalpha has been reported to influence proliferation and differentiation of chondrifying cells via modulation of MAPK-signalling [10], and calcineurin has also been described as one of the positive regulators of *in vitro* chondrogenesis in HDC by our research group [12]. Since proper function of these calcium sensitive signal transduction pathways is indispensable to *in vitro* chondrogenesis, it can be hypothesised that cytosolic [Ca^{2+}] could be tightly regulated during chondrogenic differentiation. In this chapter we are discussing several aspects of the changes in cytosolic [Ca^{2+}] of chondrogenic mesenchymal cells during their differentiation in chicken high density cultures. We also provide data on the possible roles of these changes in the regulation of cartilage formation.

CYTOSOLIC FREE [CA^{2+}] IN CELLS OF HDC SHOWS A CHARACTERISTIC AGE-DEPENDENT PATTERN

Basal level of [Ca^{2+}]$_i$ in differentiating chondroblasts measured by a fluorescent method after loading with Fura-2 Ca^{2+} sensitive fluorescent dye on various days of culturing was found to have an age-dependent pattern (Fig. 1). On day 0, [Ca^{2+}]$_i$ is low, starting with a concentration of about 75 nM, then it slightly increases in parallel with the progression of differentiation. On day 3 of culturing, however, a 140 nM peak of the cytosolic free [Ca^{2+}] was observed in chondrogenic cells. It is important to emphasise that cells of HDC differentiate into chondroblasts on this day of culturing [7]. From day 4, [Ca^{2+}]$_i$ decreases again, however, it remains slightly higher (about 100 nM) as compared to days 0–2, suggesting that mature chondrocytes are characterised by a slightly higher [Ca^{2+}]$_i$ than differentiating mesenchymal cells. It is important to consider that no enzymatic digestions were performed on cells of HDC prior to Ca^{2+} measurements, since the unique ECM surrounding chondroblasts and chondrocytes is crucial to maintain their proper function and morphology [5]. The [Ca^{2+}]$_i$ of differentiating mesenchymal cells in HDC are in the same range as that of other non-excitable cells, *e.g.* [Ca^{2+}]$_i$ of Fura-2-loaded HaCaT keratinocytes also proved to be 80–90 nM [13].

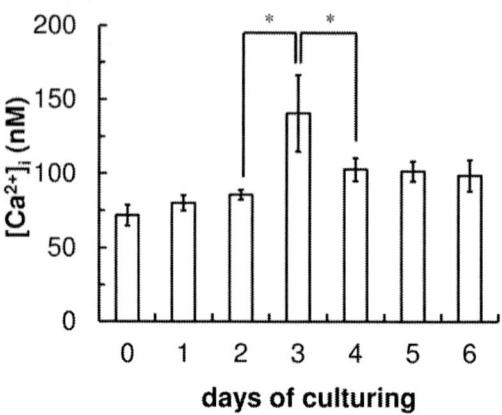

Figure 1. Basal intracellular Ca^{2+} concentration in chondrifying cells of untreated control HDC. $[Ca^{2+}]_i$ was determined in Fura-2-loaded cells. Shown are mean values of basal $[Ca^{2+}]_i$ of 30 cells ± standard error of the mean. Statistical analysis was performed by Student's t-test comparing each data to the previous culturing day; $*P < 0.01$.

CA^{2+} HOMEOSTASIS OF CELLS IN CHONDRIFYING MICROMASS CULTURES LARGELY DEPENDS ON CA^{2+} INFLUX FROM EXTRACELLULAR SPACE

When considering the source of elevated cytosolic Ca^{2+} levels on day 3, the following question obviously arises: whether a Ca^{2+} influx from the extracellular space, or rather a Ca^{2+} release from intracellular Ca-stores could be accounted for the phenomenon. To test the hypothesis if Ca^{2+} influx was of crucial importance in the Ca^{2+} peak on day 3, HDC were grown in a medium with free calcium ions unavailable for the cells. To this end, EGTA was applied in 0.8 mM concentration for 12 h on various days of culturing. EGTA treatment not only eliminated the $[Ca^{2+}]_i$ peak in cells of 3-day-old HDC but it also significantly decreased the cytosolic Ca^{2+} level to approximately 60% of that of untreated control cells on other days of culturing. The dependence of chondrifying mesenchymal cells on the availability of extracellular $[Ca^{2+}]$ is further supported by the fact that when cultures were maintained in 0.8 mM EGTA throughout the culturing period or in 5 mM EGTA even for shorter periods, cultures detached from the glass or plastic surface and died.

Besides the considerable decrease in cytosolic free $[Ca^{2+}]$, EGTA treatments also had an effect on cartilage matrix production. Cultures treated with EGTA for 12 h on day 2 or 3 of culturing showed a profound decrease in metachromatic staining by dimethyl-methylene blue (DMMB) by day 6 (Fig. 2A). It is remarkable that administration of EGTA after day 5 did not have any significant effect on matrix production suggesting that Ca^{2+} influx is only indispensible to proper cartilage formation during commitment and differentiation of chondrogenic cells. Reduced intracellular $[Ca^{2+}]$ attenuates *in vitro* cartilage matrix formation, at least in part, via inhibition of signalling pathways and factors affecting cartilage differentiation, since a considerable decrease in the mRNA levels of both aggrecan core protein and sox9 was observed under the effect of EGTA on each day of treatments (Fig. 2B). These findings were confirmed by Western blot analyses: treatments with EGTA also reduced the protein expression—and in a more pronounced manner, phosphorylation levels—of Sox9 (Fig. 2C). Although the proliferation rate of cells in micromass cultures was significantly reduced under the effect of 0.8 mM EGTA (approx. 15% of untreated control cultures), the mitochondrial activity of cells was not affected as revealed by ^{3}H-thymidine incorporation and MTT assays, respectively (Fig. 3). These data support the hypothesis that the parameter primarily affected by the proper maintenance of Ca^{2+} homeostasis at the beginning of *in vitro* chondrogenesis in HDC might be the rate of cellular proliferation, which also reaches its maximum on culturing days 1 and 2. The calcium-sensitivity of proliferation and its regulation has been reported in several cell types examined (*e.g.* epithelial cells in colon [14]). In chondrogenic cells, proliferation can be altered by a decrease of [Ca^{2+}]$_i$ via modulation of the activity of classical PKC izoenzymes, particularly PKCalpha [15].

A. Metachromasia

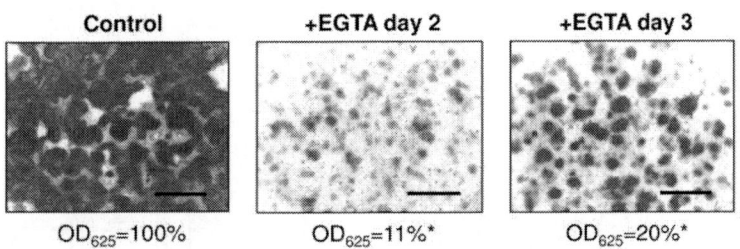

B. RT-PCR

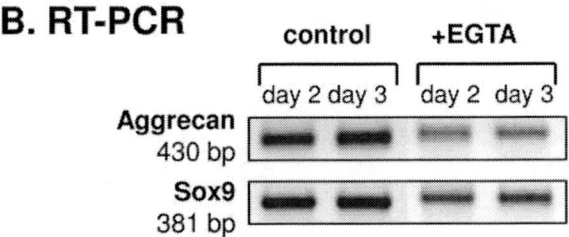

C. Western blot

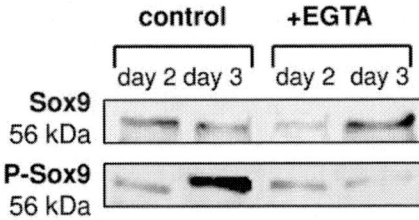

Figure 2. Effect of 0.8 mM EGTA on cartilage matrix production of chondrifying micromass cultures. (A) Metachromatic cartilage areas in 6-day-old HDC stained by acidic DMMB (pH 2). Scale bar, 500 μm. Optical density (OD$_{625}$) was determined by toluidine blue staining procedures. Data are mean values ± standard error of the mean (± 5 %). Asterisks indicate significant (*$P < 0.01$) decrease in optical density of extracted toluidine blue as compared to the respective control. (B & C). Effect of EGTA treatments on the mRNA and protein expression of cartilage-specific marker genes in 2- and 3-day-old HDC.

At this level we can conclude that decreasing $[Ca^{2+}]_i$ by EGTA treatments during the first 3 days of culturing leads to an inhibition of cartilage matrix production, however, we cannot say anything about the effects of the opposite change of $[Ca^{2+}]_i$. Therefore, the question arises: how does elevation of $[Ca^{2+}]_i$ influence *in vitro* chondrogenic differentiation and cartilage matrix production? To this end, A23187 Ca^{2+} ionophore was administered to cells of HDC on various days of culturing at 0.1 and 5 mg/L concentrations, leading to uncontrolled Ca^{2+} influx. 0.1 mg/L concentration of the Ca^{2+} ionophore A23187 elevated $[Ca^{2+}]_i$ to approximately 125% of control cells (150 nM), and 5 mg/L concentration of Ca^{2+} ionophore resulted in an even higher increase (about 250 nM). Interestingly, however, the two concentrations of the ionophore had opposite effects on cartilage matrix production. Treatment with the lower concentration of the ionophore on days 2 and 3 augmented cartilage formation, whereas the higher concentration of A23187 caused a marked inhibition of this

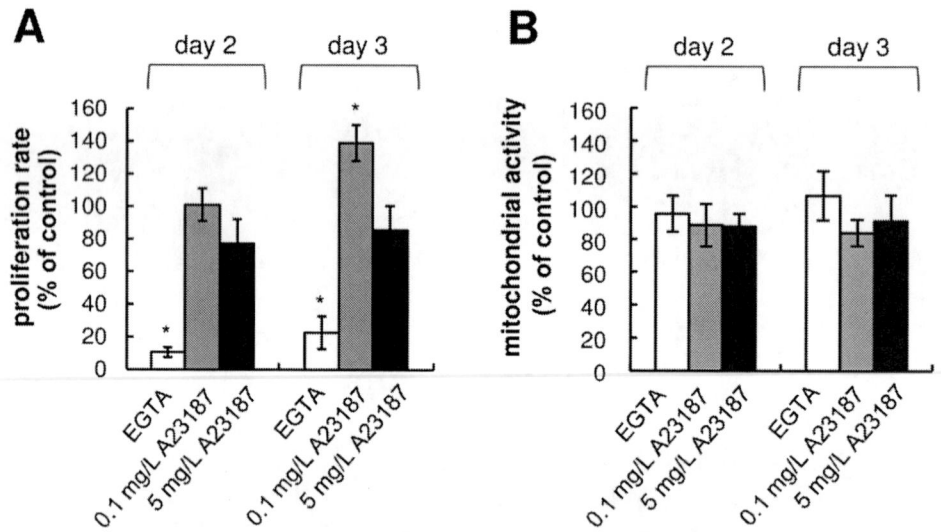

Figure 3. Effect of EGTA and A23187 Ca^{2+} ionophore on the proliferation rate and cellular viability of cells in HDC. (A) Proliferation rate was determined by monitoring the incorporation of ^{3}H-thymidine. Data are mean values ± standard error of the mean; asterisks indicate significant (*$P < 0.01$) changes compared to untreated control cultures. (B) Cellular viability was determined by MTT assay. Data are mean values ± standard error of the mean.

parameter (Fig. 4A). Cellular proliferation was slightly stimulated (130% of the control) by low concentration of the Ca^{2+} ionophore A23187, whereas in high concentration it resulted in an opposite effect (75% of untreated cultures). Neither concentration of the ionophore have caused any significant decrease in mitochondrial activity, further supporting the hypothesis that modulation of $[Ca^{2+}]_i$ exerts its effects by interfering with cell proliferation (see Fig. 3). Since the mRNA level of sox9 and aggrecan core protein, and both protein level and phosphorylation of Sox9 slightly increased rather than decreased under the effect of 5 mg/L ionophore (Figs. 4B–C), the precise mechanism underlying the marked inhibition of cartilage formation remains to be elucidated. Nonetheless, our data indicate that a slight elevation of $[Ca^{2+}]_i$ in differentiating cells of HDC promotes differentiation, but further elevation beyond a certain threshold is already detrimental and inhibits chondrogenesis.

INTRACELLULAR STORES DO NOT CONTRIBUTE SIGNIFICANTLY TO CA^{2+} HOMEOSTASIS IN CELLS OF HDC

Cyclopiazonic acid (CPA), an inhibitor of the Ca^{2+} pump of the smooth endoplasmic reticulum (SERCA) was applied to cells in a Ca^{2+}-free medium to investigate the role of intracellular Ca-stores. During fluorescent [Ca^{2+}]$_i$ measurements, soon after cells were perfused with a Ca^{2+}-free medium the basal cytosolic Ca^{2+} level decreased from 140 nM to 120 nM showing the dependence of this parameter on the extracellular Ca^{2+} concentration. Shortly after the administration of 10 μM CPA, [Ca^{2+}]$_i$ started to increase very slowly (Fig. 5), which shows that the intracellular Ca^{2+}-stores are not empty, however, the amount of stored Ca^{2+} may be relatively low. When cells were perfused again with normal Ca^{2+}-containing solution, a well-defined [Ca^{2+}]$_i$ peak was observed as a result of Ca^{2+} influx reflecting on a possible activation of store-operated Ca^{2+} entry (SOCE) channels [16]. The exact nature of the observed phenomenon, however, remains to be further investigated.

A. Metachromasia

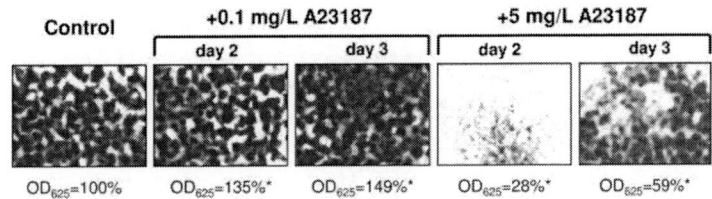

B. RT-PCR

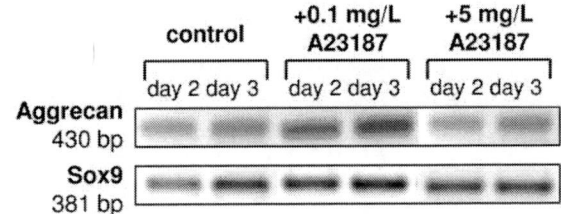

C. Western blot

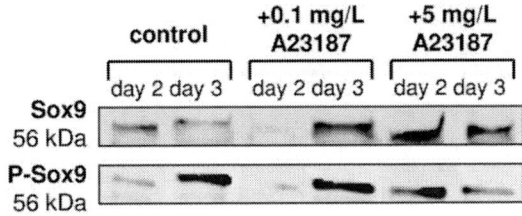

Figure 4. Effects of A23187 Ca^{2+} ionophore on cartilage matrix production of chondrifying micromass cultures. (A) Metachromatic cartilage areas in 6-day-old HDC stained by acidic DMMB (pH 2). Optical density (OD$_{625}$) was determined by toluidine blue staining procedures. Data are mean values ± standard error of the mean (± 6 %). Asterisks indicate significant (*P < 0.01) change in optical density of extracted toluidine blue as compared to the respective control. (B & C) Effect of A23187 on the mRNA and protein expression and phosphorylation of cartilage-specific marker molecules.

When 10 µM CPA was administered to the culture medium of HDC on days 2 or 3, the prolonged inhibition of the Ca-pump must have resulted in a complete depletion of intracellular stores, however, no adverse effects on chondrogenesis could be observed by day 6. Combined treatments with EGTA and CPA for 12 h (*i.e.*, inhibition of Ca^{2+} entry from both extracellular and intracellular sources) resulted in a complete loss of cartilage matrix reflecting on the highly sensitive Ca^{2+} homeostasis of differentiating cells in HDC. These findings may also suggest the insufficient capacity of intracellular Ca-stores to replenish the function of Ca^{2+} entry pathways in chondrogenic cells.

The investigation of the role of internal stores in the Ca^{2+} homeostasis of chondrifying cells also involved the study of the ryanodine receptor (RyR) and the inositol-1, 4, 5-trisphosphate (IP_3) receptors. We were unable to detect RyR from total cell lysates and only a weak expression profile was seen in separated endoplasmic reticulum fraction of HDC. Moreover, no Ca^{2+} transients were detected when caffeine, an agonist of RyR was administered during single cell measurements, which probably indicates that the low amount of RyR located in the endoplasmic reticulum of cells of HDC probably does not significantly contribute to the Ca^{2+} homeostasis of these cells. As far as the inositol-1, 4, 5-trisphosphate pathway is concerned, only the IP_3R type 1 was found to be expressed by cells of HDC.

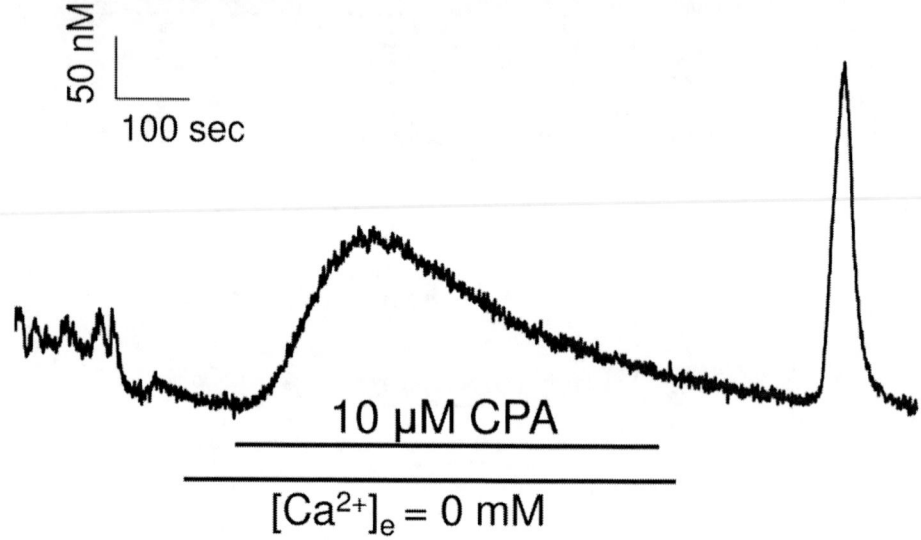

Figure 5. Effect of administration of cyclopiazonic acid (CPA) on the release of Ca^{2+} from intracellular stores. Measurements were carried out in Fura-2-loaded cells.

Taken together, we can conclude that intracellular Ca^{2+} stores are probably negligible in the modulation of the $[Ca^{2+}]_i$ peak in cells of HDC during differentiation. Nonetheless, we provide evidence that some well-known components of the Ca^{2+} toolkit of non-excitable cells (store-operated Ca^{2+} entry pathways, PMCAs, SERCA and IP_3 receptors) were also expressed by cells of HDC, and it work together to regulate the Ca^{2+} homeostasis. Based on our findings, the intracellular stores rather seem to be contributing to the maintenance of basal resting $[Ca^{2+}]_i$, rather than playing a role in its changes.

IONOTROPIC PURINERGIC RECEPTOR P2X$_4$ CAN BE ONE OF THE CHANNELS THAT ALLOW CA^{2+} INFLUX DURING DIFFERENTIATION

It can be concluded at this level that although intracellular elements of the Ca^{2+} toolkit were detected in chondrogenic cells, we failed to reveal strong evidence concerning their contribution to generate the elevation of cytosolic Ca^{2+} concentration in differentiating chondrogenic cells, and therefore we can hypothesise that the extracellular space can be the source of the elevated cytosolic Ca^{2+} concentration on day 3. To this end, we aimed to identify plasma membrane calcium channels that might be responsible for the Ca^{2+} influx. Ubiquitously expressed by various cell types, *purinergic receptors* that are activated by extracellular nucleotides, have two major types: P1 receptor families are sensitive to adenosine, while P2 receptor families can be activated by ATP, ADP, and UTP. The latter group is subdivided into two major subtypes: P2Y and P2X receptors. The metabotropic P2Y purinergic receptors are 7 transmembrane domain-containing proteins signalling through the PLC pathway leading to the release of intracellular Ca^{2+} from IP$_3$-sensitive Ca^{2+} stores. The ionotropic P2X receptors are ATP-gated ion channels allowing Ca^{2+} influx. As yet, seven P2X (P2X$_1$–P2X$_7$) and eight P2Y isoforms (P2Y$_1$, P2Y$_2$, P2Y$_4$, P2Y$_6$, P2Y$_{11}$, P2Y$_{12}$, P2Y$_{13}$ and P2Y$_{14}$) have been described in human tissues [17-19]. Since P2X receptors are sequentially expressed in embryonic rat and mouse skeletal muscle cells and osteoblasts [20-22], we presumed that such receptors might be involved in the Ca^{2+} homeostasis of differentiating chondrogenic mesenchymal cells.

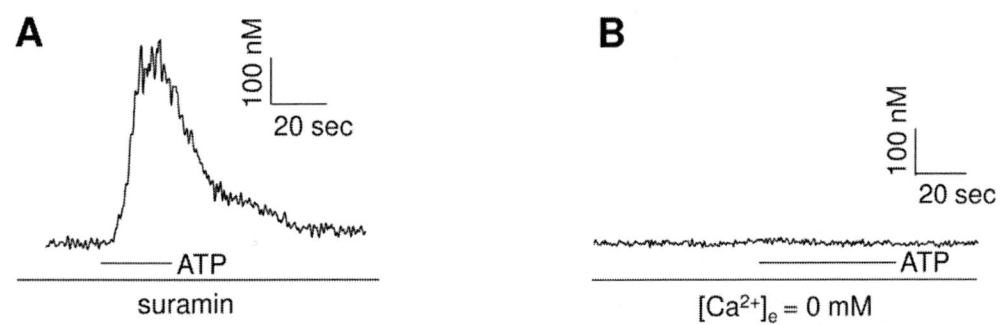

Figure 6. Response of cells in 3-day-old HDC to administration of ATP. (A) Effect of ATP (180 μM) and the P2X antagonist suramin (10 μM) on [Ca^{2+}]$_i$ in the presence of external calcium ([Ca^{2+}]$_e$=1.8 mM). (B) Representative record showing the lack of ATP-evoked Ca^{2+} transients in the absence of external calcium ([Ca^{2+}]$_e$=0 mM).

To test this hypothesis, cells were challenged by the administration of extracellular ATP during Ca^{2+} measurements on various days of culturing. Extracellular ATP only induced a transient increase of [Ca^{2+}]$_i$ when cells were perfused with a Ca^{2+}-containing solution (Fig. 6A–B). Noteworthy that the amplitude of the average response exhibited a differentiation-dependent pattern; Ca^{2+}-transients with the highest amplitude were recorded on culturing days 3 and 4. Extracellular ATP was unable to evoke changes in the [Ca^{2+}]$_i$ on days 1 and 2 of culturing, and the amplitude of the transients considerably decreased by day 6. The number of cells responding to ATP also showed variations: while almost all the cells (90%) responded to

ATP in 3-day-old cultures, the proportion of cells responded on the other culturing days was hardly comparable with that on day 3.

Next, we wanted to establish whether P2Y (metabotropic) or P2X (ionotropic) purinergic receptors were responsible for the observed effects. Since no response was detected on either days of culturing when ATP was administered in the absence of extracellular Ca^{2+}, we concluded that influx of extracellular Ca^{2+} was needed to evoke the Ca^{2+} transients and the receptor of ATP expressed by cells of HDC could be a member of the ionotropic purinergic receptor family (P2X). To find possible candidates among P2X receptors, their non-specific antagonist suramin was tested during Ca^{2+} measurements. Suramin has been reported to inhibit all P2X receptors except $P2X_4$ and $P2X_6$ [18]. Since cells treated with suramin (10 μM) exhibited the same pattern of Ca^{2+} concentration changes as the untreated control ones did following the administration of ATP (see Fig. 6A), we assumed the presence and function of the suramin-insensitive $P2X_4$ and/or $P2X_6$ receptor subtypes.

A: RT-PCR

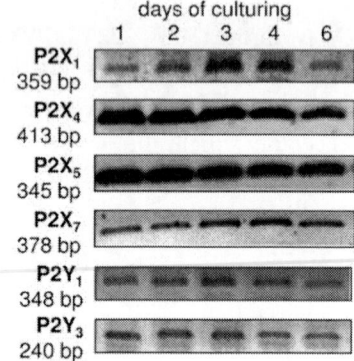

B: Western blot

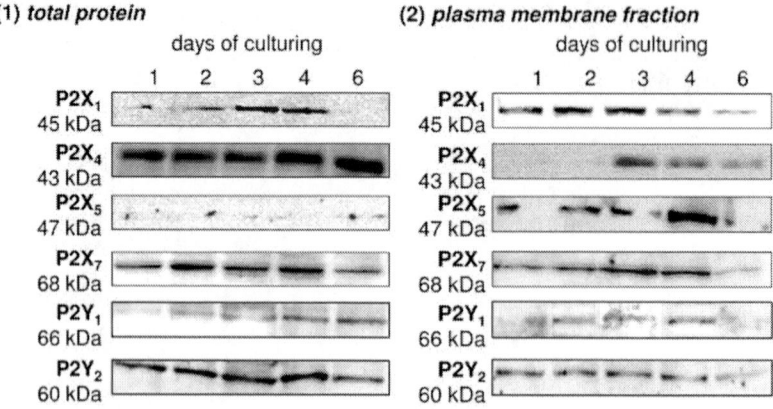

Figure 7. Expression patterns of P2X and P2Y purinergic receptor subtypes in cells of HDC on different days of culturing. (A) mRNA expression patterns of various P2X and P2Y receptors detected by RT-PCR reactions. (B) Protein expression patterns of P2X and P2Y receptor proteins in cells of HDC by Western blot analyses. For the analyses of the protein expression levels, total protein and plasma membrane fraction samples were used (50 μg in each lane).

We also aimed to identify the expression of various purinergic receptors during chondrogenic differentiation of chicken mesenchymal cells. HDC were found to express various ionotropic purinergic receptors both at the mRNA and protein levels (Fig. 7). Protein expression of $P2X_1$ subtype in total lysates followed a similar profile to the mRNA expression—that is, they exhibited a peak-like pattern with the highest level of expression on day 3—, but in the plasma membrane fraction strong signals were only detected on the first days of the culturing period. Protein expression of $P2X_5$ receptor subtype was hardly detectable in total cell lysates, but in the plasma membrane fraction a strong signal appeared on day 4. For $P2X_7$, a protein expression profile with an irregular pattern was observed in total lysates with the strongest signals on days 2 and 4, however, in the plasma membrane fractions a peak-like pattern with strongest bands on days 2, 3 and 4 was observed. The protein expression of $P2X_4$ receptor subtype proved to be the most characteristic: while it showed a rather irregular profile in total cell lysates, it first appeared in the plasma membrane fraction on day 3 with a strong signal, and by days 4 and 6 its expression rapidly diminished. Noteworthy that the vast majority of chondrogenic mesenchymal cells responded to ATP on the same day of culturing, which also coincides with the day of differentiation characterised by elevated cytosolic Ca^{2+} levels discussed earlier in this chapter. We were unable to provide evidence on the expression of P2X receptor subtypes other than $P2X_1$, $P2X_4$, $P2X_5$ and $P2X_7$ in cells of HDC. The sequential and age-dependent expression of various P2X receptor subtypes suggests the involvement of purinergic signalling in the mediation of chondrogenic differentiation similar to hematopoietic cell lines [23]. According to the above described molecular biological analyses of the receptor expression pattern and the data gained from the lack of the influence of suramin on chondrogenesis, we suggest that the $P2X_4$ receptor subtype can be primarily responsible for the Ca^{2+} influx on day 3, while the other P2X receptors may rather contribute to the maintenance of the basal cytosolic Ca^{2+} level.

METABOTROPIC PURINERGIC P2Y RECEPTORS ARE ALSO EXPRESSED BY CELLS OF HDC BUT THEY ONLY PLAY A MINOR ROLE IN THE CA^{2+} HOMEOSTASIS

Although the findings discussed above clearly suggested that P2X receptors play the decisive role in the Ca^{2+} homeostasis of differentiating chondroprogenitor cells, we also investigated the possible contribution of metabotropic P2Y receptors. First, we examined whether intracellular Ca^{2+} stores are present and contain releasable Ca^{2+} by the activation of IP_3 pathway. Since bradykinin is a known agonist of this pathway in chondrocytes [24], and the IP_3R1 was found to be present in cells of HDC as described previously in this chapter, we administered bradykinin to 3-day-old cells during Ca^{2+} measurements. As a result, a slight elevation of $[Ca^{2+}]_i$ was observed in 60% of cells proving the presence and active functioning of the IP_3 pathway. Next, 3-day-old cells were challenged with non-specific agonists of metabotropic purinergic receptors (ADP, UDP and UTP) during Ca^{2+} measurements. While administration of ADP and UDP did not result in any significant Ca^{2+} transients, a slight elevation of $[Ca^{2+}]_i$ was detected in approximately 50% of cells measured during the administration of UTP. The expression of metabotropic purinergic receptors was also investigated in HDC. Whereas $P2Y_1$ was present at a constant level both in total cell lysates and in the plasma membrane fraction until day 6, $P2Y_2$ receptor protein followed a peak-like pattern in total lysates with strongest signals on days 2–4. To the contrary, $P2Y_2$ proved to exhibit a constant level of expression in the plasma membrane fraction (see Fig. 7). No other P2Y receptor protein subtypes were identifiable in cells of HDC.

ADMINISTRATION OF EXTRACELLULAR ATP INCREASES CARTILAGE MATRIX PRODUCTION

Results obtained thus far concerning the Ca^{2+} homeostasis of chicken chondrifying mesenchymal cell cultures all seemed to support a hypothesis that the Ca^{2+} signalling in chondrogenic cells of HDC chiefly depends on entrance of extracellular Ca^{2+} mainly via P2X receptors, particularly via $P2X_4$ subtype. In order to give evidence on the physiological contribution of purinergic signalling, extracellular ATP was administered to 3-day-old HDC. Administration of ATP resulted in an extensive matrix production by day 6; and the chondrogenesis-promoting effect of this compound was further confirmed with the increased expression levels of collagen II, aggrecan core protein and Sox9 transcription factor (Fig. 8).

A. Metachromasia

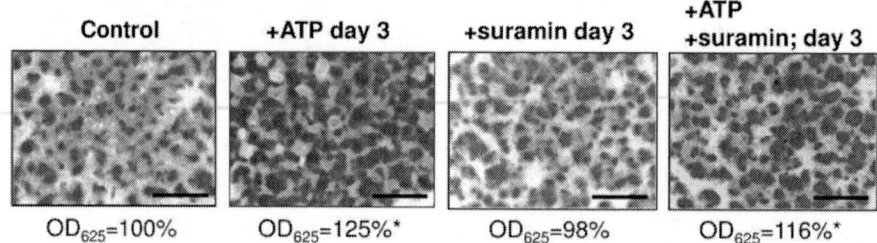

Control	+ATP day 3	+suramin day 3	+ATP +suramin; day 3
$OD_{625}=100\%$	$OD_{625}=125\%^*$	$OD_{625}=98\%$	$OD_{625}=116\%^*$

B. mRNA and protein expression

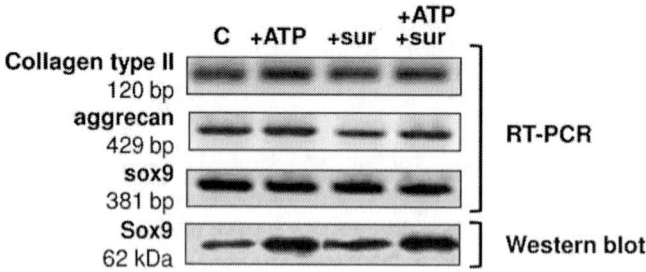

Figure 8. Effects of ATP and suramin on cartilage matrix production of chondrifying micromass cultures. Chemicals were administered on day 3 of culturing. (A) Metachromatic cartilage areas in 6-day-old HDC stained by acidic DMMB (pH 2). Scale bar, 500 µm. Optical density (OD_{625}) was determined by toluidine blue staining procedures. Data are mean values; standard error of the mean was within ± 7% in all experimental groups. Asterisks indicate significant (*$P < 0.01$) increase in optical density of extracted toluidine blue as compared to the respective control. (B) mRNA and protein expression of cartilage-specific marker molecules after treatment with ATP and/or suramin.

Interestingly, treatment of HDC with ATP on other culturing days did not alter the cartilage matrix production indicating that ATP only exerts a positive effect on both cartilage matrix production and chondroblast differentiation when it is applied at the time of differentiation of chondroprogenitor cells and does not have any effect on chondroprogenitor cells or mature chondrocytes. The non-specific P2X receptor antagonist suramin did not significantly interfere with the abovementioned effects of ATP treatments, further supporting our hypothesis that P2X$_4$ receptor could be one of the main factors that contribute to the chondrogenesis-promoting mechanism of extracellular ATP (see Fig. 8).

The results gained by experiments concerning the administration of extracellular ATP raise the question whether the chondrogenic mesenchymal cells in HDC produce and secrete ATP into the culture medium as an autocrine or paracrine mediator to promote and facilitate their own differentiation. To this end, luminescent ATP assays were performed revealing that a small amount of ATP was detectable in the culture medium in the range of 2–10 nM on each day of culturing. The observations that chondrogenic cells in HDC secreted ATP throughout the 6-day-long culturing period and they responded to the extracellularly administred ATP with Ca^{2+} transients only at the time of differentiation further support the purinergic concept in the control of *in vitro* chondrogenesis.

CALCINEURIN PLAYS A DUAL ROLE IN THE CA^{2+} HOMEOSTASIS OF CELLS IN HDC

In vitro cartilage differentiation is at least in part regulated by various Ca^{2+} sensitive protein kinases and the Ser/Thr-specific protein phosphatase PP2B or calcineurin [11,12], which possesses a calcium binding subunit and can also bind to Ca^{2+}–calmodulin. Calcineurin has been described as a regulator of T-lymphocyte activation, and its pharmacological inhibitors (*e.g.* cyclosporine A (CsA), rapamycin, etc.) are widely applied immuno-suppressants [25]. PP2B is also known to be involved in the differentiation of muscle tissue and the nervous system [26]. Since calcineurin is one of the key molecules involved in the differentiation of HDC which is regulated by the changes of [Ca^{2+}]$_i$, we measured the enzyme activity of this phosphatase in cultures with altered Ca^{2+} homeostasis. In cultures treated with EGTA, the activity of calcineurin significantly decreased on both days of treatments. Interestingly, elevation of [Ca^{2+}]$_i$ by A23187 had partly unexpected effects. On day 2, the elevation of Ca^{2+} concentration induced by the ionophore increased, while on day 3 it decreased calcineurin activity (Fig. 9A). When calcineurin activity was inhibited by its pharmacological inhibitor CsA, the cytosolic Ca-peak on day 3 of culturing observed under control conditions was eliminated, although cells had slightly higher basal [Ca^{2+}]$_i$ than those of the untreated control cultures (Fig. 9B).

Since calcineurin is one of the positive regulators of *in vitro* chondrogenesis of HDC [11,12], we can hypothesise that the maintenance of its enzymatic activity requires a very precisely set [Ca^{2+}]$_i$ when chondrogenic cells undergo the actual differentiation process. Noteworthy that when the activity of calcineurin was inhibited by treatments with CsA, differentiating cells failed to produce the peak-like [Ca^{2+}]$_i$ increase on culturing day 3, in spite of the fact that basal Ca^{2+} levels were higher in CsA-treated cultures than in the untreated ones. This observation may implicate that calcineurin plays an active regulatory role in the enhancement of Ca^{2+} influx responsible for the rapid transient elevation of $[Ca^{2+}]_i$ in cells of HDC on day 3. This hypothesis is supported by the fact that calcineurin has been reported to regulate the activity of various components of the Ca^{2+} toolkit. It has recently been described to dephosphorylate, and in this way desensitize or inactivate, several types of Ca-channels (*e.g.* store-operated Ca-channels [27], high-threshold voltage-activated Ca-channels [28] and TRPV1 receptors) in both excitable and non-excitable cell types [29,30].

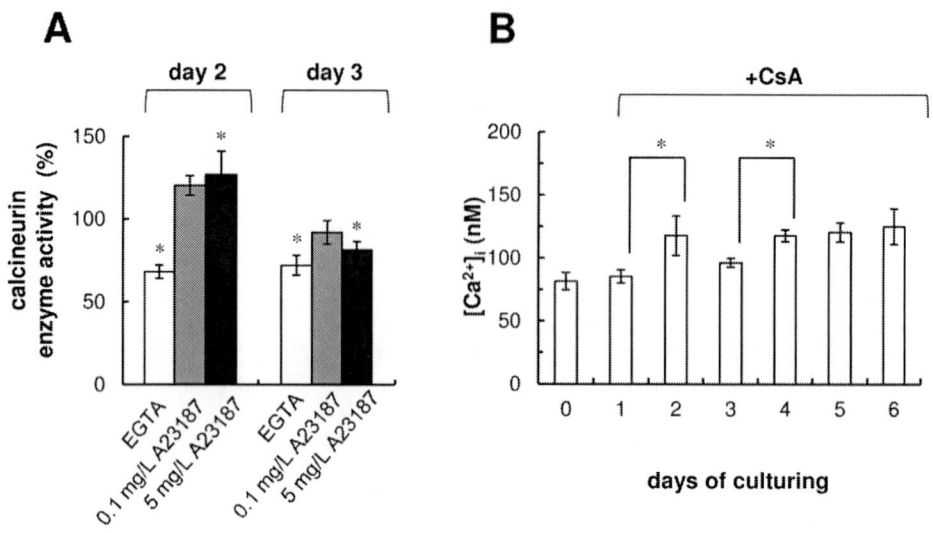

Figure 9. Dual role of calcineurin in the Ca^{2+} homeostasis of cells in HDC. (A) Effect of A23187 Ca^{2+} ionophore and EGTA on the activity of calcineurin. **(B)** Effect of the calcineurin-inhibitor CsA (10 μM) on the regulation of basal cytosolic Ca^{2+} levels of HDC. $[Ca^{2+}]_i$ were determined in Fura-2-loaded cells. Shown are mean values of intracellular Ca^{2+} levels of 30 cells ± standard error of the mean. Statistical analysis was performed by Student's *t*-test comparing each data to the previous culturing day; *$P <$ 0.01.

CONCLUSION

In this chapter we provided insights into the Ca^{2+} homeostasis of chondrifying chicken high density cultures during differentiation. We revealed a distinct pattern of the changes of $[Ca^{2+}]_i$: the day of chondrocyte differentiation was characterised by a significant and transient elevation. We provided evidence that the precise temporal pattern of the changes of $[Ca^{2+}]_i$ in chondrifying cells was indispensible to proper chondrogenesis and cartilage matrix formation, and depended on the availability of extracellular Ca^{2+} rather than the active function of intracellular Ca^{2+} stores, since in the absence of extracellular free Ca^{2+} ions, differentiating cells of HDC failed to elevate their cytosolic Ca^{2+} level on day 3, which, in turn, interfered with the differentiation process leading to an almost complete blockade of the formation of metachromatically stained cartilage areas by day 6. We can conclude that proper *in vitro* chondrogenesis requires a very tightly controlled $[Ca^{2+}]_i$ and the proliferation ability of cells in HDC is more sensitive to the decrease than to the increase of $[Ca^{2+}]_i$. Moreover, based on our data a slight elevation in $[Ca^{2+}]_i$ of chondrifying mesenchymal cells by A23187 Ca^{2+}

ionophore promoted differentiation, but further elevation above a certain threshold was already inhibitory to chondrogenesis.

Nonetheless, the extracellular Ca^{2+} is required to initiate *in vitro* differentiation of chondrogenic mesenchymal cells to chondroblasts and chondrocytes. Among the receptors that can be responsible for the elevation of Ca^{2+} concentration are members of the P2X family, and based on our results, we propose that P2X$_4$ receptors contribute to the elevation of [Ca^{2+}]$_i$ in chondrogenic cells on day 3 of culturing. Moreover, we reported that cells of HDC responded to administration of extracellular ATP by Ca^{2+} influx mainly at the time of chondroblast formation and addition of ATP to the culture medium stimulated chondrogenesis. We provided evidence that cells of HDC secreted ATP into the culturing medium, which further supports our hypothesis that a purinergic autocrine regulation is involved in the proper control of *in vitro* chondrogenesis.

We also demonstrated that the Ser/Thr phoshpatase calcineurin plays a dual (both 'active' and 'passive') regulatory role in the Ca^{2+} homeostasis of chondrogenic cells: its activity was modulated by [Ca^{2+}]$_i$ and the inhibition of calcineurin with CsA eliminated the Ca^{2+} peak of HDC resulting in a pronounced decrease in cartilage formation. These findings raise the possibility of the active role of calcineurin in the regulation of Ca^{2+} influx into chondrifying cells.

In conclusion, the regulation of *in vitro* chondrogenesis is sensitive to changes of cytosolic free [Ca^{2+}] and calcineurin is an important signalling factor in these events. Cells of HDC possess various members of the extensive Ca^{2+} toolkit, and our results provide the first evidence on the concept that a purinergic autocrine mechanism may be involved in the regulation of the elevated cytosolic Ca^{2+} levels during *in vitro* cartilage differentiation.

REFERENCES

[1]   Olsen, B. R., Reginato, A. M. & Wang, W. (2000). Bone development. *Annu.Rev.Cell Dev.Biol.* 16, 191-220.

[2]   Tavella, S., Raffo, P., Tacchetti, C., Cancedda, R. & Castagnola, P. (1994). N-CAM and N-cadherin expression during in vitro chondrogenesis. *Exp.Cell Res.* 215, 354-362.

[3]   Goldring, M. B., Tsuchimochi, K. & Ijiri, K. (2006). The control of chondrogenesis. *J.Cell Biochem.* 97, 33-44.

[4]   Dessau, W., von der M. H., von der M. K. & Fischer, S. (1980). Changes in the patterns of collagens and fibronectin during limb-bud chondrogenesis. *J.Embryol.Exp.Morphol.* 57, 51-60.

[5]   Cancedda, R., Castagnola, P., Cancedda, F. D., Dozin, B. & Quarto, R. (2000). Developmental control of chondrogenesis and osteogenesis. *Int.J.Dev.Biol.* 44, 707-714.

[6]   Lefebvre, V., Huang, W., Harley, V. R., Goodfellow, P. N. & de C.B. (1997). SOX9 is a potent activator of the chondrocyte-specific enhancer of the pro alpha1(II) collagen gene. *Mol.Cell Biol.* 17, 2336-2346.

[7]   Hadhazy, C., Lazlo, M. B. & Kostenszky, K. S. (1982). Cartilage differentiation in micro-mass cultures of chicken limb buds. *Acta Morphol.Acad.Sci.Hung.* 30, 65-78.

[8]   Ahrens, P. B., Solursh, M. & Reiter, R. S. (1977). Stage-related capacity for limb chondrogenesis in cell culture. *Dev.Biol.* 60, 69-82.

[9]   Chin, D. & Means, A. R. (2000). Calmodulin: a prototypical calcium sensor. *Trends Cell Biol.* 10, 322-328.

[10]   Han, Y. S., Bang, O. S., Jin, E. J., Park, J. H., Sonn, J. K. & Kang, S. S. (2004). High dose of glucose promotes chondrogenesis via PKCalpha and MAPK signaling pathways in chick mesenchymal cells. *Cell Tissue Res.* 318, 571-578.

[11]   Tomita, M., Reinhold, M. I., Molkentin, J. D. & Naski, M. C. (2002). Calcineurin and NFAT4 induce chondrogenesis. *J.Biol.Chem.* 277, 42214-42218.

[12]   Zakany, R., Szijgyarto, Z., Matta, C., Juhasz, T., Csortos, C., Szucs, K., Czifra, G., Biro, T., Modis, L. & Gergely, P. (2005). Hydrogen peroxide inhibits formation of cartilage in chicken micromass cultures and decreases the activity of calcineurin: implication of ERK1/2 and Sox9 pathways. *Exp.Cell Res.* 305, 190-199.

[13]   Bakondi, E., Gonczi, M., Szabo, E., Bai, P., Pacher, P., Gergely, P., Kovacs, L., Hunyadi, J., Szabo, C., Csernoch, L. & Virag, L. (2003). Role of intracellular calcium mobilization and cell-density-dependent signaling in oxidative-stress-induced cytotoxicity in HaCaT keratinocytes. *J.Invest Dermatol.* 121, 88-95.

[14]   Rey, O., Young, S. H., Jacamo, R., Moyer, M. P. & Rozengurt, E. (2010). Extracellular calcium sensing receptor stimulation in human colonic epithelial cells induces intracellular calcium oscillations and proliferation inhibition. *J.Cell Physiol* 225, 73-83.

[15]   Chang, S. H., Oh, C. D., Yang, M. S., Kang, S. S., Lee, Y. S., Sonn, J. K. & Chun, J. S. (1998). Protein kinase C regulates chondrogenesis of mesenchymes via mitogen-activated protein kinase signaling. *J.Biol.Chem.* 273, 19213-19219.

[16]   Lewis, R. S. (2007). The molecular choreography of a store-operated calcium channel. *Nature* 446, 284-287.

[17]   North, R. A. & Surprenant, A. (2000). Pharmacology of cloned P2X receptors. *Annu.Rev.Pharmacol.Toxicol.* 40, 563-580.

[18]   Ralevic, V. & Burnstock, G. (1998). Receptors for purines and pyrimidines. *Pharmacol.Rev.* 50, 413-492.

[19]   Soto, F., Garcia-Guzman, M. & Stuhmer, W. (1997). Cloned ligand-gated channels activated by extracellular ATP (P2X receptors). *J.Membr.Biol.* 160, 91-100.

[20]   Orriss, I. R., Knight, G. E., Ranasinghe, S., Burnstock, G. & Arnett, T. R. (2006). Osteoblast responses to nucleotides increase during differentiation. *Bone* 39, 300-309.

[21]   Collet, C., Strube, C., Csernoch, L., Mallouk, N., Ojeda, C., Allard, B. & Jacquemond, V. (2002). Effects of extracellular ATP on freshly isolated mouse skeletal muscle cells during pre-natal and post-natal development. *Pflugers Arch.* 443, 771-778.

[22]   Ryten, M., Hoebertz, A. & Burnstock, G. (2001). Sequential expression of three receptor subtypes for extracellular ATP in developing rat skeletal muscle. *Dev.Dyn.* 221, 331-341.

[23]   Bernhard, M. K. & Ulrich, K. (2006). Rt-PCR study of purinergic P2 receptors in hematopoietic cell lines. *Biochemistry (Mosc.)* 71, 607-611.

[24]   Koolpe, M., Rodrigo, J. J. & Benton, H. P. (1998). Adenosine 5'-triphosphate, uridine 5'-triphosphate, bradykinin, and lysophosphatidic acid induce different patterns of calcium responses by human articular chondrocytes. *J.Orthop.Res.* 16, 217-226.

[25]   Feske, S., Okamura, H., Hogan, P. G. & Rao, A. (2003). Ca2+/calcineurin signalling in cells of the immune system. *Biochem.Biophys.Res.Commun.* 311, 1117-1132.

[26] Aramburu, J., Heitman, J. & Crabtree, G. R. (2004). Calcineurin: a central controller of signalling in eukaryotes. *EMBO Rep.* 5, 343-348.

[27] Gwack, Y., Feske, S., Srikanth, S., Hogan, P. G. & Rao, A. (2007). Signalling to transcription: store-operated Ca2+ entry and NFAT activation in lymphocytes. *Cell Calcium* 42, 145-156.

[28] Lukyanetz, E. A., Piper, T. P. & Sihra, T. S. (1998). Calcineurin involvement in the regulation of high-threshold Ca2+ channels in NG108-15 (rodent neuroblastoma x glioma hybrid) cells. *J.Physiol* 510 (Pt 2), 371-385.

[29] Bultynck, G., Vermassen, E., Szlufcik, K., De, S.P., Fissore, R. A., Callewaert, G., Missiaen, L., De, S. H. & Parys, J. B. (2003). Calcineurin and intracellular Ca2+-release channels: regulation or association? *Biochem.Biophys.Res.Commun.* 311, 1181-1193.

[30] Mignen, O., Thompson, J. L. & Shuttleworth, T. J. (2003). Calcineurin directs the reciprocal regulation of calcium entry pathways in nonexcitable cells. *J.Biol.Chem.* 278, 40088-40096.

ISBN: 978-1-61324-313-8
©2012 Nova Science Publishers, Inc.

Chapter 7

CA$_V$2.1 CHANNELOPATHIES AND GENETIC APPROACHES FOR INVESTIGATINGS CA$_V$2.1 CHANNEL FUNCTION AND DYSFUNCTION

*Eiki Takahashi**

Research Resources Center, RIKEN Brain Science Institute,
Saitama, JAPAN

ABSTRACT

Voltage-gated Ca^{2+} (Ca_v) channels allow the entry of Ca^{2+} into a cell when the membrane is depolarized. In the nervous system, Ca_v channels control a broad array of functions including neurotransmitter release, neurite outgrowth, synaptogenesis, neuronal excitability, activity-dependent gene expression, and neuron survival, differentiation, and plasticity. Ca_v2.1 (P/Q-type) channels have a dominant, specific role in synaptic transmission at central excitatory synapses. In *CACNA1A* gene mutations, changes to the pore-forming α1 subunit of the Ca_v2.1 channel cause several autosomal-dominant neurological disorders in humans including familial hemiplegic migraine type 1 (FHM1), episodic ataxia type 2 (EA2), spinocerebellar ataxia type 6 (SCA6), and epilepsy. Mice with mutations in the *Cacna1a* gene are a useful tool for obtaining insights into disease processes and defining channel functions. Mouse Ca_v2.1 mutants include FHM1 model strains, including R192Q and S218L knockin mice, a SCA6 model strain carrying additional CAG repeats in the *cacna1a* locus of knockin mice, a knockout strain lacking Ca_v2.1 currents, and spontaneous strains that include *rocker*, *tottering*, *rolling Nagoya*, *leaner*, *tottering-4j*, *tottering-5j*, and *wobbly* mice. This chapter summarizes the human disease phenotypes and functional consequences of disease-causing mutations expressed in cell culture models, and overviews the results that Ca_v2.1 mutant mice have provided regarding the disease mechanisms of Ca_v2.1 channelopathy and the functions of Ca_v2.1 channels.

* Corresponding author: Eiki Takahashi, DVM, PhD, DJCLAM, Research Resources Center, RIKEN Brain Science Institute, Hirosawa 2-1, Wako, Saitama, Japan, 351-0198, E-mail: etakahashi@brain.riken.jp

Keywords: $Ca_v2.1$; channelopathy; epilepsy; episodic ataxia type 2 (EA2); familial hemiplegic migraine type 1 (FMH1); mouse model; spinocerebellar ataxia type 6 (SCA6); synaptic transmission.

INTRODUCTION

Voltage-gated Ca^{2+} (Ca_v) channels play an important role in regulating diverse neuronal functions attributed to elevated intracellular Ca^{2+} concentrations [4]. Ca_v channels exist in resting (closed), activated (open), or inactivated (desensitized) states. Resting channels open in response to membrane depolarization, and then transition to an inactivated state. Intracellular Ca^{2+} concentrations affect various cellular responses [2]. The Ca_v channel is a molecular complex consisting of $\alpha 1$, β, $\alpha 2$-δ, and γ subunits [2]. The pore-forming $\alpha 1$ subunit functions as a voltage sensor and is capable of generating channel activity in heterogeneous expression systems [3]. The $\alpha 1$ subunit has four homologous transmembrane domains (I-IV), each containing six membrane-spanning helices (S1-S6), plus a reentrant p-loop motif that lines the channel pore, enabling the passage of Ca^{2+}. The four domains are connected through cytoplasmic linkers, and both the C- and N-termini are cytoplasmic and interact with regulatory proteins. In heterogeneous expression systems, different β and $\alpha 2$-δ subunits in combination with a given $\alpha 1$ subunit give rise to Ca_v channels with different biophysical properties [11, 40]. Ca_v channels in the central nervous system have been identified and classified by their biophysical and pharmacological profiles as Ca_v1 (L-type), $Ca_v2.1$ (P/Q-type), $Ca_v2.2$ (N-type), $Ca_v2.3$ (R-type), or Ca_v3 (T-type) channels [4].

At the presynaptic terminal, three major Ca_v2 channel types are involved in the Ca^{2+}-dependent exocytotic release of neurotransmitters: $Ca_v2.1$, $Ca_v2.2$, and $Ca_v2.3$ [46]. Ca^{2+} influx via these channels triggers neurotransmitter release in a cooperative process with other components of the vesicle fusion machinery [2]. The mechanisms of Ca^{2+} influx through Ca_v2 channels linked to neurotransmitter release have been determined by identifying proteins that are involved in the trafficking, docking, and fusion of vesicles, and by characterizing their protein-protein interactions during Ca^{2+}-evoked exocytosis [15, 16, 33]. Although Ca_v2 $\alpha 1$ subunit genes are widely expressed throughout the nervous system, high expression levels of each $\alpha 1$ subunit are localized differentially [32, 43]. Given the pivotal role of Ca_v2 channels in controlling specific neurotransmitter production and release in specific regions, defects in the expression, localization, structure, or modulation of presynaptic $Ca_v2.1$, $Ca_v2.2$, and $Ca_v2.3$ channels may result in aberrant synaptic signaling, leading to various patterns of neural network dysfunction.

Mutations in the *CACNA1A* gene encoding the pore-forming $\alpha 1$ subunit of voltage-gated $Ca_v2.1$ channels have been identified in humans. These mutations cause several neurological defects, including familial hemiplegic migraine type 1 (FMH1) [27], episodic ataxia type 2 (EA2) [27], spinocerebellar ataxia type 6 (SCA6) [47], and some forms of epilepsy [12, 21]. No point mutations in the $\alpha 1$ subunit gene of voltage-gated $Ca_v2.2$ or $Ca_v2.3$ channels has been reported to cause detrimental conditions.

To examine $Ca_v2.1$ channel functions and disease processes, mouse genetic approaches can be useful. Mice with mutations in the *Cacna1a* gene include FHM1 model strains, including R192Q and S218L knockin mice, a SCA6 model strain carrying additional CAG repeats in the *cacna1a* locus of knockin mice, a knockout strain lacking Ca$_v$2.1 currents, and spontaneous strains, including *rocker*, *tottering*, *rolling Nagoya*, *leaner*, *tottering-4j*, *tottering-5j*, and *wobbly* mice.

HUMAN CHANNELOPATHIES AND CA$_v$2.1 A1 SUBUNIT MUTANT MICE

Human Ca$_v$2.1 Channelopathies

FMH1

FHM1 is an autosomal-dominant subtype of migraine with an aura and childhood onset, caused by mutations in the *CACNA1A* gene. All 21 reported FHM1 mutations (R192Q, R195K, S218L, V581M, R583Q, T666M, V714A, D715E, Y1246C, K1336E, R1347Q, C1370Y, Y1385C, V1457L, C1535S, R1668W, L1682P, W1684R, V1696I, I1710T, I1811L) are missense mutations that produce substitutions of conserved amino acids in important functional regions of the Ca$_v$2.1 channel, including the pore lining and voltage sensors [29]. The functional consequences of FHM1 mutations have been investigated by expressing recombinant Ca$_v$2.1 channel subunits in heterogeneous expression systems. The R192Q mutation is localized in S4 of the first domain and is associated with a relatively mild FHM1 phenotype [27]. The S218L mutation is localized in the S4-S5 linker of the first domain and is associated with a particularly severe clinical phenotype of FHM1 [22]. Both mutations expressed in cell culture models promote Ca$_v$2.1 current recovery from direct G-protein inhibition [42], suggesting increased neuronal network hyperexcitability, possibly as a consequence of reduced presynaptic inhibition by the gain of function of Ca$_v$2.1 activities.

EA2

EA2 is an autosomal-dominant neurological disease, typically with an onset in childhood, and is an acetazolamide-responsive disorder, characterized by transient attacks of ataxia, instability, and dyskinesia with progressive inter-episode dystonia, general weakness, and mild ataxia [18]. These episodes are triggered by diverse stressors, ranging from psychological stress and exercise to caffeine and ethanol. The symptoms last from hours to days and are primarily cerebellar in origin [18]. Many different point mutations in the *CACNA1A* gene have been shown to cause EA2. The majority of the more than 50 reported EA2 mutations are nonsense mutations or indels that disrupt the open reading frame, leading to truncation of the protein or are intronic mutations that also presumably disrupt the reading frame by exon skipping or intron inclusion of the gene product [29]. *In vitro* models show that EA2 mutations induce the loss of function of Ca$_v$2.1 channel activities [10]. A consequence of the mutations is a loss of precision of pacemaking in cerebellar Purkinje cells [30]. This reduced precision reduces the information encoded by Purkinje cells and is thought to contribute to symptoms associated with this disorder.

SCA6

SCA6 is a late-onset slowly progressive ataxic syndrome associated with marked cerebellar atrophy, especially in the superior vermis, characterized by more severe loss of Purkinje than granule cells [47]. It is an autosomal-dominant neurodegenerative disease caused by expansion of a small CAG repeat in the C-terminal cytoplasmic domain (exon 47) of the *CACNA1A* gene. In normal individuals, the CAG repeat in *CACNA1A* ranges from 4 to 20 repeats, whereas in SCA6 patients, this repeat usually ranges from 20 to 28 repeats [31]. This CAG repeat expansion in SCA6 is smaller than the normal-length CAG repeat in other CAG repeat diseases, such as Huntington's disease. The pathogenesis of SCA6 is thought to be associated with intracellular aggregations of these mutant proteins. In cell culture models of SCA6 disease, early apoptotic cell death occurs [14].

CLINICAL OVERLAP AMONG FHM1, EA2, AND SCA6

The majority of FHM1 patients have cerebellar symptoms and signs [5]. Over half of EA2 patients have migraines [17]. Some SCA6 patients have fluctuating ataxia, similar to EA2 [20]. Conversely, some EA2 patients present with prominent progressive ataxia reminiscent of SCA6 [45].

Epilepsy

Epilepsy is a disorder that results from abnormal hyperexcitability and hypersynchronous activity of neurons and has a wide variety of origins, including brain damage and genetic causes [36]. It is thought that mutations that give rise to reduced $Ca_v2.1$ channel function have a propensity to cause absence seizures. In patients with FHM1, seizures are exceedingly rare and limited to a few case reports [5]. The mutation in S2 of the first domain reduces $Ca_v2.1$ channel activity and gives rise to an ataxic, epileptic phenotype [12]. A truncation mutation in the C-terminal region of $Ca_v2.1$ that results in a non-functional channel has been associated with the occurrence of childhood episodes of absence epilepsy and primary generalized seizures in a patient with EA2 [21]. An 11-year-old girl with a missense mutation (I712V) in the *CACNA1A* gene was described to have a range of symptoms, including seizures, headache, and ataxia [9].

$Ca_v2.1\alpha1$ Subunit Mutant Mice

FHM1 model strains including R192Q and S218L knockin mice, a SCA6 model strain carrying expanded CAG repeats in the *Cacna1a* locus of knockin mice, a knockout strain lacking $Ca_v2.1$ currents, and spontaneous strains including *rocker*, *tottering*, *rolling Nagoya*, *leaner*, *tottering-4j*, *tottering-5j*, and *wobbly* mice have been reported as mice with mutations in the *Cacna1a* gene (Fig. 1).

FHM1 Knockin Mouse

A knockin model expressing the human pathogenic R192Q FHM1 mutation revealed increased Ca^{2+} influx and a decreased threshold [37, 38]. The R192Q FHM1 knockin data indicate that there was enhanced glutamate release as a gain of function of $Ca_v2.1$ channels and increased velocity of cortical spreading depression (CSD). CSD is the mechanism underlying the aura of migraine [23], and has been shown to activate headache mechanisms [1]. The S218L knockin mice exhibited mild permanent cerebellar ataxia, spontaneous attacks of hemiparesis or fatal generalized seizures, and brain edema after only a mild head impact, modeling the main features of the severe S218L clinical syndrome [38, 39]. In agreement with the lower threshold of activation of human S218L Ca$_v$2.1 channels compared with human R192Q Ca$_v$2.1 channels [34], the gain of function of the Ca$_v$2.1 current at low voltages was larger in S218L than in R192Q knockin mice. Correlated with the more severe S218L clinical phenotype and larger gain of function of the neuronal Ca$_v$2.1 current, the facilitation of both the induction and propagation of CSD was greater in S218L than in R192Q knockin mice [39]. Studies of CSD in FHM1 knockin mice indicate that CSD is a key player in the pathogenesis of migraine and that the Ca$_v$2.1 channel plays a key role in the initiation and propagation of CSD.

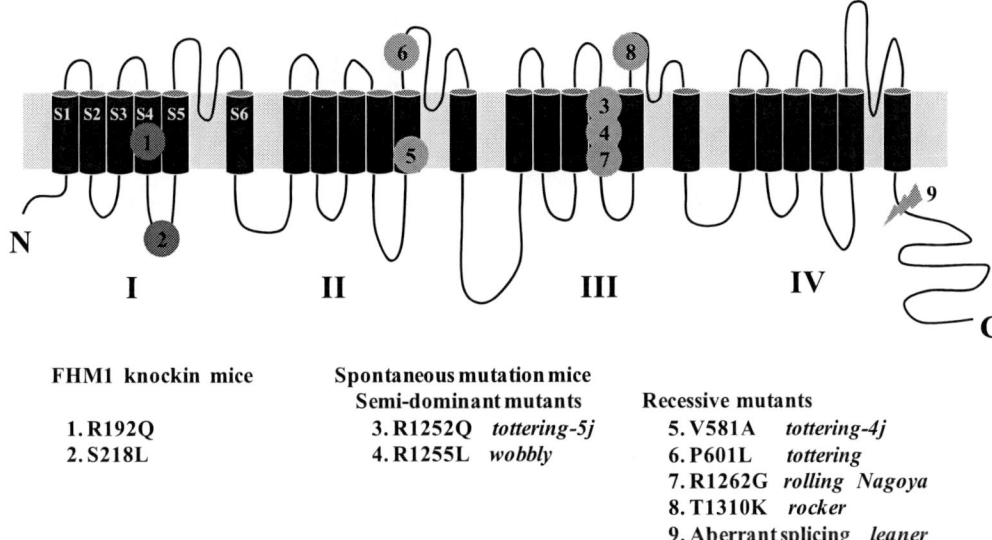

FHM1 knockin mice	Spontaneous mutation mice	
	Semi-dominant mutants	Recessive mutants
1. R192Q	3. R1252Q *tottering-5j*	5. V581A *tottering-4j*
2. S218L	4. R1255L *wobbly*	6. P601L *tottering*
		7. R1262G *rolling Nagoya*
		8. T1310K *rocker*
		9. Aberrant splicing *leaner*

Figure 1. Locations of mouse mutations in the secondary structure of the *Cacna1a* gene. In 3. R1252Q *tottering-5j, please see PPT.*

SCA6 Knockin Mice

As a model of SCA6 in mice, Watase *et al.* (2008) generated three lines of knockin mice, carrying either 14 (normal human allele), 30 (expanded SCA6 allele), or 84 (hyperexpanded allele) CAG repeats in the mouse *cacna1a* locus (*Sca6^{14Q}*, *Sca6^{30Q}*, and *Sca6^{84Q}*) [41]. The *Sca6^{84Q}* mice, but not those with an expanded CAG repeat similar to that of SCA6 patients, developed motor impairment with age; both the onset and severity of symptoms were gene-dosage dependent. The mutant Ca$_v$2.1 channels with the expanded CAG repeat formed insoluble aggregates in the cerebellum in an age- and gene-dosage-dependent manner. These findings are consistent with the hypothesis that the pathogenesis of SCA6 involves a toxic gain-of-function mechanism, associated with accumulation of the expanded CAG repeat protein, similar to that of other polyglutamine diseases [8].

Ca$_v$2.1 Knockout Mouse

Ca$_v$2.1 knockout mice exhibit dystonia and most do not survive past weaning [7]. The Ca$_v$2.1 knockout mice surviving at 4 months of age show selective degeneration of the cerebellum in a specific pattern, with extensive loss of Purkinje cells in parasagittal stripes and graded loss of granule cells that is more severe in the anterior lobe [7]. Cerebellar atrophy in EA2 and SCA6 patients also predominates in the anterior lobe [28]. At the calyx of the held presynaptic terminal of $Ca_v2.1$ knockout mice, obliteration of the $Ca_v2.1$ channel currents was partially compensated by an increase in the $Ca_v2.2$ channel, but not the $Ca_v2.3$ channel, and this replacement altered the short-term plasticity mediated by Ca^{2+}-dependent modulation of the presynaptic Ca_v2 channel [13]. In contrast, at the neuromuscular junction of $Ca_v2.1$ knockout mice, both $Ca_v2.2$ and $Ca_v2.3$ channels were upregulated and cooperated in controlling Ca^{2+} release [35]. These data indicate that Ca_v2 channels appear to differ in their capacity to interact with the release machinery in a synapse-dependent manner, and that other Ca_v2 channels do not completely compensate for the loss of $Ca_v2.1$ channels.

SPONTANEOUS MUTATION MICE

In the spontaneous *Cacna1a* mutants, semi-dominant mutations were detected in the *tottering-5j* [24] and *wobbly* [44] mice and recessive mutations were detected in the *rocker* [48], *tottering* [6], *rolling Nagoya* [26], *tottering-4j* [24], and *leaner* [6] mice. In contrast to the heterozygous *tottering-5j* and *wobbly* mice, which showed mild ataxia and had normal life spans, the homozygous *tottering-5j* and *wobbly* mice showed severely ataxia and died prematurely. All of the homozygous recessive mice developed ataxia, ranging from mild in *rocker, tottering, rolling Nagoya*, and *tottering-4j* mice to severe ataxia in *leaner* mice. The *rocker, tottering, tottering-4j*, and *leaner* mice exhibit intermittent seizures similar to human absence epilepsy, while *rolling Nagoya* mice do not have seizures. These spontaneous mutants with loss-of-function mutations in the *Cacna1a* gene that lead to a reduced $Ca_v2.1$ current density are often considered EA2 mouse models, although these mutations do not seem to produce $Ca_v2.1$ trafficking-expression defects or dominant negative defects comparable to those induced by EA2 mutations [29]. They have provided important insights into the mechanism of ataxia caused by $Ca_v2.1$ channels.

The point mutation identified in *wobbly* mutants is very close to the *rolling Nagoya* mutation structurally (Fig. 1). The *wobbly* mutation leads to the positive charge-neutralizing substitution R1255L and the *rolling Nagoya* mutation leads to the positive charge-neutralizing change R1262G. Leucine (L) and glycine (G) belong to the aliphatic R group, which is nonpolar and hydrophobic. Although these mutations lead to similar ataxic phenotypes in homozygous mice, *wobbly* mice have more severe cerebellar ataxia than *rolling Nagoya* mice. Interestingly, *wobbly* is a semi-dominant mutation, whereas *rolling Nagoya* is recessive. Transmembrane S4 segments have been implicated as voltage sensors and have a characteristic arrangement with a varying number of positively charged amino acids at every third position. The *rolling Nagoya* mutation produces a deficit in the gating properties of the $Ca_v2.1$ channel [25]. Functional and structural studies of the voltage-gated potassium channel Shaker suggest that the first four positively charged residues (counting from the extracellular side of S4) are the most relevant for voltage sensing [19]. The *wobbly* mutation leads to neutralization of the fourth positively charged arginine, and its position complies with the characteristic arrangement of the positively charged amino acids at every third position. In contrast, the *rolling Nagoya* mutation leads to the neutralization of the seventh positively charged arginine, which is an extra positively charged residue in S4 of the third domain in a position that deviates from the characteristic arrangement of positively charged amino acids at every third position. Thus, the *wobbly* mutation likely alters a functionally more important positively charged residue in S4 than that altered by *rolling Nagoya*, and may induce greater impairment in gate charging of the Ca$_v$2.1 channel than the *rolling Nagoya* mutation does, and lead to more severe cerebellar ataxia as a semi-dominant mutation. Different *Cacna1a* gene allelic variants exhibit considerable variability in their phenotypes, so a detailed comparison of the allelic variants and gene dosages may help in revealing causal relationships among the many different biophysical, structural, and site-specific synaptic abnormalities and behavioral deficits.

CONCLUSION

Ca$_v$2.1 channels have a dominant, specific role in synaptic transmission at central excitatory synapses. The *CACNA1A* gene encodes the pore-forming, voltage-sensitive subunit of the voltage-dependent calcium Ca$_v$2.1 channels. Mutations in this gene have been linked to several human disorders, including FHM1, EA2, SCA6, and epilepsy. Genetic approaches in mice have provided a wealth of information on the role of neuronal processes. These studies would have been impossible without the overwhelming amount of data gathered from biochemical and cellular studies. Conversely, without the mutants, the physiological relevance of these *in vitro* findings would not have been revealed. The *Cacna1a* gene mutants have proven to be a useful tool for unraveling the neurological defects, especially synaptic diseases, and Ca$_v$2.1 function. Studies of *Cacna1a* gene mutants have been important for investigating the molecular basis of neuronal network systems.

REFERENCES

[1]   Bolay H, Reuter U, Dunn AK, Huang Z, Boas DA, Moskowitz MA. Intrinsic brain activity triggers trigeminal meningeal afferents in a migraine model. *Nat Med.* 2002. 8:136-142.

[2]   Catterall WA. Structure and function of neuronal Ca^{2+} channels and their role in neurotransmitter release. *Cell Calcium.* 1998. **24**:307-323.

[3]   Catterall WA. Interactions of presynaptic Ca^{2+} channels and snare proteins in neurotransmitter release. *Ann. NY Acad. Sci.* 1999. 868:144-159.

[4]   Catterall WA and Few AP. Ca^{2+} channel regulation and presynaptic plasticity. *Neuron.* 2008. 59:882-901.

[5]   Ducros A, Denier C, Joutel A, Cecillon M, Lescoat C, Vahedi K, Darcel F, Vicaut E, Bousser MG, Tournier-Lasserve E. The clinical spectrum of familial hemiplegic migraine associated with mutations in a neuronal calcium channel. *N Engl J Med.* 2001. 345:17-24.

[6]   Fletcher CF, Lutz CM, O'Sullivan TN, Shaughnessy JD Jr, Hawkes R, Frankel WN, Copeland NG, Jenkins NA. Absence epilepsy in tottering mutant mice is associated with calcium channel defects. *Cell.* 1996. 87:607-617.

[7] Fletcher CF, Tottene A, Lennon VA, Wilson SM, Dubel SJ, Paylor R, Hosford DA, Tessarollo L, McEnery MW, Pietrobon D, Copeland NG, Jenkins NA. Dystonia and cerebellar atrophy in Cacna1a null mice lacking P/Q calcium channel activity. *Faseb J.* 2001. 15:1288-1290.

[8] Gatchel JR, Zoghbi HY. Diseases of unstable repeat expansion: mechanisms and common principles. *Nat Rev Genet.* 2005. 6:743-755.

[9] Guerin AA, Feigenbaum A, Donner EJ, Yoon G. Stepwise developmental regression associated with novel CACNA1A mutation. *Pediatr Neurol.* 2008. 39:363-364.

[10] Guida S, Trettel F, Pagnutti S, Mantuano E, Tottene A, Veneziano L, Fellin T, Spadaro M, Stauderman K, Williams M, Volsen S, Ophoff R, Frants R, Jodice C, Frontali M, Pietrobon D. Complete loss of P/Q calcium channel activity caused by a CACNA1A missense mutation carried by patients with episodic ataxia type 2. *Am J Hum Genet.* 2001. 68:759-764.

[11] Hobom M, Dai S, Marais E, Lacinova L, Hofmann F, Klugbauer N. Neuronal distribution and functional characterization of the calcium channel alpha2delta-2 subunit. *Eur J Neurosci.* 2000. 124:1217-1226.

[12] Imbrici P, Jaffe SL, Eunson LH, Davies NP, Herd C, Robertson R, Kullmann DM Hanna MG. Dysfunction of the brain calcium channel Cav2.1 in absence epilepsy and episodic ataxia. *Brain.* 2004. 127:2682-2692.

[13] Inchauspe CG, Forsythe ID, Uchitel OD. Changes in synaptic transmission properties due to the expression of N-type calcium channels at the calyx of Held synapse of mice lacking P/Q-type calcium channels. *J Physiol.* 2007. 584:835-851.

[14] Ishikawa K, Fujigasaki H, Saegusa H, Ohwada K, Fujita T, Iwamoto H, Komatsuzaki Y, Toru S, Toriyama H, Watanabe M, Ohkoshi N, Shoji S, Kanazawa I, Tanabe T, Mizusawa H. Abundant expression and cytoplasmic aggregations of alpha1A voltage-dependent calcium channel protein associated with neurodegeneration in spinocerebellar ataxia type 6. *Hum Mol Genet.* 1999. 8:1185-1193.

[15] Jahn R and Scheller RH. SNAREs-engines for membrane fusion. *Nat Rev Mol Cell Biol.* 2006. 7:631-643.

[16] Jarvis SE and Zamponi GW. Trafficking and regulation of neuronal voltage-gated calcium channels. *Curr Opin Cell Biol.* 2007. 19:474-482.

[17] Jen J, Kim GW, Baloh RW. Clinical spectrum of episodic ataxia type 2. *Neurology.* 2004. 62:17-22.

[18] Jen JC, Graves TD, Hess EJ, Hanna MG, Griggs RC, Baloh RW. Primary episodic ataxias: diagnosis, pathogenesis and treatment. *Brain.* 2007. 130:2484-2493

[19] Jiang Y, Ruta V, Chen J, Lee A, MacKinnon R. The principle of gating charge movement in a voltage-dependent K^+ channel. *Nature.* 2003. 423:42-48.

[20] Jodice C, Mantuano E, Veneziano L, Trettel F, Sabbadini G, Calandriello L, Francia A, Spadaro M, Pierelli F, Salvi F, Ophoff RA, Frants RR, Frontali M. Episodic ataxia type 2 (EA2) and spinocerebellar ataxia type 6 (SCA6) due to CAG repeat expansion in the CACNA1A gene on chromosome 19p. *Hum Mol Genet.* 1997. 6:1973-1978.

[21] Jouvenceau A, Eunson LH, Spauschus A, Ramesh V, Zuberi SM, Kullmann DM Hanna MG. Human epilepsy associated with dysfunction of the brain P/Q-type calcium channel. *Lancet.* 2001. 358:801-807.

[22] Kors EE, Terwindt GM, Vermeulen FL, Fitzsimons RB, Jardine PE, Heywood P, Love S, van den Maagdenberg AM, Haan J, Frants RR, Ferrari MD. Delayed cerebral edema

and fatal coma after minor head trauma: role of the CACNA1A calcium channel subunit gene and relationship with familial hemiplegic migraine. *Ann Neurol.* 2001. 49:753-760.

[23] Lauritzen M. Pathophysiology of the migraine aura. The spreading depression theory. *Brain.* 1994. 117:199-210

[24] Miki T, Zwingman TA, Wakamori M, Lutz CM, Cook SA, Hosford DA, Herrup K, Fletcher CF, Mori Y, Frankel WN, Letts VA. Two novel alleles of tottering with distinct Cav2.1 calcium channel neuropathologies. *Neuroscience.* 2008. 155:31-44.

[25] Mori Y, Wakamori M, Oda S, Fletcher CF, Sekiguchi N, Mori E, Copeland NG, Jenkins NA, Matsushita K, Matsuyama Z, Imoto K. Reduced voltage sensitivity of activation of P/Q-type Ca^{2+} channels is associated with the ataxic mouse mutation rolling Nagoya (tg(rol)). *J Neurosci.* 2000. 20:5654-5662.

[26] Oda S. The observation of rolling mouse Nagoya (*rol*), a new neurological mutant, and its maintenance. *Exp Anim.* 1973. 22:281-286.

[27] Ophoff RA, Terwindt GM, Vergouwe MN, van Eijk R, Oefner PJ, Hoffman SM, Lamerdin JE, Mohrenweiser HW, Bulman DE, Ferrari M, Haan J, Lindhout D, van Ommen GJ, Hofker MH, Ferrari MD, Frants RR. Familial hemiplegic migraine and episodic ataxia type-2 are caused by mutations in the Ca^{2+} channel gene CACNL1A4. *Cell.* 1996. 87:543-552.

[28] Pietrobon D. Calcium channels and channelopathies of the central nervous system. *Mol Neurobiol.* 2002. 25:31-50.

[29] Pietrobon D. Cav2.1 channelopathies. *Pflugers Arch.* 2010. 460:375-393.

[30] Strupp M, Zwergal A, Brandt T. Episodic ataxia type 2. *Neurotherapeutics.* 2007. 4:267-273.

[31] Takahashi H, Ishikawa K, Tsutsumi T, Fujigasaki H, Kawata A, Okiyama R, Fujita T, Yoshizawa K, Yamaguchi S, Tomiyasu H, Yoshii F, Mitani K, Shimizu N, Yamazaki M, Miyamoto T, Orimo T, Shoji S, Kitamura K, Mizusawa H. A clinical and genetic study in a large cohort of patients with spinocerebellar ataxia type 6. *J Hum Genet.* 2004. 49:256-264.

[32] Tanaka O, Sakagami H, Kondo H. Localization of mRNAs of voltage-dependent Ca^{2+} channels: four subtypes of alpha 1- and beta-subunits in developing and mature rat brain. *Brain Res Mol Brain Res.* 1995. 30:1-16.

[33] Tedford HW and Zamponi GW. Direct G protein modulation of Cav2 calcium channels. *Pharmacol Rev.* 2006. 58:837-862.

[34] Tottene A, Pivotto F, Fellin T, Cesetti T, van den Maagdenberg AM, Pietrobon D. Specific kinetic alterations of human Cav2.1 calcium channels produced by mutation S218L causing familial hemiplegic migraine and delayed cerebral edema and coma after minor head trauma. *J Biol Chem.* 2005. 280:17678-17686.

[35] Urbano FJ, Piedras-Rentería ES, Jun K, Shin HS, Uchitel OD, Tsien RW. Altered properties of quantal neurotransmitter release at endplates of mice lacking P/Q-type Ca^{2+} channels. *Proc Natl Acad Sci USA.* 2003. 100:3491-3496.

[36] Vadlamudi L, Scheffer IE, Berkovic SF. Genetics of temporal lobe epilepsy. *J Neurol Neurosurg Psychiatry.* 2003. 74:1359-1361.

[37] van de Ven RC, Kaja S, Plomp JJ, Frants RR, van den Maagdenberg AM, Ferrari MD. Genetic models of migraine. *Archives of neurology.* 2007. 64:643-646.

[38] van den Maagdenberg AM, Pietrobon D, Pizzorusso T, Kaja S, Broos LA, Cesetti T, van de Ven RC, Tottene A, van der Kaa J, Plomp JJ, Frants RR, Ferrari MD. A Cacna1a knockin migraine mouse model with increased susceptibility to cortical spreading depression. *Neuron*. 2004. 41:701-710.

[39] van den Maagdenberg AM, Pizzorusso T, Kaja S, Terpolilli N, Shapovalova M, Hoebeek FE, Barrett CF, Gherardini L, van de Ven RC, Todorov B, Broos LA, Tottene A, Gao Z, Fodor M, De Zeeuw CI, Frants RR, Plesnila N, Plomp JJ, Pietrobon D, Ferrari MD. High CSD susceptibility and migraine-associated symptoms in $Ca_V2.1$ S218L mice. *Ann Neurol*. 2010. 67:85-98.

[40] Walker D and De Waard M. Subunit interaction sites in voltage-dependent Ca^{2+} channels: role in channel function. *Trends Neurosci*. 1998. 21:148-154.

[41] Watase K, Barrett CF, Miyazaki T, Ishiguro T, Ishikawa K, Hu Y, Unno T, Sun Y, Kasai S, Watanabe M, Gomez CM, Mizusawa H, Tsien RW, Zoghbi HY. Spinocerebellar ataxia type 6 knockin mice develop a progressive neuronal dysfunction with age-dependent accumulation of mutant Cav2.1 channels. *Proc Natl Acad Sci USA*. 2008. 105:11987-1199.

[42] [42] Weiss N, Sandoval A, Felix R, Van den Maagdenberg A, De Waard M. The S218L familial hemiplegic migraine mutation promotes deinhibition of Cav2.1 calcium channels during direct G-protein regulation. *Pflugers Arch*. 2008. 457:315-326.

[43] Williams ME, Marubio LM, Deal CR, Hans M, Brust PF, Philipson LH, Miller RJ, Johnson EC, Harpold MM, Ellis SB. Structure and functional characterization of neuronal alpha 1E calcium channel subtypes. *J Biol Chem*. 1994. 269:22347-22357.

[44] Xie G, Clapcote SJ, Nieman BJ, Tallerico T, Huang Y, Vukobradovic I, Cordes SP, Osborne LR, Rossant J, Sled JG, Henderson JT, Roder JC. Forward genetic screen of mouse reveals dominant missense mutation in the P/Q-type voltage-dependent calcium channel, CACNA1A. *Genes Brain Behav*. 2007. 6:717-727.

[45] Yue Q, Jen JC, Nelson SF, Baloh RW. Progressive ataxia due to a missense mutation in a calcium-channel gene. *Am J Hum Genet*. 1997. 61:1078-1087.

[46] Yokoyama CT, Myers SJ, Fu J, Mockus SM, Scheuer T, Catterall WA. Mechanism of SNARE protein binding and regulation of Cav2 channels by phosphorylation of the synaptic protein interaction site. *Mol Cell Neurosci*. 2005. 28:1-17.

[47] Zhuchenko O, Bailey J, Bonnen P, Ashizawa T, Stockton DW, Amos C, Dobyns WB, Subramony SH, Zoghbi HY, Lee CC. Autosomal dominant cerebellar ataxia (SCA6) associated with small polyglutamine expansions in the alpha 1A-voltage-dependent calcium channel. *Nat Genet*. 1997. 15:62-69.

[48] Zwingman TA, Neumann PE, Noebels JL, Herrup K. Rocker is a new variant of the voltage-dependent calcium channel gene Cacna1a. *J Neurosci*. 2001. 21:1169-1178.

In: Calcium Signaling
Editor: Masayoshi Yamaguchi

ISBN: 978-1-61324-313-8
©2012 Nova Science Publishers, Inc.

Chapter 8

REGUCALCIN: ROLE AS A SUPPRESSOR PROTEIN IN CALCIUM SIGNALING

Masayoshi Yamaguchi[*]

Division of Endocrinology and Metabolism and Lipids, Department of Medicine,
Emory University School of Medicine, Atlanta, Georgia, US

ABSTRACT

Regucalcin was discovered by Yamaguchi in the year of 1978 as a calcium-binding protein that does not contain EF-hand motif of calcium-binding domain. The name regucalcin was proposed for this calcium-binding protein, which can regulate various Ca^{2+}-dependent enzyme activations in liver cells. The regucalcin gene is localized on the chromosome X, and the organization of the regucalcin gene consists of seven exons and six introns. AP-1, NF1-A1, and RGPR-p117 bind to the promoter region of the rat regucalcin gene and enhance transcription activity of regucalcin gene expression that is mediated through calcium signaling. Regucalcin plays a pivotal role in the keep of intracellular calcium ion (Ca^{2+}) homeostasis due to activating Ca^{2+} pump enzymes in the plasma membrane (basolateral membrane), microsomes (endoplasmic reticulum), mitochondria, and nuclei of many cell types. Regucalcin has a suppressive effect on calcium signaling from the cytoplasm to the nucleus in the proliferative cells. Regucalcin has also been demonstrated to transport to the nucleus, and it can inhibit Ca^{2+}-dependent protein kinase and protein phosphatase activities, Ca^{2+}-activated deoxyribonucleic acid (DNA) fragmentation, and DNA and ribonucleic acid synthesis in the nucleus. Overexpression of regucalcin suppresses cell death and apoptosis in the cloned rat hepatoma cells induced by various signaling factors. Regucalcin can inhibit the enhancement of cell proliferation due to hormonal stimulation. Regucalcin plays an important role as a suppressor protein in cell signaling system, and it is proposed to play a pivotal role in keep of cell homeostasis and function.

---

[*] Division of Endocrinology and Metabolism and Lipids, Department of Medicine, Emory University School of Medicine, 101 Woodruff Circle, 1305 WMRB, Atlanta, Georgia 30322-0001, USA, E-mail: yamamasa1155@ yahoo.co.jp

Keywords: Regucalcin, RGN, gene expression, calcium signaling, calmodulin, protein kinase
   C, nuclear function, apoptosis, cell proliferation

INTRODUCTION

Calcium ion (Ca^{2+}) plays an important role in the regulation of many cell functions. Ca^{2+} can regulate muscle contraction, neurotransmission, hormone secretion, cell mitosis, and gene expression. A role as second messengers of Ca^{2+} in cells for hormonal stimulation comes into notice. Calcium signal is transmitted to intracellular responses, which are mediated through a family of calcium-binding protein and protein kinase C [1, 2]. Liver metabolism is regulated by an increase in Ca^{2+} in the cytoplasm of liver cells due to hormonal stimulation [3-5]. The effect of Ca^{2+} is amplified through calmodulin and protein kinase C [1-5]. Calcium signaling is important in the regulation of liver metabolism.

Liver has been shown to participate in the regulation of calcium metabolism through hepatic bile system in rats, and bile calcium excretion is increased by hormonal stimulation [6,7]. On the basis of this finding, it was found that a novel calcium-binding protein, which differs from calmodulin and other calcium-related proteins, was present in the hepatic cytoplasm of rats [8-10]. The name regucalcin was proposed for this calcium-binding protein, which regulates various Ca^{2+}- or Ca^{2+}/calmodulin-dependent enzyme activations in liver cells [11- 15].

Regucalcin and its gene (RGN) are identified in over 15 species consisting of regucalcin family [16-21]. Comparison of the nucleotide sequences of regucalcin from vertebrate species is highly conserved in their coding region with throughout evolution. The regucalcin gene is localized on the chromosome X, and the organization of the regucalcin gene consists of seven exons and six introns [21-23].

AP-1, NF1-A1, and RGPR-p117, which is a transcription factor, bind to the promoter region of the regucalcin gene and enhance transcription activity of regucalcin gene expression that is mediated through calcium and other signalings [21]. Regucalcin mRNA expression and its protein content are pronounced in the liver and kidney cortex of rats, although it is present only slightly in other tissues (including the duodenum, testis, spleen, lung, smooth muscle, heart, and brain) [17, 25-30]. The role of regucalcin is investigated in the liver, kidney cells, heart, and brain in detail [reviewed in Ref. 31-35]. After finding of regucalcin, the identical protein to regucalcin was also reported as senesence marker protein-30 (SMP30) [36, 37].

There are growing evidences that regucalcin plays an important role as a regulatory protein for calcium signaling from the cytoplasm to the nuclei in liver cells [38]. Overexpression of regucalcin has been demonstrated to inhibit cell apoptosis and cell proliferation induced by various signaling factors. This chapter has been written to outline the recent advances that have been made concerning the role of regucalcin as a regulatory protein in cell signaling in liver, kidney, and other tissues: its role in regulation of intracellular Ca^{2+} homeostasis, inhibition of Ca^{2+}-dependent enzyme activations, regulation of nuclear calcium signaling, and inhibitory effects in cell apoptosis and cell proliferation induced by various signaling factors.

PROPERTIES OF CALCIUM BINDING IN REGUCALCIN

The molecular weight of rat regucalcin is estimated as 33,388 Da, composing of 299 amino acid residues from the cloning of rat regucalcin cDNA [16]. The regucalcin molecule does not contain the EF-hand motif as a calcium-binding domain [16]. The isoelectric point of regucalcin is 5.20 [16].

From the experimental data by Scatchard plot, the apparent association constant (K_f) for calcium ion (Ca^{2+}) of regucalcin is found to be 4.19×10^5 M^{-1} by equilibrium dialysis with a correlation coefficient of 0.99, and there are 6.52 high-affinity sites per molecule of protein [8-10]. Regucalcin appears to have six or seven high-affinity binding sites for Ca^{2+} per molecule of protein [9]. Extrapolation of the regression line to infinite Ca^{2+} concentration indicated a maximal Ca^{2+}-binding of 2.28×10^{-4} mol of Ca^{2+} per gram of protein. It is known that calmodulin exists as a monomer with a molecular weight of 17,000 and contains four Ca^{2+}-binding sites [1].

The conformational changes induced by binding of Ca^{2+} to regucalcin have been investigated by means of the ultraviolet (UV) absorption spectrum [10]. UV of regucalcin showed a maximum at 278 nm in the range from 240 to 330 nm. In the presence of Ca^{2+} (0.1 and 1.0 mM), a decrease in absorption at 278 nm was observed [10]. Such a negative UV difference is similar to the Ca^{2+}-induced absorption changes in calmodulin. The spectrum can be attributed to charges in both tyrosine and tryptophan residues. Changes in the environment of both aromatic amino acids occur upon Ca^{2+}-binding [10].

Fluorescence spectroscopy is used to study the effect of Ca^{2+} on the conformation of regucalcin [10]. The spectral emission was quenched after the addition of 1.0 mM Ca^{2+}. Known fluorescence emission properties of isolated tyrosine and tryptophan residues suggest, as indicated above, that changes in the environment of these two aromatic amino acids occur upon Ca^{2+}-binding. These observations demonstrate that Ca^{2+}-binding induces conformational changes in regucalcin. These changes may result in increasing the hydrophobicity of regucalcin [10, 16].

The conformation of the polypeptide backbone of regucalcin has been studied by circular dichroism (CD) spectroscopy [10]. The presence of 1.0 mM Ca^{2+} caused clear alterations in the CD spectrum. The apparent α-helical content of regucalcin in Ca^{2+}-free buffer was estimated to be 34%, and the presence of 1.0 mM Ca^{2+} decreased this by 4.5%. Thus, conformational changes are induced by Ca^{2+} binding to regucalcin. This binding also loosens the conformation of regucalcin. The apparent α-helical content of calmodulin is 30%, and it is increased after Ca^{2+} binding and its conformation is tightened. [10,16]. The increase in hydrophobicity after Ca^{2+} binding represents the mechanism that calmodulin activates its target proteins. However, regucalcin reverses the activation of many enzymes by Ca^{2+}/calmodulin. The role of regucalcin may be different from that of calmodulin in cells.

Regucalcin differs entirely from other calcium-binding proteins of the EF-hand type: calmodulin, calcineurin, parvalbumin, S-l00a, S-100b proteins, caligulin, calregulin, calbindin, calreticulin, and annexins. This is supported from the results of the molecular cloning and sequencing of the cDNA coding for a regucalcin from various mamarlian livers [16, 20]. The nucleotide and amino acid sequences of regucalcin does not have statistically significant homology, as compared with the registered sequences which are found in the EMBL and GenBank detabases (D14327 and D86217) [16].

The hydropathy profile of regucalcin shows that there is a hydrophobic sequence in both N-terminal and C-terminal regions of the regucalcin molecules [16]. Regucalcin shows a hydrophilic character as molecule [16, 20]. The most common EF-hand is composed of the helix-loop-helix-domain. The prototype loop consists of 12 amino acids, of which five have a carboxyl (or a hydroxyl group) in their side chain, precisely spaced so as to coordinate the Ca^{2+}. Analysis of the structure of the EF-hand from the regucalcin sequence did not give the expected pattern of amino acids conforming to the typical EF-hand structure of a calcium-binding site.

Amino acid analysis shows that regucalcin has a relatively high content of glycine and much lower amounts of glutamic acid, valine, aspartic acid, and lysine [16, 20]. Regucalcin contains about 20% (mol%) glutamic and aspartic acids and about 17% amide residues (lysine, histidine, and arginine) and thus a high proportion of charged residues: regucalcin molecule contains aspartic acid (24 residues) and glutamic acid (16 residues) [16]. Acidic amino acids comprise approximately one-fifth of the regucalcin molecule, a prevalence that appears to be characteristic of the composition of all Ca^{2+}-binding proteins. The significance of the high di-carboxylic acid content is that it permits binding of Ca^{2+} to protein. These amino acids may be related to Ca^{2+} binding.

The result of crystal structure with X-ray diffraction data shows that regucalcin contains the metal site bound with either a Ca^{2+} or a Zn^{2+} atom, suggesting that the Ca^{2+}-bound form may be physiologically relevant for stressed cells with an elevated Ca^{2+} level [39]. This supports our finding for regucalcin as a calcium-binding protein.

REGUCALCIN GENE EXPRESSION AND CA^{2+} SIGNALING

The rat regucalcin gene is localized on the proximal end of the rat chromosome Xq11.1-12 [22], and the gene are demonstrated in human, mouse, cow, monkey, dog, rabbit, and chicken but not yeast [17]. The amino acid sequence of mouse regucalcin had 94% homology as compared with that of rat regucalcin [20]. Regucalcin may be a protein that is highly differentiated. The organization of the rat regucalcin gene seems to be about 18 kb in size, and consisted of seven exons and six introns [24]. There are many regulatory elements (AP-1, NF1-A1, RGPR-p117, β-catenin, and NF-κB) in the 5'-flanking region [21, 40-48]. The promoter activity of the rat regucalcin gene is enhanced by treatment with Bay K 8644, dibutyryl cyclic AMP, phorbol esters, insulin, and dexamethasone [42]. Using gel mobility shift assays, it is found that nuclear proteins from rat liver cells and rat hepatoma H4-II-E cells specifically bind to the 5'-flanking region of the rat regucalcin gene [40-42]. Treatment with Bay K 8644, dibutyryl cyclic AMP, phorbol esters, and insulin stimulates the binding of nuclear factors to the 5'-flanking region of the rat regucalcin gene in H4-II-E cells. These factors-inducible nuclear proteins are related to enhance promoter activity of the regucalcin gene [42].

Regucalcin mRNA expression has been demonstrated to mediate through signaling pathway of Ca^{2+}/calmodulin-dependent protein kinase, protein kinase C, and tyrosine kinase in the cells [49-51]. AP-1 factor binds to the 5'-flanking region of the rat regucalcin gene that is mediated through the Ca^{2+} response [41]. AP-1 factor is complex of c-fos/c-jun that is phosphorylated by protein kinases [52, 53]. Calcium signaling system is an important pathway in the stimulation of regucalcin mRNA expression.

Regucalcin mRNA is mainly present in liver and renal cortex with a size of 1.8kb [25]. The expression of regucalcin mRNA in the liver and renal cortex is clearly stimulated through an increase in the cellular Ca^{2+} levels following an oral administration of calcium chloride in rats in vivo [54-56]. Hepatic regucalcin mRNA expression is increased with fetal development and its expression is stimulated after the intake of dietary calcium to maternal rats in vivo [57]. Liver regucalcin concentration is increased after an oral administration of calcium in rats [55].

Hepatic regucalcin mRNA expression is also stimulated after a single subcutaneous administration of calcitonin [58], insulin [59], and estrogen [60], suggesting that the expression of regucalcin mRNA is enhanced through various hormonal stimulation. Regucalcin mRNA expression is increased in regenerating rat liver, suggesting its role in the proliferation of liver cells [61]. Aging has been shown to decrease liver regucalcin mRNA expression [62].

Rat regucalcin immunoreactivity is most pronounced in the liver of rats; it is not seen in the duodenum, testicle, spleen, lung, and smooth muscle (bladder) and is barely visible in the kidney, heart, and brain [26, 27]. Regucalcin is primarily located in the rat liver. Thus, the tissue specific distribution of regucalcin is demonstrated by Northern blotting analysis or enzyme immunoassay.

ROLE OF REGUCALCIN IN INTRACELLULAR CA^{2+} HOMEOSTASIS

Intracellular Ca^{2+} homeostasis is regulated through plasma membrane $(Ca^{2+}-Mg^{2+})$-adenosine 5'-triphosphatease (ATPase), microsomal Ca^{2+}-ATPase, mitochondrial Ca^{2+} uptake, and nuclear Ca^{2+} transport in the cells. Regucalcin has been demonstrated to regulate Ca^{2+}-transporting systems in the liver, renal cortex cells, heart, and brain tissues, suggesting its role in the regulation of intracellular Ca^{2+} homeostasis.

Regucalcin Regulates Ca^{2+} Homeostasis in Liver Cells

The high-affinity $(Ca^{2+}-Mg^{2+})$-ATPase is located on the plasma membranes of liver cells [63, 64]. This enzyme acts as a Ca^{2+} pump to exclude the metal ion from the cytoplasm of liver cells. Addition of regucalcin into the reaction mixture in vitro caused an increase in $(Ca^{2+}-Mg^{2+})$-ATPase activity in the plasma membranes isolated from rat liver, suggesting a role in the regulation of Ca^{2+} pump activity [65]. Regucalcin directly activates $(Ca^{2+}-Mg^{2+})$-ATPase independently of Ca^{2+}-stimulated phosphorylation of the enzyme [65-67], and it has been shown to stimulate ATP-dependent calcium transport across the plasma membrane vesicles of rat liver after addition of $^{45}Ca^{2+}$ into the reaction mixture in vitro [68]. Regucalcin-enhanced ATP-dependent $^{45}Ca^{2+}$ uptake in the plasma membrane vesicles is completely inhibited in the presence of N-ethylmaleimide or digitonin [67]. Regucalcin has been shown to bind the lipid components of liver plasma membrane, and it acts on the sulfhydryl (SH) groups that are an active site of $(Ca^{2+}-Mg^{2+})$-ATPase [67]. The mechanism of regucalcin in activating $(Ca^{2+}-Mg^{2+})$-ATPase may be not involved on GTP-binding protein that modulates the receptor-mediated hormonal effect (including calcitonin, epinephrine, phenylephrine, and insulin) in liver plasma membranes [69].

The effect of hormonal signaling factors (inositol-glycan, dibutyryl cyclic AMP, and inositol 1,4,5-trisphosphate) on the regucalcin-increased $(Ca^{2+}-Mg^{2+})$-ATPase activity in rat liver plasma membranes is examined [70]. Inositol-glycan, which is generated by insulin [70], can directly activate the plasma membrane $(Ca^{2+}-Mg^{2+})$-ATPase, and its effect is modulated by regucalcin. Cross talk with signaling factors may be seen in the regulation of Ca^{2+} pump activity in the plasma membranes of liver cells.

The physiological role of regucalcin in the regulation of $(Ca^{2+}-Mg^{2+})$-ATPase activity in liver plasma membranes is examined after an orall administration of calcium chloride solution in rats [71]. Calcium administration caused an increase in $(Ca^{2+}-Mg^{2+})$-ATPase activity in liver plasma membranes [71]. This increase was abolished in the presence of anti-regucalcin antibody, suggesting an involvement of endogenous regucalcin that is distributed in the cytoplasm. Regenerating rat liver with a proliferative cells significantly increased liver calcium content and plasma membrane $(Ca^{2+}-Mg^{2+})$-ATPase activity between 12 and 48 h after partial hepatectomy [72]. This increase was completely abolished in the presence of anti-regucalcin antibody, indicating an involvement of endogenous regucalcin [72]. Activatory effect of regucalcin on hepatic plasma membrane $(Ca^{2+}-Mg^{2+})$-ATPase was impaired in liver injury with carbon tetrachloride administration in rats [73]. Regucalcin plays a role as an activator protein for Ca^{2+} pump enzyme in the hepatic plasma membranes.

Regucalcin binds Ca^{2+} in the cytoplasm of liver cells, and the metal is subsequently transported into the organell dependent on ATP [74]. Regucalcin may not tightly bind cytosolic Ca^{2+} of lower levels, since its calcium-binding constant is 4.19×10^5 M^{-1} [9]. Regucalcin may regulate cytoplasmic Ca^{2+} levels by activating Ca^{2+} pump enzyme in the plasma membranes of liver cells.

Regucalcin can stimulate the uptake of Ca^{2+} by rat liver mitochondria [75, 76]. The effect of regucalcin on mitochondrial Ca^{2+} uptake is inhibited in the presence of ruthenium red or lanthanum chloride [75], which is an inhibitor of mitochondrial Ca^{2+} transport. Regucalcin may have a role in the reduction of cytoplasmic Ca^{2+} levels due to activating mitochondrial Ca^{2+} uptake.

Regucalcin has also been demonstrated to activate Ca^{2+} pump enzymes $(Ca^{2+}$-ATPase) and to stimulate ATP-dependent $^{45}Ca^{2+}$ uptake by liver microsomes [77, 78], suggesting a role in the regulation of cytoplasmic Ca^{2+} levels. The effect of regucalcin in increasing Ca^{2+}-ATPase activity in the microsomes is inhibited in the presence of thapsigargin, a specific inhibitor of microsomal Ca^{2+} pump enzyme. Regucalcin acts on the SH groups of microsomal Ca^{2+}-ATPase due to the binding on the membranous lipids [78].

The components of Ca^{2+} uptake $(Ca^{2+}$-ATPase) and Ca^{2+} release $(Ca^{2+}$-channels) are located at separate sites on liver microsomes. Interestingly, regucalcin has been found to stimulate Ca^{2+} release from rat liver microsomes [79]. The mechanism is related to the inositol 1, 4, 5-triphosphate (IP3)-induced Ca^{2+} release [80]. Regucalcin may bind to IP3 receptors on the microsomes. This cell physiological significane of regucalcin is unknown. Presumably, regucalcin stimulates microsomal Ca^{2+} uptake when cytosolic Ca^{2+} concentration is raised. Also, regucalcin regulates Ca^{2+} storage in the endoplasmic reticulum of liver cells: it stimulates Ca^{2+} release from the microsomes to restore the microsomal calcium accumulation to regulate Ca^{2+}-related microsomal functions. Regucalcin has been shown to have the reversible effect on liver microsomal glucose-6-phosphatase activity increased by Ca^{2+} addition [31].

The existence of an ATP-stimulated Ca^{2+}-sequestration system is also found in liver nuclei, and it generates a net increase in nuclear matrix free Ca^{2+} concentration [81]. This system may play an important role in the regulation of intranuclear Ca^{2+}-dependent processes [82]. ATPase, which is stimulated by Ca^{2+} in the presence of Mg^{2+}, exists in the nuclei of rat liver, and the Ca^{2+}-stimulated ATPase activity is involved in the nuclear Ca^{2+} uptake [83]. Regucalcin increases Ca^{2+}-ATPase activity in rat liver nuclei [84]. Regucalcin has also been shown to stimulate Ca^{2+} release from liver nuclei [85]. Presumably, regucalcin has a role in the regulation of liver nuclear function through the effect on Ca^{2+} transporting system in the nuclei.

Regucalcin (SMP30) has been shown to lower intracellular Ca^{2+} levels by modulating plasma membrane Ca^{2+}-pumping activity in the cloned human hepatoma HepG2 cells that overexpress regucalcin [86].

Regucalcin Regulates Ca^{2+} Homeostasis in Kidney Cells

Regucalcin largely is present in the kidney cortex of rats [27]. Kidney plays a physiological role in the regulation of calcium homeostasis in blood through re-absorption of urinary calcium [87, 88]. Renal cortex cells play a role in the re-absorption of urinary calcium.

Regucalcin mRNA is expressed in the kidney cortex but not the medulla of rats [56], and its expression is stimulated after calcium administration in vivo [56]. The binding of kidney nuclear proteins to the 5'flanking region of the rat gene for regucalcin has been shown to enhance through Ca^{2+}/calmodulin signaling [89, 90]. Regucalcin mRNA expression is stimulated after the administration of dexamethasone in rats [91], and its expression is suppressed in hypertensive state in rats [92-94]. The specific nuclear factor binds to the NF1-like sequence in the promoter region of regucalcin gene in the kidney cortex of rats [95], and the nuclear factor binding and regucalcin mRNA expression are suppressed after administration of cisplatin that induces kidney damage [95, 96].

Ca^{2+}-ATPase system has been shown to exceed the capacity of the Na^{+}/Ca^{2+} exchanger, and it plays a primary role in Ca^{2+} homeostasis of rat kidney cortex cells [97, 98]. Regucalcin has been demonstrated to play a role as an activator of the ATP-dependent Ca^{2+} pumps in the basolateral membranes isolated from rat kidney cortex [99]. The effect of regucalcin in increasing Ca^{2+} pump enzyme (Ca^{2+}-ATPase) activity in the basolateral membranes is completely inhibited in the presence of N-ethylmaleimide, indicating that regucalcin may act on the SH groups of Ca^{2+}-ATPase [99].

Regucalcin has been shown to increase Ca^{2+}-ATPase activity and ATP-dependent calcium uptake in the microsomes [100] and mitochondria [101] of rat kidney cortex [100]. These increases are clearly decreased in the presence of N-ethylmaleimide, suggesting that regucalcin acts on the SH groups of Ca^{2+}-ATPase in the microsomes.

The finding, that regucalcin increases Ca^{2+}-ATPase activity in the basolateral membranes, microsomes, and mitochondria of rat renal cortex, suggests a physiological role of regucalcin in the regulation of the Ca^{2+} homeostasis in renal cells. Regucalcin may be responsible for ATP-dependent transcellular Ca^{2+} transport, and it participates in the promotion of Ca^{2+} re-absorption in the nephron tubulo of kidney cortex. Regucalcin may play a physiological role in the regulation of calcium metabolism in blood through re-absorption of urinary calcium in kidney.

Regucalcin Regulates Ca^{2+} Homeostasis in Heart Cells

The Ca^{2+} current is one of the most important components in cardiac excitation-contraction coupling [102]. This coupling mechanism is based on the regulation of intracellular Ca^{2+} concentration by Ca^{2+} pump in the sarcoplasmic reticulum of heart muscle [102]. The role of regucalcin in the regulation of heart muscle function is shown. Regucalcin mRNA is expressed in rat heart [103]. The result with Western blot analysis indicates that regucalcin is present in the cytoplasm of heart muscle cells [103]. Regucalcin concentration in the heart muscle tissues has been shown to be about 3.86×10^{-8}M [27]. Regucalcin mRNA expression in the hearts of rats is decreased with increasing age [104], and free radical stress has a suppressive effect on its gene expression [104]. Overexpression of regucalcin in transgenic rats has been found to accelerate free radical stress-induced death of rats [104].

Regucalcin has been found to increase Ca^{2+}-ATPase activity and ATP-dependent Ca^{2+} uptake, which regulates intracellular Ca^{2+} concentration related to cardiac excitation-contraction coupling, in rat heart microsomes, suggesting its role in the regulation of heart muscle function [103]. The effect of regucalcin in increasing heart microsomal Ca^{2+}-ATPase activity is inhibited in the presence of thapsigargin [103], a specific inhibitor of the sarcoplasmic reticulum Ca^{2+} pump enzyme (Ca^{2+}-ATPase) [105], indicating that regucalcin activates Ca^{2+} pump enzyme in the sarcoplasmic reticulum. It is suggested that regucalcin binds to the lipids at the close site of Ca^{2+}-ATPase in heart misrosomes, and that it acts on the SH group which may be an active site of the enzyme and stimulates Ca^{2+}-dependent phosphorylation of Ca^{2+}-ATPase [103].

Phospholamban has been known to inhibit Ca^{2+} pump enzyme (Ca^{2+}-ATPase) in the sarcoplasmic reticulum (misrosomes) of heart muscle [106]. Ca^{2+}-ATPase is activated through cAMP-dependent phosphorylation of phospholamban following hormonal stimulation [106]. The endogenous activatory protein of sarcoplasmic reticulum Ca^{2+}-ATPase is unknown. Regucalcin, which is present in the cytoplasm of heart mescle, may play an important role as an endogenous activator in the regulation of sacroplasmic reticulum Ca^{2+}-ATPase activity in rat heart muscle [103]. Regucalcin may play a physiological role in the regulation of cardiac excitation-contraction coupling.

The role of regucalcin in the regulation of Ca^{2+}-ATPase activity in the heart mitochondria of rats is examined, moreover [107]. Regucalcin is found to be present in the mitochonria of normal rat heart [107], and this localization is increased in the heart of regucalcin transgenic rats as compared with that of normal rats [107]. The addition of regucalcin (10^{-11} to 10^{-8} M) in the enzyme reaction mixture caused a significant increase in Ca^{2+}-ATPase activity in the heart mitochondria in the presence of 50 μM $CaCl_2$ [107]. Ca^{2+}-ATPase activity was also increased in the heart mitochondria of regucalcin transgenic rats [107]. Regucalcin has an activating effect on Ca^{2+}-ATPase in rat heart mitochondria, suggesting its role in the regulation of heart mitochondrial function.

Regucalcin may play a role in the regulation of cytoplasmic Ca^{2+} levels due to activating Ca^{2+} pump activity in the sarcoplasmic reticulum and mitochondria in heart cells.

Regucalcin Regulates Ca^{2+} Homeostasis in Brain

Intracellular Ca^{2+} concentration in the neuronal cells of brain is regulated by various buffering and transport systems such as the membrane Na^{+}-Ca^{2+} exchanges, the membranous Ca^{2+}-ATPase, Ca^{2+}-binding proteins and intracellular Ca^{2+} uptake systems [108-112]. The changes in the neuronal Ca^{2+} homeostasis with aging may be implicated in age-related disturbance in cognitive functions [112]. There is growing evidence that the alteration in the neuronal Ca^{2+} regulation is also implicated in the pathology of Alzheimer's disease [112].

The expression of regucalcin mRNA is demonstrated in brain tissues [29, 113]. Regucalcin concentration in the brain tissues has been shown to be about 5x10^{-9} M as measured using enzyme-linked immuoadsorbent assay, and this level is lowered with aging [27]. Regucalcin is localized in the neurons isolated from rat brain tissues [29], suggesting its role in brain function.

Brain calcium accumulation has been shown to increase after oral administration of calcium in rats, and fasting enhances its accumulation [114]. The supply of glucose may be required in the regulation of brain Ca^{2+} homeostasis in rats *in vivo* [114], suggesting a physiological significance of energy-dependent mechanism in brain calcium metabolism. Fasting caused a significant increase in brain calcium content and Ca^{2+}-ATPase activity in the microsomes and mitochondria of brain tissues in young and aged rats, and these increases were restored after the supply of glusoce, supporting a physiologic significance of energy-dependent mechanism in the regulation of brain calcium in rats with different ages [115].

Aging causes a decrease in Ca^{2+}-ATPase activity in the brain plasma membranes of rats [116], suggesting that aging enhances the entry of Ca^{2+} into brain neuronal cells across the plasma membranes. In addition, aging induces an attenuation of Ca^{2+}-seguestrating system in the brain microsomes, supporting the view that a disturbance of the neuronal Ca^{2+} regulation is brought with increasing age [117]. Protein kinase C activates brain microsomal Ca^{2+}-ATPase in aged rats [118]. It is speculated that aging-induced increase in Ca^{2+}-ATPase activity results from the translocation to the microsomes of protein kinase C in brain cytosol [118]. The development of brain disease with aging may be partly related to the toxicity of brain calcium raised by the increase in microsomal Ca^{2+}-ATPase activity with aging. The disturbance of brain Ca^{2+} homeostasis may play a pivotal role in the revelation of brain disease.

Regucalcin has been found to have an inhibitory effect on Ca^{2+}-ATPase activity in rat brain microsomes [113], suggesting that regucalcin plays a role in the regulation of microsomal Ca^{2+}-ATPase activity in rat brain. Interestingly, the concentration of regucalcin in the cerebral cortex and hippocampus of brain tissues is decreased with aging [113]. The suppressive effect of regucalcin on brain microsomal Ca^{2+}-ATPase activity was weakened in aged rats [113]. Presumably, the aging-induced elevation of brain microsomal Ca^{2+}-ATPase activity is partly resulted from an attenuation of regucalcin action on the enzyme activity with aging. There may be a possibility that the translocation of protein kinase C to the brain microsomes of aged rats [118] is a cause of the attenuation of regucalcin effect on the enzyme activity. Regucalcin may play a physiological and pahophysiological role in the regulation of intracellular Ca^{2+} concentration in the brain tissues.

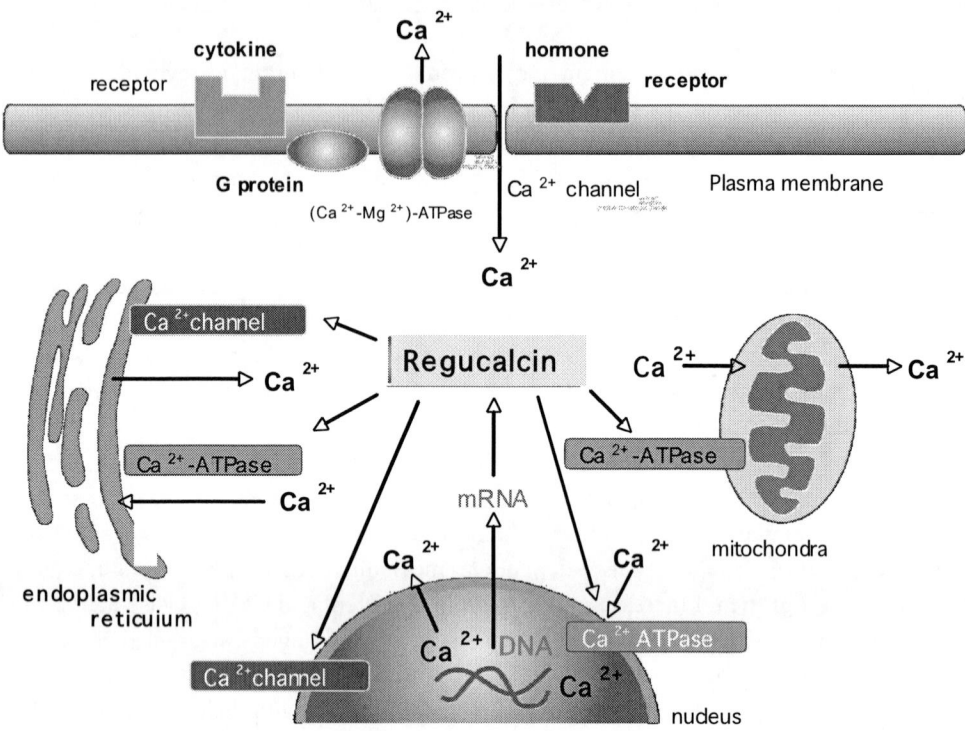

Figure 1. Regucalcin has a pivotal role in keeping intracellular Ca^{2+} homeostasis that is attenuated with various stimulating in cells. Regucalcin increases plasma membrane (Ca^{2+}-Mg^{2+})-ATPase, mitochondrial Ca^{2+}-ATPase and microsomal Ca^{2+}-ATPase activities in cells. Regucalcin also stimulates Ca^{2+} release from the microsomes (endoplasmic reticulum). Regucalcin has an inhibitory effect on nuclear Ca^{2+}-ATPase and a stimulatory effect on Ca^{2+} release from the nucleus. Through thus mechanism, regucalcin plays a part in regulating the rise of cytosolic Ca^{2+} concentration and nuclear matrix Ca^{2+} levels in cells that suppresse Ca^{2+}-dependent cellular events.

Regucalcin has an activatory role in the regulation of Ca^{2+}-ATPase activity in the mitochondria of brain tissues of rats [119]. The addition of regucalcin (10^{-10} to 10^{8} M), which is a physiological concentration in rat brain tissues, into the enzyme reaction mixture containing 25 μM calcium chlrodie caused a significant increase in Ca^{2+}-ATPase activity, while it did not significantly change in Mg^{2+}-ATPase activity [119].

Regucalcin levels are increased in the brain tissues or the mitochondria obtained from regucalcin transgenic rats. The mitochondrial Ca^{2+}-ATPase activity has been found to increase in regucalcin transgenic rats as compared with that of wild-type rats [119]. Endogenous regucalcin plays a role in the regulation of Ca^{2+}-ATPase activity in the brain mitochondria of rats.

Conclusion

Regucalcin plays a cell physiological role as a regulatory protein that is involved in the regulation of Ca^{2+} homeostasis in various cell types. The low cytoplasmic Ca^{2+} concentration of living cells is maintained through energy-requiring pumps. These pumps either remove

Ca^{2+} to the extracellular space by transport across the plasma membrane or accumulate it inside of intracellular organelles such as the mitochondria and endoplasmic reticulum (microsomes). Regucalcin stimulates the activity of these pumps to lower the cytoplasmic Ca^{2+} levels, as shown in Figure 1. This may provide the cellular mechanism of the inhibitory effect of regucalcin in the regulation of cell functions related to Ca^{2+} signaling.

REGUCALCIN REVERSES CA^{2+} EFFECT IN ENZYME REGULATION

Ca^{2+} and calmodulin systems generally activate various enzymes in many cells. There are many evidences that regucalcin has an inhibitory effect on enzyme activation by Ca^{2+}/calmodulin [25-29]. This section outlines the findings on the inhibitory effect of regucalcin on action of Ca^{2+} in cell metabolism.

Liver metabolism is regulated through an increase in Ca^{2+} level in the cytoplasm of liver cells due to hormonal stimulation [3-5]. Regucalcin has an inhibitory effect on Ca^{2+}/calmodulin-dependent enzyme activity in vitro. The hormonal effect on fructose 1,6-diphosphatase, which promotes the conversion from fructose 1,6-diphosphate to glucose-6-phosphate in the hepatic cytoplasm of rats, is mediated through Ca^{2+}. This enzyme activity is activated through Ca^{2+}/calmodulin [31]. The activation of fructose 1,6-diphosphatase by Ca^{2+}/calmodulin was completely reversed after the addition of regucalcin in the enzyme reaction mixture [11].

Phosphorylase *a* activity in the liver particulate glycogen is increased after addition of Ca^{2+} (10 μM) [13]. This increase was completely reversed after addition of regucalcin in the enzyme reaction mixture [13]. Regucalcin (1.0 μM) reversed activations of pyruvate kinase [12] and glucose-6-phosphatase [11] after addition of Ca^{2+} in the enzyme reaction mixture. These findings suggest that regucalcin regulates glycogenolysis and gluconeogenesis that is stimulated by Ca^{2+} in liver cells.

The reversible effect of regucalcin is also shown in Ca^{2+}-induced inhibition of 5'-nucleotidase activity in liver plasma membranes [15] and deoxyuridine 5'-triphosphatase activity in hepatic cytosol [120].

Thus, regucalcin has a reversible effect on the activation and inhibition of many enzymes by Ca^{2+}.

The controlled use of energy to maintain homeostasis and cellular function is a basic property of all cells. The free energy of ATP is belived to be the major cytosolic intermediate in this process. ATP is produced from the energy obtained by the oxidation of metabolic substrates in glycolysis and oxidative phosphorylation. ATPase produces the energy from ATP in the cell cytosol. The cytosolic factors regulating ATPase activity is important. Regucalcin has been shown to an inhibitory role in the regulation of ATPase activity in the brain cytosol of young and aged rats [121], suggesting a role in the regulation of energy conversion in brain tissues.

The mechanism of the reversible effect of regucalcin on various enzyme activities, which are regulated through Ca^{2+}, has not been well known. However, action of regucalcin may be partly based on Ca^{2+} binding, because the protein has 6-7 high-affinity binding sites per molecule (Kf = 4.19 x 10^5 M^{-1}) [9]. Regucalcin may directly bind to Ca^{2+} and/or calmodulin.

In addition, it is possible that regucalcin may bind to enzymes and that affects on enzyme activity.

REGUCALCIN INHIBITS CA^{2+}/CALMODULIN-DEPENDENT ENZYME ACTIVATION

Ca^{2+}/calmosulin-dependent enzymes are localized in many tissues and cells. Regucalcin has been found to have an inhibitory effect on various Ca^{2+}/calmodulin-dependent enzyme activations.

Cyclic adenosine monophosphate (AMP) is a second messenger for hormonal stimulation in many cells. Cyclic AMP is degraded by cyclic AMP phosphodiestrase in liver cytosol [122]. The enzyme activity is increased through Ca^{2+}/calmodulin [1]. Regucalcin has been found to inhibit the activation of cyclic AMP phosphodiesterase by Ca^{2+}/calmodulin in the cytosols of liver and renal cortex [123, 124], suggesting a role of regucalcin in the regulation of cyclic AMP level in the cells.

Nitric oxide (NO) may be important as a signaling factor in many cells [125]. NO, which an unpaired electron reacts with protein, targets primarily through their thiol or heme groups, and acts as a messenger or modulator molecule in many biological systems. NO is produced from L-arginine with L-citrulline as a coproduct in a reaction catalyzed by NO synthase that requires Ca^{2+}/calmodulin [125].

Regucalcin has been shown to have a suppressive effect in the enhancement of NO synthase activity in the cytosols of liver [126, 127] and kidney cortex [128] of rats. Overexpression of regucalcin did not cause a significant alteration of NO synthase activity in the kidney cortex cytosol of regucalcin transgenic rats as compared with that of wild-type rats [128]. However, the effect of calcium chloride (10 μM) in increasing NO synthase activity in the kidney cortex cytosol of wild-type rats was weakened in regucalcin transgenic rats [128]. The presence of anti-regucalcin monoclonal antibody (25 or 50 ng/ml) in the reaction mixture caused a significant increase in NO synthase activity, and this increase was completely ablolished after the addition of regucalcin (10^{-7} M). Endogenous regucalcin has a suppressive effect on NO synthatse activity in the cytosol of various tissues of rats.

Regucalcin has also been found to suppress Ca^{2+}/calmodulin-dependent NO synthase activity in the heart cytosol of rats [129]. Regucalcin had an inhibitory effect on NO synthase activity in the presence of antagonist for calmodulin [129], indicating a direct effect of regucalcin on the enzyme indepent on Ca^{2+}/calmodulin. The physiological significance of regucalcin inhibition of NO synthase in heart muscle cytosol is unknown. However, regucalcin may participate in the regulation of NO production in heart muscle cells. NO acts as a messenger or modulator molecule in heart muscle. NO production may be stimulated through Ca^{2+} signaling due to hormonal stimulation in heart muscle cells. Regucalcin may have a suppressive effect on over-production of NO due to inhibiting NO synthase in heart muscle cells.

NO acts as a messenger or modulator molecule in brain neurons. Regucalcin has also been shown to reveal a suppressive effect on NO synthase activity in the brain cytosol of young and aged rats, even though regucalcin levels are reduced with increasing age [130]. A remarkable expression of regucalcin protein was seen in the cytosol and nucleus of the brain tissues of regucalcin tranasgenic rats as compared with that of wild-type rats [131]. NO synthtase activity was decreased in the brain cytosol of transgenic rats, and the prsence of anti-regucalcin monoclonal antibody (50 ng/ml) in the enzyme reaction mixture caused a significant increase in cytosolic NO synthase activity in the cytosol of brain tissues of wild-type rats [131]. Endogenous regucalcin plays a suppressive role in the regulation of brain neuronal NO synthase activity in rats.

Thus, regucalcin has been demonstrated to have a suppressive role on Ca^{2+}/calmodulin-dependent NO synthase activity in various tissues. Regucalcin may play a role as a suppressor protein in NO production in many cell types, and it may regulate many cellular events that are involved in NO signaling.

Superoxide dismutase (SOD) plays a role in the prevention of cell death and apoptosis in the heart. The decrease in Mn-SOD activity is associated with increased mitochondrial oxidative damage as demonstrated by a decrease in the activities of iron sulfhydryl proteins sensitive to oxygene stress [132]. Cu/Zn-SOD has been shown to play the role of protector against dexorubicin-induced cardiotoxicity in mice [133]. Meanwhile, NO has a role in the suppression of myocardial O_2 consumption in rats [134].

Regucalcin has been found to increase SOD activity in the cytosol of rat liver [135] and heart [136]. Regucalcin has an inhibitory effect on NO synthase activity in the heart cytosol [136]. Production of superoxide radicals is widely accepted as the cause of the cardiodamage. Presumably, regucalcin participates in the control of production of superoxide radicals in rat heart muscle cells.

The multifunctional Ca^{2+}/calmodulin-dependent protein kinases play an important role in the response of the cells to a calcium signal [1, 137]. Regucalcin has been shown to inhibit Ca^{2+}/calmodulin-dependent protein kinase activity in the cytosol of rat liver [138], kidney cortex [139], and brain tissues of rats [140, 141]. An appreciable effect of regucalcin is seen at 0.5 μM, which is a cell physiological concentration. As Ca^{2+}/calmodulin-dependent protein kinase in the cytoplasm is activated through calcium signal, regucalcin may regulate a signal transduction for Ca^{2+}. The mechanism of action of regucalcin may be partly based on its binding to Ca^{2+}/calmodulin and/or enzyme. Regucalcin has been demonstrated to bind on calmodulin in analysis with sodium dodecyl sulfate-polyacrylamide gel electrophoresis (SDS-PAGE) using calmodulin-agarose beads [142].

Nishizuka discovered a diacylglycerol-activated Ca^{2+} and phospholipid-dependent protein kinase (protein kinase C) [2]. Protein kinase C is distributed widespread in the body, with amounts in the liver being intermediate between the high levels found in brain and spleen [2]. Protein kinase C is capable of phosphorylating cytoplasmic proteins. It is found that regucalcin inhibits protein kinase C activity in the cytoplasm of rat liver [143], kidney cortex [144], and brain cytosol and brain neurons [140, 141]. supporting the view that regucalcin plays a role in the regulation of Ca^{2+}-dependent cellular functions. The presence of anti-regucalcin monoclonal antibody in the enzyme reaction mixture caused a significant elevation of protein kinase activity, indicating that the endogenous regucalcin has an inhibitory effect on the enzyme activity [140, 141].

The regulatory effect of regucalcin in rat brain function may be attenuated with aging. Increasing age enhances protein kinase activity in rat brain cytosol [140]. This enhancement may be partly involved in aging-decreased regucalcin in rat brain tissues [113]. It is speculated that the endogenous regucalcin plays a suppressive role in the activation of Ca^{2+}-dependent protein kinase in the brain cytosol, and that aging may weaken the effect of regucalcin. Regucalcin may play a pivotal role in the regulation of phosphorylation of the cytosolic proteins in brain tissues. Interestingly, it has been reported that regucalcin gene is localized on human chromosome X that encompasses the map location for a growing number of diseases with a genetic basis; these include syndromic and non-syndromic forms of X-linked mental retardation and X-linked neurpmuscler diseases [23]. Regucalcin may have a pathophysiological role in brain disease with aging.

As mentioned above, regucalcin plays an inhibitory role in signaling pathway that is mediated through cyclic AMP, NO, and Ca^{2+}-dependent protein kinases in many tissues and cell types.

Protein phosphorylation-dephosphorylation is a universal mechanism by which numerous cellular events are regulated [145]. It has become apparent that there may exist many phosphatases that, like the kinases, are just elaborately and rigorously controlled [145, 146]. Protein phosphatase plays an important role in intracellular signal transduction due to hormonal stimulation [145].

Calcineurin, a calmodulin-binding protein, has been shown to possess a Ca^{2+}-dependent and calmodulin-stimulated protein phosphatase activity [147]. Regucalcin has been demonstrated to inhibit calcineurin activity in the cytosol of liver and renal cortex after its binding to calmodulin [148, 149]. Protein phosphatases, which endogenous regucalcin acts in liver cytoplasm, may be insensitive to okadaic acid [150]. Protein phosphatase activity toward phosphotyrosine, phosphoserine, and phosphothreonine in the cytosol of rat liver was elevated in the presence of anti-regucalcin monoclonal antibody in the enzyme reaction mixture in vitro, suggesting its role of endogenous regucalcin [148]. Regucalcin may be a unique protein, which has inhibitory effects on protein tyrosine phosphatase and protein serine/threonine phosphatase. Regucalcin, which was localized in rat liver nuclei, has been shown to inhibit nuclear protein phosphatase activity [151].

Endogenous regucalcin plays a role in the regulation of protein phosphatase activity in the cytosol and nuclei of rat renal cortex [152-154]. Regucalcin has been found to be present in the cytosol and nuclei of rat kidney cortex using Western blot analysis [154]. The addition of regucalcin (50-250 nM) in the enzyme reaction mixture obtained from the cytoplasm and nuclei from rat kidney cortex caused a decrease in protein phosphatase activity toward phosphotyrosin, phosphoserine, and phosphothreonine [152, 153]. The effect of calcium (25 µM) and calmodulin (2.5 µg/ml) in increasing protein phosphatase activity was decreased after the addition of regucalcin. Protein phosphatase activity in the cytosol and nuclei was increased in the presence of anti-regucalcin monoclonal antibody (10-50 ng/ml) in the enzyme reaction mixture [152, 153]. Regucalcin plays a suppressive role in the regulation of protein phosphatase activity in the cytoplasm and nucleus of rat kidney cortex.

Kidney cortex calcium content and the cytosolic and nuclear regucalcin levels were increased at 0.5 - 5 h after a single intraperitoneal administration of calcium chloride solution (10 mg Ca/100 g body weight) in rats [154]. The cytosolic and nuclear protein phosphatase activity, which is raised in calcium-administered rats, was found to enhance when anti-regucalcin monoclonal antibody was added in the enzyme reaction mixture [154]. The effect of antibody was completely abolished after the addition of regucalcin in the enzyme reaction mixture. Thus, endogenous regucalcin has a suppressive effect on the enhancement of protein phosphatase activity in the cytosol and nucleus of kidney cortex in calcium-administered rats.

Cardiac hypertrophy is induced by calcineurin, which dephosphorylates the teanscription factor NF-A3, enabling it to translocate to the nucleus [155]. Transgenic mice, that express activated forms of calcineurin or NF-AT3 in the heart, develop cardiac hypertrophy and heart failure that mimic human heart diaease [155], suggesting a hypertrophic signaling pathway. If regucalcin has a suppressive effect on calcineurin activity in the heart cytosol of normal and transgenic rats [156], overexpression of regucalcin may have a pathophysiologic role in the prevention of development of cardiac hypertrophy and heart faliure.

Regucalcin has also been shown to have an inhibitory effect on Ca^{2+}/calmodulin-dependent protein phosphatase activity toward phosphotyrosine, phosphoserine, and phosphothreonine in rat brain cytosol [157] and neurons [158]. The presence of anti-regucalcin monoclonal antibody in the enzyme reaction mixture caused a significant elevation of protein phosphatase activity in the brain cytosol, indicating that the endogenous regucalcin has a suppressive effect on the cytosolic enzyme activity [157]. Regucalcin has been shown to have an inhibitory role in the regulation of protein phosphatase activity in rat brain cytosol.

Regucalcin is localized in the microsomes of rat brain, and aging causes a decrease in its protein levels [159]. Regucalcin has been also shown to have a suppressive effect on protein tyrosine phosphatase activity in rat brain microsomes [159]. Aging caused an increase in protein tyrosine phosphatase activity in rat brain microsomes and the suppressive effect of regucalcin on the enzyme activity was weakened in aged rats [159], suggesting that the decrease in microsomal regucalcin with aging is partly involved in the enhancement of microsomal protein tyrosine phosphatase activity with aging [159].

Regucalcin has been found to be present in the nucleus of rat brain and the endogenpous regucalcin has a suppressive effect on the nuclear protein tyrosine phosphatase activity [160]. Increasing age has also been found to induce a reduction in rat brain nucleus and may lead to attenuation of the suppressive effect of regucalcin on the nuclear protein tyrosine phosphatase activity [160]. Regucalcin may play a pivotal role as a regulatory protein in the regulation of brain function that relates to protein phosphorylation-dephosphorylation.

Thus, regucalcin may play a physiological role in the intracellular control of the hormonal stimulation for phosphorylation and dephosphorylation of many proteins in various cell types.

As mentioned above, regucalcincan has been demonstrated to reverse the activity of many Ca^{2+}-activated enzymes (phosphorylase *a,* glucose-6-phosphatase, fructose 1,6-bisphosphatase, pyruvate kinase, protein kinase C, Ca^{2+}/calmodulin-dependent protein kinase, protein phosphatase, and Ca^{2+}/calmodulin-dependent cyclic AMP phosphodiesterase) and of Ca^{2+}-inhibited enzymes (5'-nucleotidase and dUTPase).

The first action is that regucalcin binds Ca^{2+} and that inhibits the metal's effect on many enzymes. The effect of regucalcin, that reverses Ca^{2+} action on many enzymes, may be based on its binding of Ca^{2+}, since the protein has 6 to 7 high-affinity binding sites per molecule, and a Ca^{2+}-binding constant of $4.l9x10^5$ M^{-1} [9]. The intrinsic significance of regucalcin action may be the binding of Ca^{2+} by its protein.

The second is that Ca^{2+}-binding regucalcin directly inhibits the function of enzymes and somewhat stimulates enzyme function. The direct action of Ca^{2+}-binding regucalcin or the protein itself may be decided by the protein structure of enzymes. Spectroscopical studies have clearly demonstrates that Ca^{2+}-binding induces conformational changes in regucalcin, which may then result in increased hydrophobicity of the protein, and loosening of the conformation of regucalcin [10].

Regucalcin can directly inhibit the enzymes that are activated through Ca^{2+}-calmodulin [11]; this also results from regucalcin that affects the binding of Ca^{2+} to calmodulin. Which of the two proteins binds Ca^{2+} may be decided by their relative concentrations of Ca^{2+} in hepatic cytosol, since the Ca^{2+}-binding constant of regucalcin is greater than that of calmodulin [9]. Calmodulin exists as a monomer of molecular weight 17,000 and contains 4 Ca^{2+}-binding sites [1]. In the enzyme assay system of Ca^{2+}/calmodulin-dependent cyclic AMP phosphodiesterase activity, the inhibitory effect of regucalcin on the enzyme activation through Ca^{2+}-calmodulin is completely blocked with the addition with increasing concentrations of Ca^{2+} [120, 121]. This further supports the view that the mechanism by which regucalcin inhibits Ca^{2+} action is based on the binding of Ca^{2+}.

Moreover, in the enzyme assay system of protein kinases (protein kinase C and Ca^{2+}/calmodulin-dependent protein kinase), the effect of regucalcin, which inhibits the activation of enzymes by Ca^{2+}, is seen with increasing concentrations of Ca^{2+}. This suggests that regucalcin directly inhibits the enzyme activation in addition to Ca^{2+}-binding. Also, it is possible that Ca^{2+}-binding regucalcin or the protein itself can directly inhibit the enzyme activity.

There may be many enzymes that are regulated by regucalcin and/or Ca^{2+}-binding regucalcin in various cell-types. This remains to be elucidated.

REGUCALCIN REGULATES PROTEIN SYNTHESIS AND DEGRADATION

Regucalcin has been shown to have a regulatory effect on prptein synthesis and protein degradation, suggesting that regucalcin play a role in the regulation of protein turnover in cells.

Role of Regucalcin in Protein Synthesis

Protein synthesis is depressed in a variety of eukaryoticx cell types exposed to conditions depleting Ca^{2+} but not Mg^{2+} [161]. It has been also proposed that hormones (vasopressin and α-adrenergic agonist), which are known to mobilize sequestered Ca^{2+} within liver cells, inhibit amino acid incoporation by influencing a Ca^{2+} requirement associated with protein synthesis [162]. Moreover, vasopressin inhibits the rate of protein synthesis in isolated hepatocytes partially depleted of Ca^{2+} [163]. These investigations propose the hypothesis that a sequestered pool of intracellular Ca^{2+} is required for the maintenance of high rates of protein synthesis in liver cells [162, 163]. On the other hand, vasopressin and α-adrenergic agonist causes an increase in the intracellular free Ca^{2+} concentration of hepatocytes that are not depleted of Ca^{2+} [164, 165]. Whether the increase in intracellular Ca^{2+} influences hepatic protein synthesis is well undefined, however. It may be important to clarify the effect of Ca^{2+} addition on hepatic protein synthesis in vitro.

It has been demonstrated that Ca^{2+}, of various metals, can uniquely inhibit in vitro protein synthesis using the 5500 g supernatant fraction (the microsomes and cytosol) of rat liver homogenate [166]. Its inhibition was seen after addition of 1.0 μM Ca^{2+}. Ca^{2+} addition caused a remarkable decrease of the activity of aminoacyl (leucyl)-tRNA synthetase, which is a rate-limiting enzyme of protein synthesis at translational process, in hepatic cytosol [165]. Ca^{2+} may directly inhibit hepatic protein synthesis in subcellular fraction of liver cells. Ca^{2+} is required for protein synthesis in hepatocytes exposed to conditions depleting the cation [166]. It is not clarified whether protein synthesis in hepatocytes not depleted of Ca^{2+} requires exogenous Ca^{2+}. The mechanism by which Ca^{2+} is required in protein synthesis of hepatocytes depleted of Ca^{2+} may be complex.

Calmodulin, which can amplify Ca^{2+} effects on enzymes [1], did not have an appreciable effect on in vitro protein synthesis using the 5500 g supernatant fraction of liver homogenate in the presence of Ca^{2+} (10 μM) [166]. The activity of aminoacyl-tRNA synthetase in hepatic cytosol was not altered through calmodulin [167]. Presumably, the protein synthesis is inhibited by Ca^{2+}, which is not bound to calmodulin.

The role of regucalcin in the regulation of in vitro protein synthesis using the 5500 g supernatant fraction of rat liver homogenate is investigated [166]. Regucalcin caused a remarkable inhibition of hepatic protein synthesis in vitro [166]. Regucalcin could not reverse the Ca^{2+}-induced inhibition of protein synthesis [166], although it has been shown that regucalcin can reverse the Ca^{2+} effect on many enzymes in liver cells. Since regucalcin can bind to liver cytosolic proteins and the binding is slightly enhanced with the coexistence of 0.1 μM Ca^{2+} [163], Ca^{2+}-binding regucalcin and/or Ca^{2+} free regucalcin may be able to inhibit hepatic protein synthesis. In fact, the presence of regucalcin (1 and 2 μM) could fairly decrease hepatic protein synthesis that was reduced after the addition of 10 μM Ca^{2+}[166]. Regucalcin itself may play a role in the regulation of protein synthesis in liver cells.

Regucalcin has been shown to inhibit hepatic aminoacyl-tRNA synthase activity [167]. The inhibitory effect of regucalcin was seen in the presence of Ca^{2+} (10 μM). The inhibitory effect of regucalcin on hepatic protein synthesis may be partly based on a remarkable decrease of aminoacyl-tRNA synthetase activity caused by regucalcin. Regucalcin may bind aminoacyl (leucyl)-tRNA synthetase in hepatic cytosol, since iodinated regucalcin can bind the proteins in hepatic cytosol [65]. Regucalcin may be able to regulate liver cell function that is not affected by cellular Ca^{2+}.

The role of endogenous regucalcin on protein synthesis is examined using anti-regucalcin monoclonal antibody, moreover [167]. The presence of anti-regucalcin monoclonal antibody in the reaction mixture caused a significant increase in protein synthesis and [³H] leucyl-tRNA synthetase activity in normal rat liver. These increases were completely prevented in the addition of exogenous regucalcin (1.0 μM). Liver cytosol contained about 16 μg of regucalcin per 1 mg of the cytosolic protein; the reaction mixture contained about 0.17-0.19 μM of endogenous regucalcin, because the cytosolic protein in the range 360 – 390 μg were added into the mixture of 1.0 ml. Endogenous regucalcinin may have a suppressive effect on protein synthesis in liver cells.

Hepatic protein synthesis has been shown to enhance in regenerating rat liver, which induces a proliferation of liver cells after a partial hepatectomy [167]. This enhancement was reamarkable at 24 and 48 h after partial hepatectomy. Hepatic protein synthesis in regenerating liver was further enhanced in the presence of anti-regucalcin monoclonal antibody in the reaction mixture. Endogenous regucalcin has a suppressive role on the enhancement of protein synthesis in regenerating liver.

Role of Regucalcin in Protein Degradation

Evidence for the role of Ca^{2+}-activated protease (calpains) is implicated in signal transduction [168]. Two neutral Ca^{2+}-requiring proteinases, differing in molecular size, have been isolated from rabbit liver cytosol [169]. Both are recovered as inactive proenzymes that can be converted to the active forms by high (0.1-1.0 mM) concentrations of Ca^{2+} in the absence of substrate or, in the presence of a protein substrate, by low (1-5 μM) concentrations of Ca^{2+} [169]. The activated proteinases required only 1-5 μM Ca^{2+} for maximal activity [169].

The addition of regucalcin (0.25 – 2.0 μM) into the enzyme reaction mixture has been found to induce a remarkable increase of neutral proteinase activity in the presence of 5.0 μM Ca^{2+} [170]. The effect of regucalcin was seen at 0.25 μM. Regucalcin may activate Ca^{2+}-requiring proteinase in rat liver cytosol. The effect of regucalcin in increasing liver cytosolic proteinase activity is also seen in the absence of Ca^{2+} [170]. This increase was remarkable as compared with that of Ca^{2+} addition. Regucalcin may activate Ca^{2+}-not requiring neutral proteinase in rat liver cytosol. Regucalcin has a reversible effect on the activation of various enzymes by Ca^{2+} and/or calmodulin [123, 124, 150]. The finding, that regucalcin can activate neutral proteinase in rat liver cytosol in the presence or absence of Ca^{2+}, was novel.

The activatory effect of regucalcin on liver cytosolic proteinase activity was not seen in the presence of anti-regucalcin antiserum in the enzyme reaction mixture [170], suggesting a role of endogenous regucalcin in the activation of cytosolic proteinase.

The activatory effect of regucalcin on neutral proteases in the liver cytosol is characterized in the entatic nature of their active site [171, 172]. Leupeptin is a potent inhibitor of SH-proteinase. The regucalcin-increased protenase activity was inhibited in the presence of leupeptin in the enzyme reaction mixture [172]. Regucalcin may activate neutral cysteinyl-proteinase in the liver cytosol. The effect of regucalcin in increasing proteolytic activity in rat liver cytosol was not abolished in the presence of diisopropylfluorophosphate (DFP), an inhibitor of serine-protease, although DFP alone had an inhibitory effect on the proteolytic activity [172]. Regucalcin does not act on serine-proteases in liver cytosol. The effect of regucalcin was also seen in the presence of a chelator of metal ions, suggesting that regucalcin does not activate metal-related proteases in liver cytosol.

The proteolytic activity in liver cytosol was markedly increased after the addition of dithiothreitol (DTT), a protecting reagent for SH group, in the enzyme reaction mixture and this increase was completely inhibited in the presence of N-ethylmaleiimide (NEM), a SH group-modifying agent [172]. The effect of regucalcin in increasing proteolytic acitivity was seen in the presence of DTT in liver cytosol, although it was completely inhibited in the presence of NEM [172]. Regucalcin may act on the SH group of cysteinyl-proteases in liver cytosol.

Two forms of Ca^{2+}-activated neutral proteases (calpines) have been identified in hepatocytes and other cells, m-calpaine and μ-calpaine [169]. Isolated μ-calpain requires micromolar concentrations of Ca^{2+} for activation, while m-calpain requires millimolar concentrations [173]. The activation of proteases in rabbit liver cytosol is required only 1-5 μM Ca^{2+} for maximal activity [169]. Ca^{2+} (10 μM)-increased proteolytic activity in rat liver cytosol was inhibited in the presence of NEM, indicating that neutral proteases (including calpains), which are activated by Ca^{2+}, exist in liver cytoplasm [172].

The activity of calpaines, which are thiol protease, increases in hepatocytes following addition of ATP [174]. The proteolytic activity in liver cytosol was decreased after the addition of calpastatin, a specific inhibitory of calpains, indicating the existence of calpains in the cytosol [172]. Regucalcin-increased proteolytic activity was abolished in the presence of calpastatin. Regucalcin may be an activator for calpains. m-Calpain isolated from rabbit skeletal muscle was activated by regucalcin independent on Ca^{2+}. This activation was inhibited after the addition of NEM [172]. Regucalcin acts on the SH group in m-calpain. Presumably, regucalcin may be able to activate both m- and μ-calpains.

The role of regucalcin in the regulation of neutral proteolytic activity in rat kidney cortex cytosol is also examined [175, 176]. Regucalcin had an activatory effect on neutral proteolytic activity in the kidney cortex cytosol [176]. This increase was abolished in the presence of anti-regucalcin monoclonal antibody, supporting the view that endogenous regucalcin plays a role as activator of proteases in the renal cortex ctrtosol [176]. The effect of regucalcin on proteolytic activity was not altered in the presence of calcium chloride (0.01 and 1.0 mM) or EGTA (1.0 mM), indicating that the effect of regucalcin was independent on Ca^{2+}. Regucalcin may activate both calpaines and other proteases in the kidney cortex cytosol. Regucalcin has been shown to activate SH proteases in rat renal cortex cytosol. Presumably, regucalcin acts on the SH-groups of protease in rat renal cortex cytosol.

The activatory effect of regucalcin on proteases is seen in the concentrations of 0.01 – 0.25 μM [175]. The concentration of regucalcin in rat kidney tissue is found to be present about 5.3 μM. Regucalcin plays a physiological role in the activation of thiol proteases in renal cortex cells.

Regucalcin uniquely activates thiol proteases independent on Ca^{2+} in the liver cytosol, whereas it has no effect on serine proteases and metalloproteases [170, 172]. Such an effect of regucalcin was also found in the renal cortex cytosol [176]. Regucalcin may be an activator on thiol proteases in liver and kidney cells. Regucalcin also plays a role as an activator in other tissues that express regucalcin.

Calpains are ubiquitous, non-lysosomal, calcium-dependent proteases that may play important roles in Ca^{2+}-mediated intracellular processes [168, 169, 174]. The ability of calpain to alter the limited proteolysis, the activity or function of numerous cytoskeletal proteins, protein kinases, receptors and transcription factors suggests the involvement of the protease in various Ca^{2+}-regulated cellular functions [178]. Calpains may also play an integral role in modulating the activity of protein kinase C, a key protein in many signal transduction processes [179], since they convert the native Ca^{2+}/phospholipids-dependent kinase to a soluble form that does not require Ca^{2+} or phospholipids for activity [2]. Regucalcin can increase the activity of thiol proteases including calpain in rat liver and renal cortex cytosols. Regucalcin may play a pivotal role in the regulation of cellular functions related to Ca^{2+} mediated through thiol proteases. Presumably, regucalcin plays an important role in the regulation of signal transduction that is involved in proteases.

REGULATORY ROLE OF REGUCALCIN IN CA^{2+} SIGNALING-RELATED NUCLEAR FUNCTION

Mounting evidence suggests that Ca^{2+} is active in liver nuclear function [81-83]. Calmodulin exists in rat liver nuclei [82]. The existence of an ATP-stimulated Ca^{2+}-sequestration system in rat liver nuclei that requires calmodulin and generates a net increase in nuclear matrix free Ca^{2+} concentration has been reported [81]. Calmodulin stimulates DNA synthesis in liver cells [180], and the effect of calmodulin is mediated through α-adrenergic stimulation [181, 182]. There are many evidences that regucalcin plays an important role in the regulation of nuclear functions.

Nuclear Localization of Regucalcin

Regucalcin has been shown to localize in the nuclei of rat liver [38, 151]. Regucalcin can bind on calmodulin-agarose beads [142]. Liver nuclear extract were incubated with calmodulin-agarose beads, and calmodulin-agarose beads were applied to SDS-PAGE. Band that coincides with regucalcin was found on SDS-PAGE [142], suggesting that regucalcin is present in the nucleus [151].

Exogenous regucalcin has been shown to transport into the nucleus isolated from normal rat liver [38]. Endogenous regucalcin is present in the nuclei of rat liver using Western blotting analysis [38]. When isolated liver nuclei were incubated in the presence of exogenous regucalcin (50 μg/ml; 1.5 μM), potent band for regucalcin was found in the nucleus [38], supporting that the protein is translocated into the nucleus. This translocation was seemed to be an early event, since potent band for regucalcin was seen with only 10 min of incubation. A part of regucalcin, which is localized in the cytoplasm of liver cells, is translocated to the nucleus [38].

Nuclear regucalcin translocation was not appreciably changed in the presence of ATP (2 mM), guanosine 5'-triphosphate (GTP, 2 mM), and calcium chloride (0.1 mM), suggesting that its translocation is not mediated through nuclear localization signal [38]. ATP and GTP are required for nuclear import of proteins that are localized in the nuclei. ATP or GTP does not regulate the translocation of regucalcin into liver nuclei. Regucalcin is translocated independently of Ca^{2+}.

Nuclear protein transport is blocked in the presence of the lectin wheat germ agglutinin (WGA) [83]. Nuclear regucalcin translocation was not appreciably changed in the presence of WGA in the reaction mixture [38]. This finding suggests that the nuclear translocation of regucalcin is not related to nuclear localization signal that is responsible for selection for intranuclear active transport. Presumably, regucalcin is passively transported to the nucleus through nuclear pore in liver cells, since a molecular weight of regucalcin is about 33 kDa [16].

Regucalcin has been also shown to localize in the nuclei of the cloned normal rat kidney proximal tubular epithelial NRK52E cells with immunocytochemical analysis [183]. The nuclear localization of regucalcin is enhanced through hormonal signaling process that is involved in protein kinase C [183].

Regucalcin has been shown to bind proteins in isolated rat liver nuclei using Far-Western blot analysis [184]. The results of the Far-Western analysis showed the existence of protein components that bind to regucalcin in the nucleus isolated from rat liver [184]. Regucalcin has been also demonstrated to bind DNA using Western blot analysis for regucalcin with DNA cellulose-binding assay [184]. These findings show that regucalcin binds proteins and DNA in liver nucleus. Regucalcin may have a regulatory effect on signaling pathways that modulate transcriptional activity in liver cells.

Regucalcin Inhibits Ca^{2+}-Stimulated Nuclear DNA Fragmentation

Isolated rat liver nucleus contains a DNA endonuclease activity dependent upon Ca^{2+} and Ca^{2+} results in extensive DNA hydrolysis [185]. The Ca^{2+} dependence of this endogenous DNA fragmentation process is based on DNA endonuclease activity dependent upon sub-micromolar Ca^{2+} when the nucleus is reconstituted with NAD^+ and ATP [185]. This endogenous endonuclease activity may be responsible for the DNA fragmentation occurring during programmed cell death (apoptosis) and certain forms of chemically induced cell killing [185, 186].

To explore the regulatory role of regucalcin in liver nuclear function, it was first examined whether the protein has an effect on Ca^{2+}-activated DNA fragmentation in isolated rat liver nuclei [187]. Among various metals, Ca^{2+} has been shown to stimulate uniquely in vitro DNA fragmentation in isolated rat liver nuclei [187]. This increase was seen after the addition of 1.0 μM Ca^{2+}, in agreement with previous work [187]. The presence of regucalcin (0.5-2.0 μM) completely inhibited the activation of liver nuclear DNA fragmentation when 10 μM Ca^{2+} was added. This inhibition was not seen in the presence of Ca^{2+} at 25 or 50 μM. Thus, regucalcin had an inhibitory effect on DNA fragmentation when a comparatively lower concentration of Ca^{2+} (5.0 and 10 μM) was added [187]. The inhibitory effect of regucalcin on DNA fragmentation may be partly based on binding of Ca^{2+} [9].

DNA fragmentation in rat liver nucleus has been reported to be stimulated through Ca^{2+}-calmodulin [185], which exists in liver nuclei [82]. Addition of calmodulin (10 and 20 μg/ml) did not enhance Ca^{2+} (10 μM)-activated DNA fragmentation in liver nuclei [187]; however, nuclear endogenous calmodulin may be able to enhance Ca^{2+}-activated DNA fragmentation in the nucleus. Regucalcin has been found to inhibit Ca^{2+}-activated DNA fragmentation after Ca^{2+} addition [187]. Such an inhibition was also seen in the presence of exogenous calmodulin [187]. Presumably, regucalcin can inhibit Ca^{2+}/calmodulin-dependent DNA fragmentation in liver nuclei, since radio-iodinated regucalcin has been found in the nuclei isolated from rat liver in the absence or presence of 1.0 mM Ca^{2+} [65].

Several studies have shown that Ca^{2+} plays an important role in the regulation of nuclear functions [81, 82]. Also, it has been found that a sustained increase in cytosolic Ca^{2+} level precedes the activation of DNA fragmentation that is characteristic of programmed cell death (apoptosis) and in certain forms of chemically induced cell killing [185, 186]. The finding, that regucalcin inhibits the activation of DNA fragmentation by Ca^{2+}, was the first time for a role of regucalcin in liver nuclear functions.

Regucalcin Suppresses Ca^{2+}-Related Nuclear Enzyme Activation

Smal GTPase Ran (ras-related nuclear protein) is required for protein export from the nucleus and protein import into the nucleus [188]. The role of regucalcin in the regulation of GTPase activity in the nuclei of rat liver is shown [189]. We found the existence of GTPase activity in the nuclei isolated from rat liver [189]. Liver nuclear GTPase activity was increased after calcium addition with a comparatively higher concentration in the enzyme reaction mixture, and this increase was not seen in the presence of TFP, an antagonist of calmodulin [189]. The effect of calcium in increasing nuclear GTPase activity may be related to endogenous calmodulin. Calcium in liver cytoplasm is transported through an energy-dependent mechanism to the nucleus, and the comparatively higher concentration of calcium is found in the nucleus [189]. Calmodulin is shown to be present in liver nucleus [82]. GTPase, which is activated by Ca^{2+}/calmodulin, may also be localized in liver nucleus.

The presence of exogenous regucalcin (0.5 μM) used in the enzyme reaction mixture caused an inhibitory effect on GTPase activity in liver nucleus [189]. This effect was also seen in the presence of EGTA, a chelator of Ca^{2+}. Presumably, the inhibitory effect of regucalcin on liver nuclear GTPase activity is revealed independent of Ca^{2+}/calmodulin in the nucleus. Regucalcin has been shown to have an inhibitory effect on the activation of enzymes by Ca^{2+}/calmodulin due to binding Ca^{2+} and/or calmodulin [138-141]. Regucalcin can inhibit the activity of various enzymes through the mechanism by which it binds directly to the enzyme [31, 33]. Regucalcin may directly inhibit GTPase activity in liver nucleus.

The physiological significance of the inhibitory effect of regucalcin on GTPase activity in liver nucleus is unknown. However, the presence of anti-regucalcin monoclonal antibody in the enzyme reaction mixture caused a significant increase in this activity in liver nucleus [189]. This increase was completely blocked after regucalcin addition, suggesting that endogenous regucalcin has a suppressive effect on GTPase activity in liver nucleus. Endogenous regucalcin, which is localized in liver nucleus, may participate in the regulation of nuclear functions that are related to hydrolysis of GTP.

Regucalcin may be able to regulate a process of signal transduction from the cytoplasm to nucleus in liver cells. This process is mediated through various protein kinases. The role of regucalcin in the regulation of Ca^{2+}-dependent protein kinase and protein tyrosine phosphatase activities in isolated liver nucleus is examined [38, 150, 151]. Nuclear Ca^{2+}-dependent protein kinase and protein tyrosine phosphatase activities were increased in the presence of anti-regucalcin monoclonal antibody in the enzyme reaction mixture, and these increases were completely abolished after the addition of regucalcin [38]. The translocation of regucalcin to the nucleus may play a suppressive role in the regulation of protein kinase and protein tyrosine phosphatase in liver nucleus [38].

Endogenous regucalcin has been demonstrated to have a suppressive effect on the enhancement of protein kinase activity with a proliferation of liver cells [190]. Protein kinase activity is enhanced in the cytosol and nucleus of regenerating rat liver [190]. Regucalcin had a suppressive effect on tyrosine kinase, protein kinase C, and Ca^{2+}/calmodulin-dependent protein kinase in the cytoplasm and nucleus of regenerating rat liver [123, 138, 139, 190].

Phosphatase activity toward phosphotyrosine, phosphoserine, and phosphothreonine is found in the liver nucleus [151]. Nuclear phosphotyrosine phosphatase activity was increased after Ca^{2+} addition in the enzyme reaction mixture, although the enzyme activity was not altered by TFP, an inhibitor of calmodulin, or cyclosporine A, an inhibitor of calcineurin

[151]. Nuclear phosphatase activity toward phosphotyrosine may be independent of calmodulin. Meanwhile, nuclear phosphatase activity toward phosphoserine was elevated after the addition of Ca^{2+}, while the enzyme activity was appreciably decreased by TFP and cyclosporine A, suggesting that the enzyme activity is partly involved in Ca^{2+}/calmodulin-dependent protein phosphatase (calcineurin). Nuclear phosphatase activity toward phosphothreonine was not altered after the addition of Ca^{2+}, TFP, and cyclosporine A in the enzyme reaction mixture. Vanadate caused an inhibition of nuclear phosphatase activity toward phosphotyrosine and phosphoserine but not phosphothreonin. Thus different protein phosphatases toward phosphotyrosine, phosphoserine, and phosphothreonine have been shown to be present in liver nucleus.

The addition of regucalcin in the enzyme reaction mixture caused a significant decrease in phosphatase activity toward phosphotyrosine, phosphoserine, and phosphothreonine in the liver nuclei [151]. Liver nuclear phosphatase activity toward phosphoamino acids was assayed using 5-6 mg of nuclear protein per milliliter of reaction mixture; it contained 275-330 ng of nuclear regucalcin [151]. The concentration of endogenous nuclear regucalcin was estimated about 82.4-98.8 nM. Further addition of exogenous regucalcin (0.25 µM) caused a significant decrease in nuclear phosphatase activity, although the effect was saturated with increasing concentrations of regucalcin (0.5 µM) [151].

Nuclear phosphatase activity was elevated in the presence of anti-regucalcin monoclonal antibody (25 and 50 ng/ml of reaction mixture) [151]. This elevation was completely abolished after addition of regucalcin. Endogenous regucalcin may regulate protein phosphatase activity toward phosphotyrosine, phosphoserine, and phosphothreonine in liver nucleus. Regucalcin may have an inhibitory effect on various protein phosphatases in liver nucleus.

Regucalcin Suppresses Nuclear DNA and RNA Synthesis

Regucalcin has been shown to have an inhibitory effect on DNA synthesis activity in the nuclei of normal rat liver [191]. The inhibitory effect of regucalcin was seen in the presence of EGTA, a chelator of Ca^{2+}, in the reaction mixture. Ca^{2+} is present in liver nucleus [191]. Liver nuclear DNA synthesis activity was increased in the presence of EGTA in the reaction mixture, suggesting that Ca^{2+} suppresses DNA synthesis activity in the nucleus. The effect of regucalcin in inhibiting nuclear DNA synthesis may be not related to Ca^{2+} in liver nucleus.

Liver nuclear DNA synthesis has been shown to stimulate in regenerating rat liver [192]. Nuclear DNA synthesis was markedly increased at 1 day after hepatectomy, and this increase was also seen at 3 days [192]. Nuclear DNA synthesis was enhanced in the presence of EGTA (0.4 mM) in the incubation mixture. The presence of Ca^{2+} (1.0 − 25 µM) caused a significant decrease in the nuclear DNA synthesis of normal rat liver. Regucalcin (0.25 and 0.5 µM) caused an inhibition of nuclear DNA synthesis of normal rat liver [192]. This inhibition was also seen in the presence of Ca^{2+} (1.0 µM). The inhibitory effect of regucalcin was remarkable in regenerating rat liver nuclei in comparison with that of normal rat liver. Regucalcin has been shown to have a suppressive effect on nuclear DNA synthesis in regenerating rat liver. Regucalcin may have a suppressive role in the enhancement of nuclear DNA synthesis in liver cell proliferation.

Regucalcin has been shown to have a suppressive effect on DNA synthesis activity in the nuclei isolated from rat renal cortex [193]. The addition of regucalcin ($0.1 - 0.5$ μM) in the reaction mixture containing either EGTA (1 mM) or calcium chloride (50 μM) had an inhibitory effect on nuclear DNA synthesis activity [193]. The presence of anti-regucalcin monoclonal antibody (10-50 ng/ml) in the reaction mixture caused a significant increase in nuclear DNA synthesis activity [193]. This increase was completely abolished in the presence of regucalcin (0.5 μM). Endogenous regucalcin has been found to have a suppressive effect on DNA synthesis in the nuclei of rat renal cortex [193].

Regucalcin has been shown to have an inhibitory effect on RNA synthesis in the nuclei isolated from control rat liver [194] and regenerating rat liver [195]. RNA symthesis in rat liver nuclei was stimulated after Ca^{2+} addition with a comparatively lower concentration, although the sub-micromolar concentration of Ca^{2+} evoked an inhibition of nuclear RNA synthesis [194]. The addition of Ca^{2+} with higher concentrations has been reported to have an inhibitory effect of nuclear RNA synthesis in rat liver cells [196]. Since liver nuclei contain a DNA endonuclease activity dependent upon Ca^{2+} in the submicromolar range and Ca^{2+} causes extensive DNA hydrolysis [185, 187], nuclear RNA synthesis may be suppressed with higher Ca^{2+} concentrations. Inactivation of RNA polymerase III transcription has been shown to be calcium dependent; the changes in Ca^{2+} concentration, the activation of calpains, and the consequent proteolytic degradation of RNA of transcription factors has been suggested to be involved in the regulation of RNA polymerase III transcription in the presence of 1 mM Ca^{2+} [197].

The effect of regucalcin in decreasing nuclear RNA synthesis activity in normal rat liver was not seen in the presence of α-amanitin, an inhibitor of RNA polymerase II and III [194], suggesting that the suppressive effect of regucalcin on nuclear RNA synthesis activity is partly resulted from its inhibitory action on RNA polymerase II and III [194, 195]. Meanwhile, it has been reported that Ca^{2+} has a stimulatory effect on RNA synthesis in liver nucleus [196, 197]. This effect may be partly mediated through Ca^{2+}-dependent protein kinase [196, 197]. The stimulatory effect of Ca^{2+} on nuclear RNA synthesis activity was completely blocked in the presence of regucalcin [194, 195]. Regucalcin has been shown to inhibit Ca^{2+}-dependent protein kinases in rat liver nucleus [190]. Presumably, the effect of regucalcin in decreasing RNA synthesis activity in liver nucleus is partly involved in its inhibitory action on the activities of both RNA polymerase II and III and Ca^{2+}-dependent protein kinases. Further mechanism remains to be elucidated. Regucalcin has been proposed to have a role as a transcriptional factor in liver nucleus.

Regucalcin has been found to have a suppressive effect on liver nuclear DNA and RNA synthesis [191-195]. The mechanism by which regucalcin inhibits nuclear DNA and RNA synthesis is not well known. It is speculated that regucalcin has an inhibitory effect on DNA and RNA polymerase activity. However, the effect of regucalcin in inhibiting nuclear RNA synthesis activity is observed in the presence of α-amanitin, an inhibitor of RNA polymerase II. Regucalcin can directly bind DNA [184]. Which base pairs of DNA bind regucalcin remains to be elucidated. It is possible that regucalcin binds DNA and it has an inhibitory effect on nuclear DNA and RNA synthesis activity.

REGUCALCIN SUPPRESSES APOPTOSIS MEDIATED THROUGH CA²⁺ SIGNALING

Regucalcin Inhibits NO Synthase Activity Related to Apoptosis

NO may be important as a signaling factor in many cells [125], and it plays a role in apoptosis of hepatoma cells [198]. NO mediates apoptosis by D-galactosamine in a primary culture of rat hepatocytes [199]. Regucalcin has been shown to inhibit NO synthase that is related to cell apoptosis [126], suggesting that regucalcin has a suppressive role in apoptosis.

Regucalcin has a suppressive effect on Ca^{2+}/calmodulin-dependent NO synthase in the cloned rat hepatoma H4-II-E cells [126]. The effect of regucalcin in decreasing NO synthase activity was also seen in the presence of TFP or EGTA. Presumably, regucalcin has an inhbitiory effect on NO synthase activity due to binding to calmodulin and/or the enzyme independently of Ca^{2+} in proliferative cells.

Overexpressing of regucalcin has been also shown to have a suppressive effect on NO synthase activity in H4-II-E cells (transfectants) [126]. This decrease was completely abolished in the presence of anti-regucalcin monoclonal antibody in the reaction mixture. Moreover, the effect of Ca^{2+}/calmodulin addition in increasing NO synthase activity in H4-II-E cells cells (wild type) was completely prevented in transfectants. Endogenous regucalcin had a suppressive effect on NO synthase activity in the cloned rat hepatoma H4-II-E cells.

NO synthase activity was enhanced in H4-II-E cells cultured with 10% FBS as compared with that of 1% FBS [126], suggesting that the enzyme is induced in proliferative cells. The enhancement of NO synthase activity in H4-II-E cells cultured with 10% FBS was abolished in the presence of anti-regucalcin monoclonal antibody [126]. Regucalcin levels were elevated in H4-II-E cells with 10% FBS-culture [126]. Endogenous regucalcin may have a suppressive effect on the enhancement of NO synthase activity with proliferation of H4-II-E cells.

A high concentration of NO, which is produced from inducible NO synthase, has been shown to inhibit cell proliferation [200] and to induce cell apoptosis [201]. It is reported that a low concentration of NO, which is produced from endothelial NO synthase, protects against the cytotoxic effects of reaction oxygene species in cells [202]. Whether endogenous regucalcin suppresses NO production in H4-II-E cells is unknown at present. It is speculated, however, that regucalcin may inhibit NO production in H4-II-E cells, since regucalcin can decrease NO synthase activity in the cells [126]. Endogenous regucalcin may have an inhibitory effect on inducible and endothelial NO synthetases in hepatoma cells. Alternatively, regucalcin may have a physiological role in the regulation of NO-related cell functions.

Regucalcin Suppresses Various Factors-Induced Apoptosis in Liver Cells

Tumor necrosis factor α (TNF-α) and NO mediate apoptosis by D-galctosamine in a primary culture of rat hepatocytes [199, 200]. TNF-α induces apoptosis in mammary adenocarcinoma cells by an increase in intranuclear free Ca^{2+} concentration and DNA fragmentation [200]. H4-II-E cells with subconfluent monolayer cells were cultured in a medium without FBS in the presence of TNF-α. TNF-α (0.1-10 ng/ml) caused a significant decrease in the number of H4-II-E cells (wild type), inducing cell death. Overexpressing of regucalcin in H4-II-E cells (transfectants) has been found to prevent the effect of TNF-α in decreasing cell number [203]. Overexpressing of regucalcin had a preventive effect on cell death induced with the higher concentration of TNF-α (10 ng/ml). This finding demonstrates that overexpression of regucalcin has a suppressive effect on cell death induced by stimulation of TNF-α [203].

Culture with NAME, an inhibitor of NO synthase, had a significant preventive effect on TNF-α-induced cell death. Regucalcin inhibits Ca^{2+}/calmodulin-dependent NO synthase activity in H4-II-E cells [126]. The suppressive effect of regucalcin on cell death may be partly resulted from the inhibition of NO production, which can induce apoptosis, stimulated after TNF-α stimulation in H4-II-E cells.

The effect of caspase inhibitor on TNF-α-mediated cell death in H4-II-E cells is examined [203]. TNF-α-induced cell death was prevented in culture with caspase inhibitor in wild-type cells and transfectants, suggesting that TNF-α-induced cell death is partly involved activation of caspases in H4-II-E cells. Regucalcin may have a inhibitory effect on activation of caspases in the cells.

Lypopolysacharide (LPS) has been shown to induce cell apoptosis [204, 205]. H4-II-E cells with the subconfluent monolayer were cultured in a medium without FBS in the presence of LPS. LPS caused a decrease in the number of H4-II-E cells (wild-type), inducing cell death and apoptosis [206]. This decrease was completely prevented in the regucalcin cDNA-transfected hepatoma cells overexpressing regucalcin with culture for 12-48 h [206]. Overexpression of regucalcin has a suppressive effect on LPS-stimulated cell death and apoptosis.

LPS acts to modulate the expression of a large number of genes that favor apoptosis of fibroblastic cells that are dependent upon activation of caspase-8 [205]. There is evidence that LPS-induced cell death is mediated through accumulation of reactive oxygene species and activation of p38 in rat cortex and hippocampus [205]. Culture with LPS caused a significant decrease in Ca^{2+}/calmodulin-dependent NO synthase activity in H4-II-E (wild type) cells [206]. LPS-induced decrease in NO synthase activity was found to prevent in the transfectants overexpressing regucalcin [206]. LPS-induced cell death may be not resulted from NO production in hepatoma cells, and the suppressive effect of regucalcin on LPS-induced cell death is not involved in NO in the cells. Moreover, LPS-induced cell death was prevemted in culture with caspase 3 inhibitor [206]. The effect of regucalcin in suppressing LPS-induced cell death is partly related to the inhibitory effect on caspase-3 in hepatoma cells.

An induction of apoptosis is partly mediated through pathway of protein kinase. The death of H4-II-E cells (wild type) has been found to be induced in culture with PD 98059, a ERK inhibitor, dibucaine, an inhibitor of in Ca^{2+}-dependent protein kinase, or staurosporine, a potent inhibitor of protein serine/threonin kinases (protein kinase C), suggesting that cell death is partly involved in the inhibition of protein kinases [206]. Overexpression of regucalcin in H4-II-E cells rescued cell death with PD 98059 or dibucaine [206]. Such an effect was not observed with staurosporine. PD 98059 induces apoptosis, which is in part due to the inactivation of Bcl-2 by increasing phosphorylated Bcl-2 in human prostate cancer cells [207]. Dibucaine has been shown to activate various caspases, such as caspase-3, -6, -8, and -9 (-like) activities, but not caspase-1 (-like) activity, and to induce mitochondrial membrane depolarization and the release of cytochrome C from mitochondria into the cytosol in leukemia cells (HL-60) [208]. Staurosporine induces apoptosis in Chang liver cells by a mitochondria-caspase-dependent pathway, which is closely correlated with a decrease in Bcl-2 and Bcl-XL levels in cancer cells [209]. Regucalcin may partly act the activation of Bcl-2 or the inhibition of caspases in signaling mechanism that PD 98059 or dibucaine induces apoptosis.

Calcium channel blockers, the endoplasmic reticulum Ca^{2+}-ATPase inghibitor thapsigargin and calcium ionophores are potent to lead several cell types to apoptosis [211, 212]. Thapasigargin is an inhibitor of Ca^{2+}-ATPase in the endoplasmic reticulum (Ca^{2+} store) in cells, and treatment with thapsigargin causes an elevation of sustained Ca^{2+} concentration in cells and induces apoptosis in the hepatoma cells [213]. Experiments on nucleus isolated from cells clearly demonstrate the induction of Ca^{2+}-dependent endonuclease activity during triggering apoptosis events [213]. Rises in intracellular Ca^{2+} concentration are believed to activate this nuclease and to mediate DNA cleavages into oligonucleosome fragments [214]. Regucalcin has been shown to have an inhibitory effect on Ca^{2+}-activated DNA fragmentation in isolated rat liver nucleus [187], suggesting that the protein has an inhibitory effect on apoptosis in liver cells. Thapsigargin induces cell death and apoptosis causing DNA fragmentation [203]. Thapsigargin-induced DNA fragmentation in the hepatoma cells is not altered in culture with caspase inhibitor, suggesting that thapsigargin-mediated apoptosis is independent of activation of caspases [203]. Overexpression of regucalcin in the hepatoma cells has been found to suppress DNA fragmentation induced by thapsigargin [203]. This effect is further enhanced in culture with caspase inhibitor [203]. Presumably, regucalcin has a suppressive effect on thapsigargin-mediated cell death due to preventing the rise in intracellular Ca^{2+} concentration in the hepatoma cells, since the protein can keep intracellular Ca^{2+} homeostasis due to activating Ca^{2+} pum enzymes in the plasma membranes, mitochondria, and endoplasmic reticulum of rat liver cells [31-34].

Calcium entry into cells induces cell death [206, 215]. Culture with Bay K 8644, an antagonist of Ca^{2+} entry in cells, caused a significant increase in the death of hepatoma H4-II-E cells (wild-type) [206]. Culture with Bay K 8644 did not induce cell death of transfectants (H4-II-E cells) overexpressing regucalcin [206]. Overexpression of regucalcin in H4-II-E cells was found to suppress DNA fragmentation induced by Bay K 8644. Regucalcin may have a suppressive effect on Ca^{2+} entry-induced stimulation of apoptosis in the hepatoma cells [206]. Regucalcin has a suppressive effect on Ca^{2+} entry-mediated cell death due to preventing the rise in intracellular Ca^{2+} concentration in the hepatoma cells. In addition, regucalcin may suppresse the effect of Ca^{2+} on DNA fragmentation in the nucleus of H4-II-E cells.

The effect of insulin or IGF-I on cell death and apoptosis in H4-II-E cells has not been well known. H4-II-E cells were cultured in a medium containing, either vehicle, insulin, insulin-like growth factor-I (IGF-I), epinephrine, or transforming growth factor-β in the absence of FBS [216]. The number of wild-type cells was decreased in the presence of insulin or IGF-I [216]. Agarose gel electrophoresis showed the presence of low-molecular-weight DNA fragments of adherent wild-type cells cultured with insulin or IGF-I [216]. The effect of insulin or IGF-I in stimulating cell death and DNA fragmentation H4-II-E cells (wild type) was prevented in transfectants overexpressing regucalcin [216].

The effect of insulin in decreasing the number of H4-II-E cells was prevented in the presence of caspase-3 inhibitor [216]. The effect of IGF-I on cell death, however, was also observed in the presence of caspase-3 inhibitor [216]. These observations suggest that the effect of insulin on cell death is involved in activation of caspase-3, and that the effect of IGF-I is not dependent on caspase-3 in H4-II-E cells. The effect of IGF-I in inducing cell death in the presence of caspase-3 inhibitor was completely blocked in transfectants overexpressing regucalcin [216], suggesting that regucalcin inhibits signaling pathway of IGF-I-induced cell death that is not mediated through caspase-3 in H4-II-E cells.

The effect of insulin or IGF-I in inducing cell death and apoptosis of H4-II-E cells was not observed in the presence of NAME, an inhibitor of NO synthase [216], suggesting that insulin- or IGF-induced cell death is partly involved in production of NO in H4-II-E cells. Overexpression of regucalcin has been shown to have a suppressive effect on activation of Ca^{2+}/calmodulin-dependent NO synthase in H4-II-E cells [126].

The effect of IGF-I in inducing cell death of H4-II-E cells was also observed in the presence of Bay K 8644 [216]. Such an effect was not seen in the case of insulin [216]. The mode of IGF-I action differs from that of insulin. It is assumed that insulin induces cell death that is partly mediated through intracellular calcium-dependent signaling pathway in H4-II-E cells, and that IGF-I may not be mediathed through calcium-dependent signaling pathway in H4-II-E cells. The effect of IGF-I in inducing cell death in the presence of Bay K 8644 was not observed in transfectants overexpressing regucalcin [216].

Genistein has an inhibitory effect on protein tyrosine kinases and it can produce cell cycle arrest and apoptosis in leukemic cells [217]. Genistein was found to induce cell death of H4-II-E cells, and the effect was not seen in the transfectants overexpressing regucalcin [216]. Genistein-induced cell death is partly mediated through inhibition of protein tyrosine kinase in H4-II-E cells. Regucalcin has an inhibitory effect on protein tyrosine kinase activity in the cytoplasm and nucleus of rat liver [190].

The effect of insulin in inducing cell death of H4-II-E cells was not seen in the presence of genistein [216], although such an effect was not seen in the case of IGF-I. The effect of IGF-I on cell death in the presence of genistein was prevented in transfectants overexpressing regucalcin [216]. Regucalcin has a suppressive effect on cell apoptosis that is mediated through signaling pathways with dependent or independent on protein tyrosine kinase.

Vanadate is an inhibitor of protein tyrosine phosphatase in cells [145]. Regucalcin has been shown to have an inhibitory effect on protein tyrosine phosphatase activity in the cytoplasm and nucleus of rat liver [150]. Vanadate was found to induce cell death of H4-II-E cells [216], suggesting that cell death is not caused by mechanism that is mediated through inhibition of protein tyrosine phosphatase activity. Vanadate induced cell death for transfectants overexpressing regucalcin [216], suggesting that the suppressive effect of regucalcin on cell death of H4-II-E cells is independent on protein phosphatase. IGF-I had a stimulatory effect on cell death of H4-II-E cells overexpressing regucalcin in the presence of vanadate [216], suggesting that the effect of IGF-I is not mediated through protein tyrosine phosphatase in the transfectants.

The effect of insulin in inducing cell death may be partly mediated through signaling pathway which is involved in caspase-3, calcium, NO, protein tyrosine kinase, or protein tyrosine phosphatase in H4-II-E cells. The effect of IGF-I on cell death of H4-II-E cells may be mediated through NO and other molecules. Overexpression of regucalcin may have a suppressive effect on signaling mechanism by which insulin or IGF-I induces cell death of H4-II-E cells.

Sulforaphane is an isothiocyanate that is present naturally in widely consumed vegetables and has a particularly high concentration in broccoli. This compound has been shown to block the formation of tumors initiated by chemicals in the rat [218]. Sulforaphane has been shown to induce a cell cycle arrest, followed by cell death in HT29 human colon cancer cells [218]. Sulforaphane increases expression of the pro-apoptotic protein Bax, the release of cytochrome C from the mitochondria to the cytosol, and the proteolytic cleavage of poly (ADP-ribose) polymerase in HT29 human colon cancer cells [218]. In human T-cell leukemia, sulforaphane induces apoptosis due to increased p53 and Bax protein expression, and slightly affected Bcl-2 expression [219]. In cultured PC-3 human prostate cancer cells, moreover, sulforaphane-induced apoptosis is associated with up-regulation of Bax, down-regulation of Bcl-2 and activation of caspase-3, -9, and -8 [220]. Sulforaphane has been found to induce cell death and apoptosis in H4-II-E cells [221]. Caspase-3 inhibitor prevented the effect of sulforaphane, while it was not inhibited by NAME, an inhibitor of NO synthase, in H4-II-E cells [221]. Sulforaphane-induced cell death and apoptosis partly result from activation of caspase-3 in the hepatoma cells.

Overexpression of regucalcin had a suppressive effect on cell death and apoptosis induced by sulforaphane in H4-II-E cells [221]. The suppressive effect of regucalcin on sulforaphane-induced cell death and apoptosis in H4-II-E cells may be partly involved in the molecules of Bax, cytichrome C, caspase, and Bcl-2. In addition, regucalcin may have an inhibitory effect on NO synthase and Ca^{2+}-dependent endonuclease activities in H4-II-E cells [126]. Regucalcin has a suppressive effect on many signaling pathways that mediate apoptotic cell death.

As mentioned above, regucalcin has been shown to play a role in the regulation of cell death and apoptosis in H4-II-E cells [203, 206, 216]. Overexpression of regucalcin has been demonstrated to have a suppressive effect on cell death and apoptosis induced by TNF-α, LPS, thapsigargin, Bay K 8644, dibucaine, or PD98059, an inhibitor of protein thyrosine kinase, insulin, or IGF-I in H4-II-E cells [203, 206, 216]. The signaling mechanisms that TNF-α, LPS, or other factors mediate cell death and apoptosis may be different. The suppressive effect of regucalcin on apoptotic cell death is related to its inhibitory effect on the activities of various protein kinases, NO synthase, caspase-3, or Ca^{2+}-dependent endonuclease, and its activatory effect on Bcl-2. Regucalcin has a suppressive effect on many signaling pathways that mediate cell death and apoptosis, and the protein suppresses cell death and apoptosis mediated through many different signaling pathways in H4-II-E cells.

Regucalcin Suppresses Apoptosis in Kidney Cells

Regucalcin has been shown to express in the cloned normal rat kidney proximal tubular epithelial NRK52E cells and its expression is enhanced after hormonal stimulation [222]. Nuclear localizetion of regucalcin is enhanced after hormone stimulation in NRK52E cells [183]. The role of regucalcin in cell death and apoptosis is examined using NRK52E cells overexpressing regucalcin [223]. The number of wild-type cells was decreased with culture for 42-72 h in the presence of TNF-α, LPS, Bay K 8644, or thapsigargin [223]. These effects were prevented in transfectants overexpressing regucalcin. DNA fragmentation induced after culture with LPS, Bay K 8644, or thapsigargin were prevented in transfectants overexpressing regucalcin [223]. Thus, overexpression of regucalcin has a suppressive effect on apoptotic cell death induced by TNF-α, LPS, Bay K 8644, or thapsigargin in kidney NRK52E cells. The effect of regucalcin in suppressing apoptotic cell death may be mediated through its action on many intracellular signaling pathways in NRK52E cells.

Bcl-2 is a suppressor in apoptotic cell death [224]. Apaf-1 participates in activation of caspase-3 [225]. Akt-1 involves in survival signaling pathway for cell death [226]. Overexpression of regucalcin caused a remarkable elevation of Bcl-2 mRNA expression in NRK52E cells, and it slightly stimulated Akt-1 mRNA expression in the cells. Apaf-1, caspase-3, or G3PDH mRNA expressions were notsignificantly altered in transfectants [223]. Presumably, the enhancement of Bcl-2 mRNA expression contributes to the suppression of apoptotic cell death in NRK52E cells overexpressing regucalcin. Regucalcin may play a role in the regulation of Bcl-2 gene expression in NRK52E cells.

TNF-α enhanced the expression of caspase-3 mRNA in NRK52E cells [223]. This enhancement was found to suppress in transfectants [223], suggesting that the mechanism by which regucalcin suppresses TNF-α-induced cell death is partly related to the decrease in caspase-3 mRNA expression in transfectants.

The presence of LPS caused a significant decrease in Bcl-2 mRNA levels in NRK52E cells, suggesting that this decrease is partly related to LPS-induced cell death [223]. The enhancement of Bcl-2 mRNA expression induced by overexpression of regucalcin was also seen in the presence of LPS [223]. LPS-stimulated expression of Apaf-1 mRNA was suppressed after overexpression of regucalcin [223]. This may partly involve in the suppression of LPS-induced cell death in NRK52E cells overexpressing regucalcin.

Culture with Bay K 8644 or thapsigargin was found to cause a significant increase in caspase-3 mRNA levels in wild-type cells, indicating that the increased gene expression partly contributes to inducing apoptotic cell death [223]. This increase was completely prevented in transfectants. Regucalcin may have a suppressive effect on caspase-3 mRNA expression enhanced by Bay K 8644 or thapsigargin in NRK52E cells. Thus, regucalcin was found to regulate the expression of Bcl-2, caspase-3, and Akt-1 mRNAs in the cloned normal rat kidney NRK52E cells. The change in protein levels, however, remains to be elucidated.

Toxic factors have been reported to induce renal failure due to stimulating apoptotic cell death [227]. Overexpression of regucalcin was found to have a suppressive effect on apoptotic cell death induced by various factors (including TNF-a, LPS, Bay K 8644, or thapsigargin) in NRK52E cells. Regucalcin may have a role as a suppressor in the development of apoptotic cell death in kidney proximal tubular epithelial cells. Presumably, regucalcin plays a physiological role in the maintenance of homeostasis of cellular response for cell stimulation.

As described above, overexpression of regucalcin rescues cell death and apoptosis induced in culture with various factors in the hepatoma cells and normal kidney cells. The cell signaling mechanisms, that these factors mediate cell death and apoptosis, may be different. Regucalcin may have a suppressive effect on many siganaling pathways that mediate cell death and apoptosis. The suppressive effect of regucalcin on cell death and apoptosis may be related to the inhibitory effect on the activities of NO synthase, caspase-3, or Ca^{2+}-dependent endonuclease and its activatory effect on Bcl-2. Moreover, regucalcin has regulatory effects on gene expression of many molecules that is related to cell apoptosis. Regucalcin plays an important role as a regulatory protein in intracellular signaling pathway which is related to cell death and apoptosis.

REGUCALCIN SUPPRESSES CELL PROLIFERATION

Regucalcin mRNA and its protein are expressed in the cloned rat hepatoma H4-II-E cells, although these expressions show low levels as compared to that in the cytosol of normal rat liver [50, 51]. The expression of regucalcin mRNA is stimulated in H4-II-E cells after culture with addition of serum (10% FBS) [50, 51], suugesting that regucalcin plays a role in the proliferation of cells. Regucalcin has been demonstrated to have a role as suppressor in the enhancemence of proliferation of liver cells in vitro. This section is described the mechanism by which regucalcin has a suppressive effect on cell proliferation using H4-II-E cells.

Regucalcin Suppresses the Enhancemet of Protein Kinase Activity in Cell Proliferation

The role of endogenous regucalcin in the regulation of protein kinase activity in the proliferation of H4-II-E cells is examined [228]. H4-II-E cells were cultured for 6-72 h in the presence of FBS (1 or 10%). The number of cells and protein kinase activity in the 5500 g supernatant of cell homogenate was increased 24 and 48 h after the culture with FBS (1 or 10%); the culture with 10% FBS had potent effect as compared with that of 1% FBS [228]. The culture with FBS produced an increase in protein kinase activity and a corresponding elevation of cell number in H4-II-E cells [228]. The increase in protein kinase activity of the 5500 g supernatant of cell homogenate preceded a significant elevation of cell number, suggesting that serum factors (including growth factors and hormones) stimulate cell proliferation that is partly mediated through cascades of protein kinases.

Serum-stimulated protein kinase activity in H4-II-E cells was further enhanced in the presence of calmodulin or dioctanoylglycerol in the presence of calcium chloride, and this increase was inhibited in the presence of trifluoperazine, staurosporine, or genistein in the enzyme reaction mixture [228], indicating that Ca^{2+}/calmodulin-dependent protein kinase, protein kinase C, and protein tyrosine kinase are present in H4-II-E cells [228]. Various protein kinases may be involved in the enhancement of cell proliferation with serum stimulation.

The presence of anti-regucalcin monoclonal antibody in the enzme reaction mixture containing the 5500 g supernatant of cell homogenate of H4-II-E cells with FBS caused a significant increase in protein kinase activity [228]. This effect was completely abolished after the addition of exogenous regucalcin, which has an inhibitory effect on the enzyme activity [228]. This finding indicates that endogenous regucalcin plays a suppressive role in the enhancement of protein kinase activity in the cytoplasm in H4-II-E cells with cell proliferation. The anti-regucalcin monoclonal antibody-increased protein kinase activity in H4-II-E cells was inhibited in the presence of trifluoperzine, staurosporine, or genistein, suggesting that endogenous regucalcin inhibits Ca^{2+}/calmodulin-dependent protein kinase, protein kinase C, or protein tyrosine kinase activities [228].

Serum stimulation may lead to an increase in cell proliferation that is partly mediated through cascade for various protein kinases in in H4-II-E cells [228]. Regucalcin may have a suppressive effect for overexpression of cell proliferation due to inhibiting various protein kinases in the cytoplasm and nucleus of H4-II-E cells. Presumably, regucalcin plays a role as suppressor protein in cell proliferation of normal liver cells and hepatoma cells.

Regucalcin Suppresses the Enhancement of Protein Phosphatase Activity in Cell Proliferation

Regucalcin has been shown to have an inhibitory effect on protein phosphatase activity in the cytoplasm of rat liver [150, 151]. H4-II-E cells were cultured for 3 days in a medium containing serum (10% FBS). After subconfluency, the cells were used for the assay of protein phosphatase activity toward phosphotyrosine. Ca^{2+}/calmodulin-dependent protein tyrosine phosphatases were present in H4-II-E cells [229]. Regucalcin had an inhibitory effect on Ca^{2+}/calmodulin-dependent protein tyrosine phosphatase activity in the cells [229]. The culture with Bay K 8644, an agonist of Ca^{2+} channels, caused an elevation of protein tyrosine phosphatase in H4-II-E cells, whereas dibutyryl cyclic AMP had no effect [229]. Bay K 8644-induced increase in protein phosphatase activity was inhibited in the presence of TFP, indicating that this increase is mediated through Ca^{2+}/calmodulin in hepatoma cells [229]. The effect of antibody was enhanced in the presence of TFP [229]. This enhancement may result from an increase in endogenous regucalcin in H4-II-E cells, since the expression of regucalcin mRNA in H4-II-E cells is stimulated after the culture with Bay K 8644 [229]. This finding may support the view that endogenous regucalcin, which is enhanced through Ca^{2+} signaling, has a suppressive effect on Ca^{2+}/calmodulin-activated protein phosphatase activity in proliferative cells.

The presence of anti-regucalcin monoclonal antibody in the enzyme reaction mixture caused a remarkanle elevation of protein tyrosine phosphatase activity in the cell homogenate (5,500 g supernatant) of H4-II-E cells cultured with serum addition (1 or 10% FBS) [230]. This elevation was completely prevented after the addition of regucalcin [230]. Endogenous regucalcin may have a suppressive effect on the enhancement of protein tyrosine phosphatase activity in the proliferative cells.

There may be many protein phosphatases in hepatoma H4-II-E cells. The antibody-increased protein tyrosine phosphatase activity was also inhibited by okadaic acid or vanadate, which is an inhibitor of protein phosphatases [145], although cyclosporine A, an inhitor of calcineurin (protein phosphatase) [232], had no effect [230]. Endogenous regucalcin may also act on okadaic acid or vanadate-sensitive protein tyrosine phosphatases in H4-II-E cells.

The culture with serum addition (10% FBS) caused an increase in proliferation of H4-II-E cells and a corresponding elevation of protein tyrosine phosphatase activity in the cells [230]. This finding suggests that the augmentation of protein tyrosine phosphatase activity is partly involved in the proliferation of H4-II-E cells, although protein phosphatase activity toward phosphoserine and phosphothreonine was not raised in the proliferative cells cultured with serum addition [230].

Processes that are reversibly controlled by protein phosphorylation require not only a protein kinase but also a protein phosphatase [145, 231]. Target proteins are phosphorylated at specific sites by one or more protein kinases and these phosphoproteins are removed by specific protein phosphatase [232].

As mentioned above, regucalcin may plays an important role as a suppressor for the enhancement of cell proliferation due to inhibiting various protein kinases and protein phosphatases activities that are raised in the proliferation of H4-II-E cells.

Regucalcin Suppresses the Enhancement of DNA Synthesis in Cell Proliferation

Endogenous regucalcin has a suppressive effect on the enhancement of DNA synthesis in the nuclei of H4-II-E with cell proliferation [233]. Cells were cultured for 6-96 h in medium containing FBS (1 or 10%). Cell number was increased between 24 and 96 h; cell proliferation was markedly stimulated after culture with 10% FBS as compared with that of 1% FBS [233]. Nuclear DNA synthesis activity in vitro was elevated 6 h after culture with 10% FBS and its elevation was remarkable at 12 and 24 h after the culture [233]. The increase in nuclear DNA synthesis activity preceded an elevation of the number of the cloned rat hepatoma cells H4-II-E cells cultured with FBS (1 or 10%) [233].

The increase in nuclear DNA synthesis activity at 12 and 24 h after culture with FBS was inhibited in the presence of PD98059, an inhibitor of MAP kinase, staurosporine, an inhibitor of protein kinase C, and TFP, an antagonist of Ca^{2+}/calmodulin-dependent protein kinase, in the reaction mixture [233]. The increase in nuclear DNA synthesis activity after serum stimulation may be partly mediated through action of various protein kinases in the nuclei of H4-II-E cells. However, serum-stimulated increase in nuclear DNA synthesis activity was not related to various protein phosphatases [233].

Regucalcin has been shown to transport in the nucleus of rat liver [38, 151], and it can inhibit nuclear DNA synthesis of normal rat liver [191, 192]. The presence of regucalcin in the reaction mixture caused a decrease in DNA synthesis activity in the nuclei of H4-II-E cells cultured with FBS [233]. This effect was not altered in the presence of various protein kinase inhibitions. Regucalcin has been demonstrated to inhibit the activity of various protein kinases in the nuclei of liver cells [190, 228]. The effect of regucalcin in decreasing nuclear DNA synthesis activity may be partly mediated through the pathway of various protein kinases in H4-II-E cells.

Nuclear DNA synthesis activity was increased in the presence of anti-regucalcin monoclonal antibody in the reaction mixture containing the nucleus of H4-II-E cells cultured for 24 h with 10% FBS [233]. This elevation was inhibited after addition of various protein kinase inhibitors in the reaction mixture [233]. These findings support the view that endogenous regucalcin suppresses DNA synthesis activity through mechanism by which inhibits nuclear protein kinases, and that it has a suppressive effect on DNA synthesis activity in the nuclei of H4-II-E cells with proliferation.

The transcriptional activity for regucalcin gene has been shown to enhance in H4-II-E cells cultured with Bay K 8644, an agonist of Ca^{2+} entry in cells, in the presence of 10% FBS [44, 51]. Culture with Bay K 8644 caused an increase in regucalcin levels in H4-II-E cells that were cultured in the presence of FBS (1 and 10%) [233]. In this case, nuclear DNA synthesis activity was not changed after culture with Bay K 8644 [233]. The presence of anti-regucalcin monoclonal antibody in the reaction mixture containing the nuclei of H4-II-E cells cultured with Bay K 8644 resulted in an elevation of nuclear DNA synthesis activity [233], suggesting that nuclear DNA synthesis activity is suppressed through endogenous regucalcin that is increased in the nuclei of H4-II-E cells cultured with Bay K 8644.

As mentioned above, regucalcin may have a suppressive effect on the enhancement of nuclear DNA synthesis activity in proliferative cells, and it may play a suppressive role for overexpression of cell proliferation. This was further supported in H4-II-E cells overexpressing regucalcin stably [234].

The regucalcin content of regucalcin/pCXN2-transfected cells used in this study was 19.7-fold as compared with that of the parental wild-type H4-II-E cells and pCXN2 vector-transfected cells (mock type) [234]. Regucalcin/pCXN2 vector-transfected cells (transfectants) were cultured for 72 h in the presence of FBS (10%) [234]. Cell numbers and DNA synthesis activity in the transfectants were found to suppress as compared with those of wild and mock type, suggesting that overexpression of regucalcin has a suppressive effect on cell proliferation [234].

The presence of anti-regucalcin monoclonal antibody in the reaction mixture caused an increase in DNA synthesis activity in the nuclei obtained from wild-type H4-II-E cells, mock-type cells, and transfectants with overexpression of regucalcin [234]. However, the augmentation of nuclear DNA synthesis activity was remarkable in the transfectants [234]. This may support the view that endogenous regucalcin has a great suppressive effect on nuclear DNA synthesis activity.

The expression of regucalcin mRNA has been shown to stimulate through a Ca^{2+}-signaling mechanism in H4-II-E cells [41, 51]. Regucalcin is translocated to the nucleus of liver cells [38,152]. Regucalcin inhibits nuclear protein kinase and protein phosphatase activities [228, 229], which are involved in siganal transduction to the nucleus, and it causes an inhibition in nuclear DNA synthesis in proliferative liver cells [233]. Regucalcin may play a suppressive role for the overexpression of proliferation of liver cells.

Regucalcin Suppresses Cell Cycle-Realted Gene Expression in Proliferative Cells

Regucalcin has a suppressive effect on liver cell proliferation [228-230]. Whether regucalcin suppress cell cycle-realted genes is examined in proliferative cells [235]. H4-II-E cells (wild type) and stable regucalcin/pCXN2 tranafectants were cultured for 72 h in a medium containing 10% FBS to obtain subconfluent monolayers. The proliferation of cells was suppressed in transfectants cultured for 24 -72 h. The proliferation of wild-type cells was inhibited when the cells were cultured for 72 h in a medium containing an inhibitor of transcriptional activity or protein synthesis. Such an effect was not seen in transfectants [235]. Regucalcin has a suppressive effect on cytosolic protein synthesis [166, 167] and nuclear RNA synthesis [194, 195] in rat liver. If overexpression of regucalcin inhibits protein and RNA synthesis in transfectants, the inhibitors of transcriptional activity or protein synthesis may not have additional effect. This suggests that the effect of regucalcin in suppressing cell prpliferation is partly mediated through its suppressive effect on protein and RNA synthesis in the cells.

Regucalcin has an inhibitory effect on various protein kinases in rat liver cytosol and nucleus [138,143, 190, 228]. The proliferation of H4-II-E cells (wild type and transfectants) was inhibited in the presence of PD98059, dibucaine, staurosporine, or genistein, which is an inhibitor of various protein kinases [235]. The effect of regucalcin in suppressing cell proliferation may be partly related to its inhibitory effect on MAP kinase, Ca^{2+}/calmodulin-dependent kinase, and protein tyrosine kinase in H4-II-E cells.

Wortmannin is known to have an inhibitory effect on PI3-kinase. The proliferation of H4-II-E cells (wild type) was inhibited in the presence of wortmannin, an inhibitor of PI3-kinase, or vanadate, an inhibitor of protein tyrosine phosphatase [235]. These effects were not observed in transfectants [236]. Regucalcin may inhibit PI3-kinase and protein tyrosine phosphatase activities, and those inhibitory effects may partly contribute to the suppression of proliferation in H4-II-E cells.

Bay K 8644 is an agonist of calcium entry into cells. The proliferation of H4-II-E (wild type) was inhibited in the presence of Bay K 8644 [235]. This effect of Bay K 8644 was not seen in transfectants [235], because regucalcin has a role in the maintenance of intracellular calcium homeostasis in many cell types [Review in Ref. 236].

Overexpression of regucalcin has been found to suppress the inhibitory effect of various factors, which induce cell-cycle arrest, on the proliferatuon of H4-II-E cells (wild type) [235]. The effect of roscovitine, a potent and selective inhibitor of the cyclin-dependent kinase cdc2, cdk2m, and cdk5 [237], or sulforaphane, which can induce G2/M phase cell cycle arrest [238], in inhibiting the proliferation of wild-type cells was not observed in transfectants [235]. Sulforaphane with a higher concentration caused a decrease in cell number of transfectant, suggesting that the chemical induces cell death and apoptosis (unpublished data). Butyrate induced an inhibiton of the proliferation of wild-type cells and transfectants. Roscovitine can arrest in G1 and accumulate in G2 of cell cycle [237]. Butyrate induces an inhibition of G1 progression [241]. The inhibitory effect of roscovitine or sulforaphane on cell proliferation may not be seen in transfectants, if regucalcin has a suppressive effect on the same pathway which roscovitine or sulforaphane has an inhibitory effect on cell proliferation. Presumably, regucalcin induces G1 and G2/M phase cell cycle arrest in H4-II-E cells.

The expression of p21 mRNA has been found to enhance in transfectant overexpressing regucalcin, although cdc2a and chk2 (checkpoint-kinase 2) mRNA levels are not changed in transfectants [236]. P21 is an inhibitor of cyclin-dependent kinases (cdk) [239]. Regucalcin may enhance p21 expression and it inhibits G1 progression in H4-II-E cells. It cannot exclude the possibility, however, that regucalcin directly inhibits cdk activity in the cells.

Overexpression of regucalcin suppressed the expression of IGF-I mRNA in H4-II-E cells [235]. IGF-I is a growth factor in cell proliferation. Regucalcin may have a suppressive effect on IGF-I expression in H4-II-E cells, and theis suppression of IGF-I expression leads to retardation of cell proliferation.

Regucalcin Modulates Tumor-Related Gene Expression in Proliferative Cells

It is known that c-*myc,* c-*fos*, c-*jun,* and Ha-*ras* are tumor stimulator genes [240]. *p53* and *Rb* are tumor suppressor genes and *c-src* is oncogene [241]. The expression of c-*myc*, Ha-*ras,* or c-*src* mRNAs was found to suppresse in transfectant overexpressing regucalcin [242, 243]. The expression of *p53* and *Rb* mRNAs was markedly enhanced in transfectants overexpressing regucalcin [243]. Presumably, the suppression of c-*myc*, Ha-*ras* and *c-src* mRNAs expressions and the enhancement of *p53* and *Rb* mRNAs expression in transfectants overexpressing regucalcin is partly involved in the retardation of proliferation of hepatoma H4-II-E cells.

The mechanism by which regucalcin regulates the expression of genes related to tumor is unknown. Regucalcin has been shown to translocate into the nucleus of rat liver [38, 151], and the protein has an inhibitory effect on RNA synthesis in isolated rat liver nucleus [194, 195]. Regucalcin can bind DNA and modulates nuclear transcriptional activity [184]. Regucalcin may bind to promoter region of tumor-related genes and it may suppress the expression of tumor stimulator gene or stimulate the expression of tumor suppressor gene in H4-II-E cells overexpressing regucalcin. As the result, the proliferation of hepatoma cells overexpressing regucalcin may be suppressed.

The expression of regucalcin is reduced in the cloned rat hepatica H4-II-E cells as compared with that of normal rat liver interestingly [228], suggesting an involvement of regucalcin in the suppression of carcinogenesis. The down-regulation of regucalcin expression in liver cells may lead to stimulation of cell proliferation with alteration of various tumor-related gene expressions. Regucalcin may play an important role as suppressor in cell proliferation and tumorigenesis of liver cells.

Regucalcin Suppresses Cell Proliferation in Normal Kidney Cells

Regucalcin has been also shown to have a role in the regulation of the proliferation of rat kidney proximal tubular epithelial NRK52E cells [244]. The regucalcin content of regucalcin/pCXN2-transfected cells was about 21-fold as comoared with that of the parental wild-type cells. The enhancement in cell proliferation was suppressed in the trensfectants overexpressing regucalcin [244]. The decrease in cell number of NRK52E (wild type) cells after culture with butyrate, rescovitine, and sulforaphane, which is an inhibitor of cell cycle, was not observed in transfectants, suggesting that regucalcin induces G1 and G2/M phase cell cycle arrest in NRK52E cells [244].

The inhibition of the proliferation of NRK52E cells induced by PD98059, staurosporine, or dibucaine that is an inhibitor of various protein kinase inhibitors was not seen in the transfectants [245-247]. The suppressive effect of regucalcin on cell proliferation may result from the inhibitory effect of regucalcin on various protein kinases that are involved in stimulation of cell proliferation. The inhibition in proliferation of NRK52E cells by wortmannin, an inhibitor of PI3-kinase, was not observed in transfectants [248], suggesting that regucalcin inhibits PI3-kinase and that it partly contributes suppression of cell proliferation in NRK52E cells.

Bay K 8644 is an agonist of calcium entry into cells. The proliferation of NRK52E cells (wild type) was inhibited in the presence of Bay K 8644. Overexpression of regucalcin had a preventive effect on Bay K 8644-induced inhibition of cell proliferation [244]. Regucalcin has a role in the maintenance of intracellular calcium homeostasis in many cell types.

The effect of regucalcin on the gene expression of proteins that are related to cell proliferation and cell cycle is examined. The expression of c-jun and chk2 (checkpoint-kinase 2) mRNAs was found to suppress in the transfectants [244]. The expression of p53 mRNA was enhanced in transfectants, while the expression of c-myc, c-fos, cdc2, and p21mRNA was not changed in the transfectants [244]. The decrease in c-jun and chk2 mRNA expressions may partly contribute to suppress cell proliferation induced in NRK52E cells overexpressing regucalcin. The expression of the tumor suppressor gene p53 mRNA, which was enhanced with overexpression of regucalcin, may have a partial role in the retardation of proliferation of NRK52E cells. Regucalcin has been shown to localize in the nucleus of NRK52E cells [249]. Regucalcin may have a suppressive effect on cell proliferation due to regulating many gene expressions that is related to cell proliferation in normal kidney NRK52E cells.

Regucalcin, moreover, has been shown to have a regulatory effect on gene expression of proteins that regulate calcium transport system in kidney cells. Overexpression of regucalcin has been found to have suppressive effects on the gene expression of L-type Ca^{2+} channel and calcium-sensing receptor (CaR), which regulate intracellular Ca^{2+} signaling in NRK52E cells [249]. Overexpression of regucalcin caused an increase in rat outer medullary K^+ channel (ROMK) mRNA expression in NRK52E cells, while it did not have an effect on Na, K-ATPase and epithelial sodium channel (ENaC) mRNA expressions [249]. The expression of Type II Na-Pi cotransporter (NaPi-IIa) and angiotensinogen mRNAs was not changed in NRK52E cells overexpressing regucalcin [249], suggesting that regucalcin does not have effects on NaPi-IIa and angiotensinogen mRNA expressions in kidney NRK52E cells.

The blockade of calcium influx through L-type calcium channels has been shown to attenuate mitochondrial injury and apoptosis in hypoxia renal tubular cells [250]. The entry of calcium through L-type Ca^{2+} channels induces mitochondrial disruption and cell death [250]. CaR participates in the regulation of renal Ca^{2+} transport [251]. It is speculated that regucalcin regulates intracellular Ca^{2+}-signaling pathway through its suppressive effect on L-type Ca^{2+} channel or CaR mRNA expression in the kidney proximal tubular epithelial cells.

The expression of regucalcin mRNA in NRK52E cells has been shown to enhance after the treatment of parathyroid hormone (PTH) [222], suggesting that regucalcin partly mediates cellular response for PTH in kidney cells. Overexpression of regucalcin did not attenuate the expression of L-type Ca^{2+} channel or CaR mRNAs, which is decreased after PTH treatment, in NRK52E cells. Regucalcin decreased L-type Ca^{2+} channel or CaR mRNA expressions in NRK52E cells. Regucalcin may partly contribute as a mediator in cellular response for stimulation of PTH in NRK52E cells.

Whether handling of calcium in NRK52E cells is changed in transfectants overexpressing regucalcin is unknown. Overexpression of regucalcin has been shown to have suppressive effects on apoptosis with culture of Bay K 8644 in NRK52E cells [223]. Presumably, the effects of regucalcin on gene expression are not mediated through change in calcium handling in transfectants. Regucalcin has been shown to play a role in the regulation of intracellular Ca^{2+} transport; the protein activates Ca^{2+}-pumping enzymes (Ca^{2+}-ATPase) in the basolateral membranes, mitochondria, and microsomes in rat kidney cortex [99-101]. Regucalcin may regulate intracellular Ca^{2+} homeostasis in kidney proximal tubular epithelial cells that Ca^{2+} is passed through transcellular transport. Moreover, regucalcin was found to suppress the expression of L-type Ca^{2+} channel or CaR mRNAs in NRK52E cells, supporting the view that regucalcin plays a physiological role in the regulation of intracellular Ca^{2+} homeostasis in kidney proximal tubular epithelial cells.

Overexpression of regucalcin has a suppressive effect on cell responses that are mediated through signaling process following stimulation with TNF-α or transforming growth factor-β 1 (TGF-β1) in NRK52E cells [252]. Overexpression of regucalcin had a suppressive effect on apoptotic cell death induced by TNF-α or TGF-β1 that is mediated through caspase-3 in NRK52E cells [252]. Culture with TNF-α or TGF-β1 caused a remarkable increase in α-smooth muscle actin level in NRK52E cells. Such an increase was not seen in transfectants. In addition, the expression of α-smooth muscle actin was markedly suppressed in transfectants cultured without TNF-α or TGF-β1. These findings demonstrate that overexpression of regucalcin has a suppressive effect on the expression of α-smooth muscle actin in NRK52E cells cultured with TNF-α or TGF-β1, and that regucalcin regulates signaling pathway that is mediated through TNF-α or TGF-β1 to stimulate the expression of α-smooth muscle actin in NRK52E cells. TGF-β1 is a key mediator that regulates transdifferentiation of NRK52E cells into myofibroblasts expressing α-smooth muscle actin [253]. This may contribute to renal fibrosis associated with overexpression of TGF-β1 within the diseased kidney [253]. Regucalcin may regulate transdifferentiation to renal fibrosis in NRK52E cells with TGF-β1 or TNF-α.

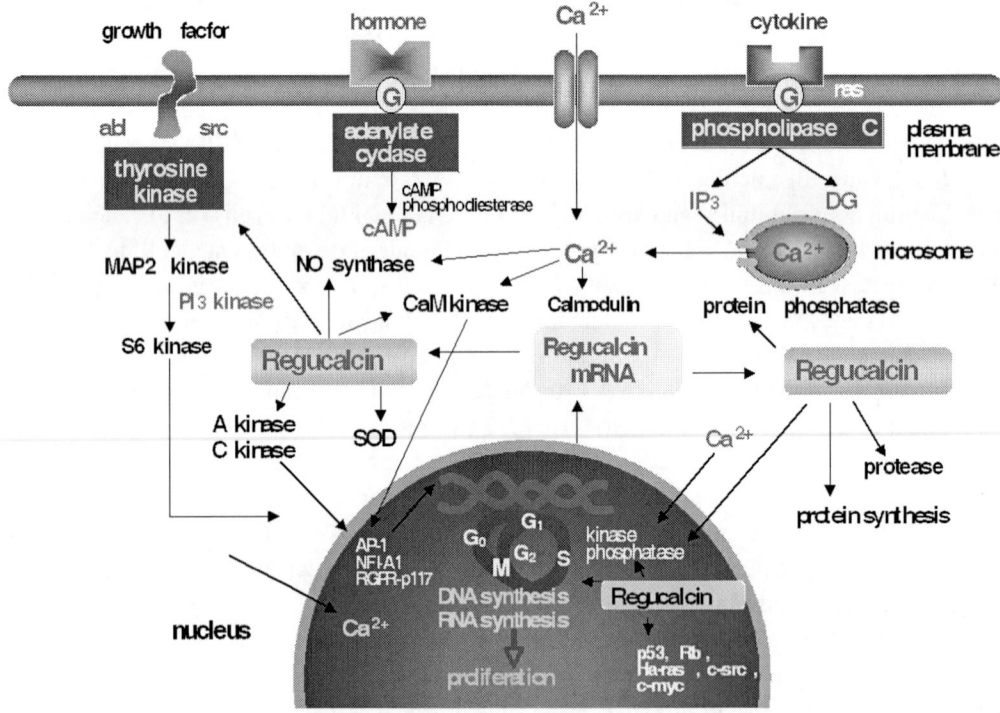

Figure 2. Regucalcin has a suppressive effect on the enhancement of cell proliferation. Regucalcin mRNA expression is stimulated through the pathway of signaling mechanism concerning Ca^{2+}/calmodulin (CaM)-dependent protein kinase (CaM kinase), protein kinase C (C kinase), protein kinase A (A kinase), and thyrosine kinase due to hormal stimulation. Regucalcin inhibits the activities of various protein kinases and protein phosphatases in the cytoplasm and nucleus of cells, and it also can inhibit Ca^{2+}/calmodulin-dependent enzyme activity (including cyclic AMP phosphodiesterase, NO synthase, superoxide dismutase (SOD), and others. Cytoplasmic regucalcin translocates into nucleus. Regucalcin inhibits nuclear DNA and RNA synthesis. Regucalcin has an inhibitory effect on the expression of c-myc, Ha-ras, and c-src mRNAs, which are tumor-stimulator genes. Regucalcin also stimulates the expression of p53 and Rb mRNAs that are tumor-suppressor genes. Moreore, regucalcin can inhibit protein synthesis and it can stimulate protein degradation. Regucalcin induces G1 and G2/M phase cell cycle arrest in cells. The suppressive effect of regucalcin on cell proliferation is mediated through regulating many signaling systems.

Overexpression of regucalcin caused a remarkable increase in the expression of mRNA of Smad 2, which is involved in signal transduction of TGF-β1 [258], or NF-κB, which is related to signaling of TNF-α [259], in NRK52E cells. Such an increase was not seen in Smad 3 mRNA expression in transfectants. This finding suggests that regucalcin stimulates the gene expression of Smad 2 or NF-κB, which is related to signaling mechanism of TNF-α or TGF-β1. Regucalcin may have a suppressive effect on signaling pathway by which TNF-α or TGF-β1 stimulates gene expression of NF-κB or Smad 2 in NRK52E cells. These cytokines may not have enhancing effects on gene expression of NF-κB or Smad 2 in transfectants. Presumably, the suppressive effects of regucalcin on apoptotic cell death and α-smooth muscle actin expression may be not involved in the expressions of NF-κB or Smad 2 that is stimulated by TNF-α or TGF-β1 in NRK52E cells.

MECHANISM BY WHICH REGUCALCIN REGULATES NUCLEAR GENE EXPRESSION

As mentioned above, regucalcin has been demonstrated to have a suppressive effect on hepatoma H4-II-E cella and normal kidney NRK52E cells. The suppressive effect of regucalcin on cell proliferation is related to its inhibitory effect on the activities of various protein kinases and protein phosphatases, calcium-dependent signaling factors, protein synthesis, nuclear DNA and RNA synthesis, and IGF-I expression and its activatory effect on p21, an inhibitor of cell cycle-related protein kinases. Moreover, regucalcin has been shown to suppress the expressions of c-*myc*, Ha-*ras* and *c-src* mRNAs and to enhance the expressions of of *p53* and *Rb* mRNAs, which are relted tumorigenesis of liver cells. *p53* is also known to stimulate p21 mRNA expression to induce cell-cycle arrest. Regucalcin has a suppressive effect on many intracellular signaling pathways that is related to cell proliferation due to hormonal stimulation, as summarized in Fig. 2. The suppressive effect of regucalcin on cell apoptosis and cell proliferation is based on the mechanism by which the protein inhibits cellular events that are mediated through many intracellular signaling factors. Regucalcin may play an important role as suppressor protein in the maintenance of homeostasis of cellular response for cell stimulation.

Regucalcin has been demonstrated to have a suppressive effect on cell death and apoptosis and cell proliferation induced by stimulation of various factors in hepatoma H4-II-E cells and normal kidney NRK52E cells. Presumably, regucalcin plays a physiological role in maintaining cell homeostasis of cellular response for various stimulating factors. Regucalcin may be a key molecule as a suppressor protein in cell regulation.

PROSPECTS

Regucalcin, which its gene is localized on the chromosome X in human and rats, is thought as a protein that is highly differentiated because of a great conservation of the regucalcin genes throughout evolution in vertebrates specifies. Regucalcin plays a multifunctional role in cellular regulation; a role in keeping intracellular Ca^{2+} homeostasis, an inhibitory role on various Ca^{2+}-dependent enzyme activations, protein kinases and protein phosphatases. Regucalcin regulates due to suppressing protein synthesis and stimulating protein degradation, and it has a suppressive effect on DNA and RNA synthesis that are mediated through various signaling systems in the nucleus. Regucalcin suppresses an enhancement of cell proliferation and apoptotic cell death that are induced by various signaling factors. Regucalcin has been proposed to play a pivotal role as a suppressor protein of intracellular signaling system in maintaining cell homeostasis. Regucalcin is the first time finding in protein molecule that has a role as a suppressor protein in cell signaling, although it is well known that many proteins enhance cell signal transduction so far.

There are growing evidences that regucalcin may be a key molecule in metabolic disease. The overexpression of regucalcin gene has been known to induce osteoporosis and hyperlipidemia in the transgenic rats [256-263]. The deficiency of regucalcin is induced a decrease in vitamin C (it is not synthesized in human), which is related to regulation of oxidative stress in mice and to inducing of cell apoptosis [264-266]. The expression of regucalcin gene is downregulated in the development of carcinogenesis in liver cells, suggesting that its suppression induces promotion of tumor cells.

The gene therapy, that targets the regucalcin gene, may be useful as a therapeutic tool for disease with the attenuation of regucalcin gene expression. Development of drug, which modulates regucalcin molecule, may have a clinical significance in the restoration of metabolic disorder that is implicated to regucalcin. Clinical studies of regucalcin for disease are expected.

REFERENCES

[1]   Cheung WY (1980) Calmodulin plays a pivotal role in cellular regulation. *Science* 202:19-27

[2]   Nishizuka Y (1986) Studies and perspectives of protein kinase C. *Science* 233: 305-31

[3]   Williamson JR, Cooper RK, Hoek JB (1981) Role of calcium in the hormonal regulation of liver metabolism. *Biochim Biophys Acta* 639: 243-295

[4]   Reinhart PH, Taylor WM, Bygrave FL (1984) The role of calcium ions in the mechanisms of action of α-adrenergic agonists in rat liver. *Biochem J* 223: 1-13

[5]   Kraus-Friedman N, Feng L (1996) The role of intracellular Ca^{2+} in the regulation of gluconeogenesis. *Metabolism* 48: 389-403

[6]   Yamaguchi M, Takei Y, Yamamoto Y (1975) Effect of thyrocalcitonin on calcium concentration in liver of intact and thyroparathyroidectomized rats. *Endocrinology* 96: 1004-1008

[7]   Yamaguchi M, Yoshida H (1985) Participation of Ca^{2+} channel in liver calcium regulation by calcitoninn in rats. *Acta Endocrinol* 110: 239-243

[8]   Yamaguchi M, Yamamoto T (1978) Purification of calcium binding substance from soluble fraction of normal rat liver. *Chem Pharm Bull* 26: 1915-1918

[9]   Yamaguchi M, Sugii K (1981) Properties of calcium-binding protein isolated from the soluble fraction of normal rat liver. *Chem Pharm Bull* 29: 567-570

[10]   Yamaguchi M (1988) Physicochemical properties of calcium-binding protein isolated from rat liver cytosol: Ca^{2+}-induced conformational changes. *Chem Pharm Bull* 36: 286-290

[11] Yamaguchi M, Yoshida H (1985) Regulatory effect of calcium-binding protein isolated from rat liver cytosol on activation of fructose 1,6-diphosphatase by Ca^{2+}-calmodulin. *Chem Pharm Bull* 33: 4489-4493

[12] Yamaguchi M, Shibano H (1987) Calcium-binding protein isolated from rat liver cytosol reverses activation of pyruvate kinase by Ca^{2+}. *Chem Pharm Bull* 35: 2025-2029

[13] Yamaguchi M, Shibano H (1987) Effect of calcium-binding protein on the activation of phosphorylase *a* in rat hepatic particulate glycogen by Ca^{2+}. *Chem Pharm Bull* 35: 2581-2584

[14] Yamaguchi M, Shibano H (1987) Reversible effect of calcium-binding protein on the Ca^{2+}-induced activation of succinate dehydrogenase in rat liver mitochondria. *Chem Pharm Bull* 35: 3766-3770

[15] Yamaguchi M, Mori S (1988) Effect of Ca^{2+} and Zn^{2+} on 5'-nucleotidase activity in rat liver plasma membranes: Hepatic calcium-binding protein (regucalcin) reverses the Ca^{2+} effect. *Chem Pharm Bull* 36: 321-325

[16] Shimokawa N, Yamaguchi M (1993) Molecular cloning and sequencing of the cDNA coding for a calcium-binding protein regucalcin from rat liver. *FEBS Lett* 327: 251-255

[17] Shimokawa N, Isogai M, Yamaguchi M (1995) Specific species and tissue differences for the gene expression of calcium-binding protein regucalcin. *Mol Cell Biochem* 143: 67-71

[18] Misawa H, Yamaguchi M (2000) Transcript heterogeneity of the human gene for Ca^{2+}-binding protein regucalcin. *Int J Mol Med* 5: 283-287

[19] Murata T, Yamaguchi M (1997) Molecular cloning of the cDNA coding for regucalcin and its mRNA expression in mouse liver: The expression is stimulated by calcium administration. *Mol Cell Biochem* 173: 127-133

[20] Misawa H, Yamaguchi M (2000) The gene of Ca^{2+}-binding protein regucalcin is highly conserved in vertebrate species. *Int J Mol Med* 6: 191-196

[21] Yamaguchi M (2011) The transcriptional regulation of regucalcin gene expression. *Mol Cell Biochem* 346:147-171

[22] Shimokawa N, Matsuda Y, Yamaguchi M (1995) Genomic cloning and chromosomal assignment of rat regucalcin gene. *Mol Cell Biochem* 151:157-163

[23] Thiselton DL, McDowall J, Brandau O, Ramser J, d'Esposito F, Bhattacharga SS, Ross MT, Hardcastle AJ, Meindl A (2002) An integrated, functionally annotated gene map of the DXS8026-ELK1 internal on human Xp11.3-Xp11.23: Potential hotspot for neurogenetic disorders. *Genomics* 79: 560-572

[24] Yamaguchi M, Makino R, Shimokawa N (1996) The 5'end seguences and exon organization in rat regucalcin gene. *Mol Cell Biochem* 165: 145-150

[25] Shimokawa N, Yamaguchi M (1992) Calcium administration stimulates the expression of calcium-binding protein regucalcin mRNA in rat liver. *FEBS Lett* 305: 151-154

[26] Yamaguchi M, Isogai M, Kato S, Mori S (1991) Immunohistochemical demonstration of calcium-binding protein regucalcin in the tissues of rats: The protein localizes in liver and brain. *Chem Pharm Bell* 36: 1601-1603

[27] Yamaguchi M, Isogai M (1993) Tissue concentration of calcium-binding protein regucalcin in rats by enzyme-linked immunoadsorbent assay. *Mol Cell Biochem* 122: 65-68

[28] Yamaguchi M, Nakajima R (2002) Role of regucalcin as an activator of sarcoplasmic reticulum Ca^{2+}-ATPase activity in rat heart muscle. *J Cell Biochem* 86: 184-193

[29] Yamaguchi M, Hamano T, Misawa H (2000) Expression of Ca^{2+}-binding protein regucalcin in rat brain neurons: Inhibitory effect on protein phosphatase activity. *Brain Res Bull* 52: 343-348

[30] Yamaguchi M, Misawa Y, Uchiyama S, Morooka Y, Tsurusaki Y (2002) Role of endogenous regucalcin in bone metabolism: Bone loss is induced in regucalcin transgenic rats. *Int J Mol Med* 10: 377-383

[31] Yamaguchi M (1992) A novel Ca^{2+}-binding protein regucalcin and calcium inhibition. Regulatory role in liver cell function. In: *Calcium Inhibition.* K Kohama (ed.) Japan Sci Soc Press, Tokyo and CRC Press, Boca Raton pp19-41

[32] Yamaguchi M (1998) Role of calcium-binding protein regucalcin in regenerating rat liver. *J Gastroen Hepatol* 13 (Suppl.): S106-S112

[33] Yamaguchi M (2000) Role of regucalcin in calcium signaling. *Life Sci* 66: 1769-1780

[34] Yamaguchi M (2000) The role of regucalcin in nuclear regulation of regenerating liver. *Biochem Biophys Res Commun* 276: 1-6

[35] Yamaguchi M (2002) Impact of aging on calcium channels and pumps. In: *Calcium Homeostasis and Signaling in Aging.* MP Mattson (ed.) Elsevier, Amsterdam pp 47-65

[36] Fujita T, Uchida K, Maruyama N (1992) Purification of senescence marker protein-30 (SMP30) and its androgen-independent decrease with age in the rat liver. *Biochim Biophys Acta* 1116: 122-128

[37] Fujita T, Shirasawa T, Uchida K, Maruyama N (1992) Isolation of cDNA clone encoding rat senescence marker protein-30 (SMP30) and its tissue distribution. *Biochim Biophys Acta* 1132: 297-305

[38] Tsurusaki Y, Misawa H, Yamaguchi M (2000) Translocation of regucalcin to rat liver nucleus: Involvement of nuclear protein kinase and protein phosphatase regulation. *Int J Mol Med* 6: 655-660

[39] Chakraborti S, Bahnson BJ (2010) Crystal structure of human senescence marker protein 30: insights linking structural, enzymatic, and physiological functions. *Biochemistry* 49: 3436-3444

[40] Murata T, Yamaguchi M (1998) Tissue-specific binding of nuclear factors to the 5'-flanking region of the rat gene for calcium-binding protein regucalcin. *Mol Cell Biochem* 178: 305-310

[41] Murata T, Yamaguchi M (1998) Ca^{2+} administration stimulates the binding of AP-1 factor to the 5'-flanking region of the rat gene for the Ca^{2+}-binding protein regucalcin. *Biochem J* 329: 157-163

[42] Murata T, Yamaguchi M (1999) Promoter characterization of the rat gene for Ca^{2+}-binding protein regucalcin. Transcriptional regulation by signaling factors. *J Biol Chem* 274: 1277-1285

[43] Misawa H, Yamaguchi M (2000) Involvement of hepatic nuclear factor I binding motif in transcriptional regulation of Ca^{2+}-binding protein regucalcin gene. *Biochem Biophys Res Commun* 269: 270-278

[44] Misawa H, Yamaguchi M (2000) Intracellular signaling factors-enhanced hepatic nuclear protein binding to TTGGC sequence in the rat regucalcin gene promoter: Involvement of protein phosphorylation. *Biochem Biophys Res Commun* 279: 275-281

[45] Misawa H, Yamaguchi M (2002) Indentification of transcription factor in the promoter region of rat regucalcin gene: Binding of nuclear factor I-A1 to TTGGC motif. *J Cell Biochem* 84: 795-802

[46] Misawa H, Yamaguchi M (2001) Molecular cloning and sequencing of the cDNA coding for a novel regucalcin gene promoter region-related protein in rat, mouse and human liver. *Int J Mol Med* 8: 513-520

[47] Misawa H, Yamaguchi M (2002) Gene expression for a novel protein RGPR-p117 in various species: The stimulation by intracellular signaling factors. *J Cell Biochem* 87: 188-193

[48] Yamaguchi M, Misawa H, Ma ZJ (2003) Novel protein RGPR-p117: The gene expression in physiologic state and the binding activity to regucalcin gene promoter region in rat liver. *J Cell Biochem* 88: 1092-1100

[49] Murata T, Shinya N, Yamaguchi M (1997) Expression of calcium-binding protein regucalcin mRNA in the cloned human hepatoma cells (Hep G2): Stimulation by insulin. *Mol Cell Biochem* 175: 163-168

[50] Nakajima M, Murata T, Yamaguchi M (1999) Expression of calcium- binding protein regucalcin mRNA in the cloned rat hepatoma cells (H4-II-E) is stimulated through Ca^{2+} signaling factors: Involvement of protein kinase C. *Mol Cell Biochem* 198: 101-107

[51] Yamaguchi M, Nakajima M (1999) Involvement of intracellular signaling factors in the serum-enhanced Ca^{2+}binding protein regucalcin mRNA expression in the cloned rat hepatoma cells (H4-II-E). *J Cell Bichem* 74: 81-89

[52] Vinson CR, Sigler PB, McKnight SL (1989) Scissors-grip model for DNA recognition by a family of leucine zipper proeins. *Science* 246: 911-916

[53] O'Shea EK, Rutkowski R, Stafford WF III, Kim PS (1989) Preferential heterodimer formation by isolated leucine zippers from fos and jun. *Science* 245: 646-648

[54] Shimokawa N, Yamaguchi M (1993) Expresion of hepatic calcium- binding protein regucalcin mRNA is mediated through Ca^{2+}/calmoclulin in rat liver. *FEBS Lett* 316: 79-84

[55] Isogai M, Yamaguchi M (1995) Calcium administration increases calcium-binding protein regucalcin concentration in the liver of rats. *Mol Cell Biochem* 143: 53-58

[56] Yamaguchi M, Kurota H (1995) Expression of calcium-binding protein regucalcin mRNA in the kidney cortex of rats: The stimulation by calcium administration. *Mol Cell Biochem* 146: 71-77

[57] Yamaguchi M, Ueoka S (1998) Expression of calcium-binding protein regucalcin mRNA in fetal rat liver is stimulated by calcium administration. *Mol Cell Biochem* 178: 283-287

[58] Yamaguchi M, Kanayama Y, Shimokawa N (1994) Expression of calcium-binding protein regucalcin mRNA in rat liver is stimulated by calcitonin: the hormonal effect is mediated through calcium. *Mol Cell Biochem* 136: 43-48

[59] Yamaguchi M, Oishi K, Isogai M (1995) Expression of hepatic calcium-binding protein regucalcin mRNA is elevated by refeeding of fasted rats: Involvement of glucose, insulin and calcium as stimulating factors. *Mol Cell Biochem* 142: 35-41

[60] Yamaguchi M, Oishi K (1995) 17β-Estradiol stimulates the expression of hepatic calcium-binding protein regucalcin mRMA in rats. *Mol Cell Biochem* 143: 137-141

[61] Yamaguchi M, Kanayama Y (1995) Enhanced expression of calcium- binding protein regucalcin mRNA in regenerating rat liver. *J Cell Biochem* 57: 185-190

[62] Ueoka S, Yamaguchi M (1998) Sexual difference of hepatic calcium-binding protein regucalcin mRNA expression in rats with different ages: Effect of ovarian hormone. *Biol Pharm Bull* 21: 405-407

[63] Lotersztajn S, Hanoune J, Pecker F (1981) A high affinity calcium-stimulated magnesium-dependent ATPase in rat liver plasma membranes. Dependence on an endogenous protein activator distinct from calmodulin. *J Biol Chem* 256:11209-11215

[64] Chen K-M, Junger KD (1983) Calcium transport and phosphorylated intermediate of $(Ca^{2+}-Mg^{2+})$-ATPase in plasma membranes of rat liver. *J Biol Chem* 258: 4404-4410

[65] Yamaguchi M, Mori S, Kato S (1988) Calcium-binding protein regucalcin is an activator of $(Ca^{2+}-Mg^{2+})$-adenosine triphosphatase in the plasma membranes of rat liver. *Chem Pharm Bull* 36: 3532-3539

[66] Takahashi H, Yamaguchi M (1993) Regulatory effect of regucalcin on $(Ca^{2+}-Mg^{2+})$-ATPase in rat liver plasma membranes: comparison with the activation by Mn^{2+} and Co^{2+}. *Mol Cell Biochem* 124: 169-174

[67] Takahashi H, Yamaguchi M (1994) Activatory effect of regucalcin on $(Ca^{2+}-Mg^{2+})$-ATPase in rat liver plasma membranes: relation to sulfhydryl group. *Mol Cell Biochem* 136:71-76

[68] Takahashi H, Yamaguchi M (1997) Stimulatory effect of regucalcin on ATP-dependent calcium transport in rat liver plasma membranes. *Mol Cell Biochem* 168: 149-153

[69] Takahashi H, Yamaguchi M (1993) Regucalcin modulates hormonal effect on $(Ca^{2+}-Mg^{2+})$-ATPase activity in rat liver plasma membranes. *Mol Cell Biochem* 125: 171-177

[70] Takahashi H, Suzuki S, Yamaguchi M (1995) Stimulatory effect of hormonal signaling factors on $(Ca^{2+}-Mg^{2+})$-ATPase activity in rat liver plasma membranes: Cross talk with regucalcin. *Mol Cell Biochem* 151: 1-7

[71] Takahashi H, Yamaguchi M (1995) Increase of $(Ca^{2+}-Mg^{2+})$-ATPase activity in hepatic plasma membranes of rats administered orally calcium: The endogenous role of regucalcin. *Mol Cell Biochem* 144: 1-6

[72] Takahashi H, Yamaguchi M (1996) Enhancement of plasma membrane $(Ca^{2+}-Mg^{2+})$-ATPase activity in regenerating rat liver: Involvement of endogenous activating protein regucalcin. *Mol Cell Biochem* 162:133-138

[73] Takahashi H, Yamaguchi M (1996) Activatory effect of regucalcin on hepatic plasma membrane $(Ca^{2+}-Mg^{2+})$-ATPase is impaired by liver injury with carbon tetrachloride administration in rats. *Mol Cell Biochem* 158: 9-16

[74] Yamaguchi M (1985) Mitocondrial uptake of $^{45}Ca^{2+}$ bound to calcium-binding protein isolated from rat liver cytosol. *Chem Pharm Bull* 33: 3390-3394

[75] Mori S, Yamaguchi M (1991) Calcium-binding protein regucalcin stimulates the uptake of Ca^{2+} by rat liver mitochondria. *Chem. Pharm Bull* 39: 224-226

[76] Takahashi H, Yamaguchi M (2000) Stimulatory effect of regucalcin on ATP-dependent Ca^{2+} uptake activity in rat liver mitochondria. *J Cell Biochem* 78: 121-130

[77] Yamaguchi M, Mori S (1989) Activation of hepatic microsomal Ca^{2+}-adenosine triphophatase by calcium-binding protein regucalcin. *Chem Pharm Bull* 37: 1031-1034

[78] Takahashi H, Yamaguchi M (1999) Role of regucalcin as an activator of Ca^{2+}-ATPase activity in rat liver microsomes. *J Cell Biochem* 74: 663-669

[79] Yamaguchi M, Mori S (1989) Effect of the calcium-binding protein regucalcin on the Ca^{2+} transport system in rat liver microsomes: The protein stimulates Ca^{2+} release. *Chem Pharm Bull* 37: 3037-3041

[80] Yamaguchi M, Mori S (1990) Regucalcin-induced Ca^{2+} release from rat liver microsomes: The effect is inhibited by heparin. *Chem Pharm Bull* 38: 2305-2307

[81] Nicotera P, McConkey DJ, Jones DP, Orrenius S (1989) ATP stimulates Ca^{2+} uptake and increases the free Ca^{2+} concentration in isolated rat liver nuclei. *Proc Natl Acad Sci USA* 86: 453-457

[82] Bachs O, Carafolli E (1995) Calmodulin and calmodulin-binding proteins in liver cell nuclei. *J Biol Chem* 262, 10786-10790

[83] Csermely P, Schnaider T, Szanto I (1995) Signaling and transport through the nuclear membrane. *Biochim Biophys Acta* 1241: 425-452

[84] Tsurusaki Y, Yamaguchi M (2000) Role of endogenous regucalcin in the regulation of Ca^{2+}-ATPase activity in rat liver nuclei. *J Cell Biochem* 78: 541-549

[85] Yamaguchi M (1992) Effect of calcium-binding protein regucalcin on Ca^{2+} transport system in rat liver nuclei: stimulation of Ca^{2+} release. *Mol Cell Biochem* 113: 63-70

[86] Fujita T, Inoue H, Kitamura T, Sato N, Shimosawa T, Maruyama N (1998) Senescence marker protein-30 (SMP30) rescues cell death by enhancing plasma membrane Ca^{2+}-pumping activity in Hep G2 cells. *Biochem Biophys Res Commun* 250:374-380

[87] Van Os CH (1985) Transcellular calcium transport in intestinal and renal epithelial cells. *Biochim Biophys Acta* 906: 195-222

[88] Taylor CW (1985) Calcium regulation in vertebrates: An overview. *Comp Biochem Physiol* 82A: 249-255

[89] Murata T, Yamaguchi M (1999) Binding of kidney nuclear proteins to the 5'flanking region of the rat gene for Ca^{2+}-binding protein regucalcin: Involvement of Ca^{2+}/calmodulin signaling. *Mol Cell Biochem* 199: 35-40

[90] Misawa H, Yamaguchi M (2001) Involvement of nuclear factor-1 (NF1) binding motif in the regucalcin gene expression of rat kidney cortex: The expression is suppressed by cisplatin administration. *Mol Cell Biochem* 219: 29-37

[91] Kurota H, Yamaguchi M (1996) Steroid hormonal regulation of calcium-binding protein regucalcin mRNA expression in the kidney cortex of rats. *Mol Cell Biochem* 155: 105-111

[92] Shinya N, Kurota H, Yamaguchi M (1996) Calcium-binding protein regucalcin mRNA expression in the kidney cortex is suppressed by saline ingestion in rats. *Mol Cell Biochem* 162: 139-144

[93] Shinya N, Yamaguchi M (1997) Alteration in Ca^{2+}-ATPase activity and calcium-binding protein regucalcin mRNA expression in the kidney cortex of rats with saline ingestion. *Mol Cell Biochem* 170: 17-22

[94] Shinya N, Yamaguchi M (1998) Stimulatory effect of calcium administration on regucalcin mRNA expression is attenuated in the kidney cortex of rats ingested with saline. *Mol Cell Biochem* 178: 275-281

[95] Misawa H, Yamaguchi M (2001) Involvement of nuclear factor-I (NFI) binding motif in the regucalcin gene expression of rat kidney cortex: The expression is suppressed by cisplatin administration. *Mol Cell Biochem* 219:29-37

[96] Kurota H, Yamaguchi M (1995) Suppressed expression of calcium-binding protein regucalcin mRNA in the renal cortex of rats with chemically induced kidney damage. *Mol Cell Biochem* 151: 55-60

[97] Agus ZS, Chiu PJS, Goldberg M (1997) Regulation of urinary calcium excretion in the rat. *Am J Physiol* 232: F545-F549

[98] Kennedy MB, Bennett MK, Erondu NE, Miller SG (1987) Calcium/calmodulin-dependent protein kinases. In: *Calcium and Cell Function*. Cheung WY (ed). Acadamic Press Inc, New York, pp 61- 107

[99] Kurota H, Yamaguchi M (1997) Activatory effect of calcium-binding protein regucalcin on ATP-dependent calcium transport in the basolateral membranes of rat kidney cortex. *Mol Cell Biochem* 169: 149-156

[100] Kurota H, Yamaguchi M (1997) Regucalcin increases Ca^{2+}-ATPase activity and ATP-dependent calcium uptake in the microsomes of rat kidney cortex. *Mol Cell Biochem* 177: 201-207

[101] Xue JH, Takahashi H, Yamaguchi M (2000) Stimulatory effect of regucalcin on mitochondrial ATP-depemdemt calcium uptake activity in rat kidney cortex. *J Cell Biochem* 80: 285-292

[102] Langer GA (1992) Calcium and the heart: Exchange at the tissue, cell, and organelle levels. *FASEB J* 6: 893-902

[103] Yamaguchi M, Nakajima R (2002) Role of regucalcin as an activator of sarcoplasmic reticulum Ca^{2+}-ATPase activity in rat heart muscle. *J Cell Biochem* 86: 184-193

[104] Akhter T, Nakagawa T, Kobayashi A, Yamaguchi M (2007) Suppression of regucalcin mRNA expression in the hearts of rats administered with free radical compound: The administration-induced death is accelerated in regucalcin transgenic rats. *Int J Mol Med* 19:653-658

[105] Thastrup O, Culler PJ, Drbbak BK, Hanley MR, Dawson AP (1990) Thapsigargin, a tumor promoter, discharges intracellular Ca^{2+} stores by specific inhibition of the endoplasmic reticulum Ca^{2+}-ATPase. *Proc Natl Acad Sci USA* 87: 2466-2470

[106] Tada M, Kadoma M (1989) Regulation of the Ca^{2+} pump ATPase by cAMP-dependent phosphorylation of phospholamban. *Bioessays* 10: 157-163

[107] Akther T, Sawada N, Yamaguchi M (2006) Regucalcin increases Ca^{2+}-ATPase activity in the heart mitochondria of normal and regucalcin transgenic rats. *Int J Mol Med* 18:171-176

[108] Dahan D, Spanier R, Rahaaminoff H (1991) The modulation of rat brain Na^{+}-Ca^{2+} exchange by K^{+}. *J Biol Chem* 266: 2067-2075

[109] MacDermott AB, Dale N (1986) Receptors, ion channels and synaptic potential underlying the intergrative actions of excitatory amino acids. *Trends Neurosic* 10: 280-284

[110] Cambray-Deakin MA, Burgoyne RD (1992) Intracellular Ca^{2+} and N- methyl-D-aspartate-stimulated neuritogenesis in rat cerebellar granule cell cultures. *Dev Brain Res* 66: 25-32

[111] Treves S, De Mattei M, Lanfred M, Villa A, Green NM, MacLennan DH, Meldolesi J, Pozzan T (1990) Calreticulin is a candidate for a calsequestrin-like function in Ca^{2+}-storage compartment (calcitosomes) of liver and brain. *Biochem J* 271: 473-480 Hartman H, Eckert A, Muller WE (1994) Disturbances of the neuronal calcium-homeostasis in the aging nervous system. *Life Sci* 55: 2011- 2018

[112] Yamaguchi M, Hanahisa Y, Murata T (1999) Expression of calcium-binding protein regucalcin and microsomal Ca^{2+}-ATPase regulation in rat brain: Attenuation with increasing age. *Mol Cell Biochem* 200: 43-49

[113] Hanahisa Y, Yamaguchi M (1996) Characterrization of calcium accumulation in the brain of rats administered orally calcium: The significance of energy-dependent mechanism. *Mol Cell Biochem* 158: 1-7

[114] Hanahisa Y, Yamaguchi M (1997) Increase in calcium content and Ca^{2+}-ATPase activity in the brain of fasted rats: Comparison with different ages. *Mol Cell Biochem* 171: 127-132

[115] Hanahisa Y, Yamaguchi M (2001) Decrease in Ca^{2+}-ATPase activity in the brain plasma membrane of rats with increasing age: Involvement of brain calcium accumulation. *Int J Mol Med* 7: 407-411

[116] Hanahisa Y, Yamaguchi M (1999) Brain microsomal calcium accumulation in rats with increasing age: Involvement of thapsigargin sensitive Ca^{2+}-ATPase. *Int J Mol Med* 4: 627-631

[117] Hanahisa Y, Yamaguchi M (1998) Increase of Ca^{2+}-ATPase activity in the brain microsomes of rats with increasing ages: Involvement of protein kinase C. Brain *Res Bull* 46: 329-332

[118] Yamaguchi M, Takakura Y, Nakagawa T (2008) Regucalcin increases Ca^{2+}-ATPase activity in the mitochondria of brain tissues of normal and transgenic rats. *J Cell Biochem* 104:795-804

[119] Yamaguchi M, Sakurai T (1992) Reversible effect of calcium-binding protein regucalcin on the Ca^{2+}-induced inhibition of deoxyuridine 5'-triphosphatase activity in rat liver cytosol. *Mol Cell Biochem* 110: 25-29

[120] Hanahisa Y, Yamaguchi M (1999) Effect of calcium-binding protein on adenosine 5'-triphosphatase activity in the brain cytosol of rats of different ages: The inhibitory role of regucalcin. *Biol Pharm Bull* 22: 313-316

[121] Rasmussen J (1970) Cell communication, calcium ion, and cyclic adenosine monophosphate. *Science* 170: 404-412

[122] Yamaguchi M, Tai H (1991) Inhibitory effect of calcium-binding protein regucalcin on Ca^{2+}/calmodulin-dependent cyclic nucleotide phosphodiesterase activity in rat liver cytosol. *Mol Cell Biochem* 106: 25-30

[123] Yamaguchi M, Kurota H (1997) Inhibitory effect of regucalcin on Ca^{2+}/calmodulin-dependent cyclic AMP phosphodiesterase activity in rat kidney cytosol. *Mol Cell Biochem* 177: 209-214

[124] Lowenstein CJ, Dinerman JL, Snyder SH (1994) Nitric oxide: A physiologic messenger. *Ann Intern Med* 120: 227-237.

[125] Izumi T, Tsurusaki Y, Yamaguchi M (2003) Suppressive effect of endogenous regucalcin on nitric oxide synthase activity in cloned rat hepatoma H4-II-E cells overexpressing regucalcin. *J Cell Biochem* 89: 800-807

[126] Yamaguchi M, Takahashi H, Tsurusaki Y (2003) Suppressive role of endogenous regucalcin in the enhancement of nitric oxide synthase activity in liver cytosol of normal and regucalcin transgenic rats. *J Cell Biochem* 88: 1226-1234

[127] Ma ZJ, Yamaguchi M (2003) Regulatory effect of regucalcin on nitric oxide synthase activity in rat kidney cortex cytosol: Role of endogenous regucalcin in transgenic rats. *Int J Mol Med* 12:201-206

[128] Ma ZJ, Yamaguchi M (2002) Suppressive role of endogenous regucalcin in the regulation of nitric oxide synthase activity in heart muscle cytosol of normal and regucalcin transgenic rats. *Int J Mol Med* 10:761-766

[129] Tobisawa M, Yamaguchi M (2003) Inhibitory role of regucalcin in the regulation of nitric oxide synthase activity in rat brain cytosol: involvement of aging. *J Neurol Sci* 209: 47-54

[130] Tobisawa M, Yamaguchi M (2003) Role of endogenous regucalcin in brain function: Suppression of cytosolic nitric oxide synthase and nuclear protein tyrosine phosphatase activities in brain tissue of transgenic rats. *Int J Mol Med* 12:581-585

[131] Van Remmen H, Williams MD, Guo Z, Estlack L, Yang H, Carlson EJ, Epstein CJ, Huang TT, Richaardson A (2001) Knockout mice heterozygous for Sod2 show alterations in cardiac mitocondrial function and apoptosis. *Am J Physol Heart Cric Physiol* 281: H1422- 1432

[132] Den Hartog GJ, Haenen GR, Boven E, van der Vijgh WJ, Bast A (2004) Lecithinized copper, zinc-superoxide dismutase as a protector against dexorubicin-induced cardiotoxicity in mice. *Toxicol Appl Pharmacol* 194: 180-188

[133] Adler A, Messina E, Sherman B, Wang Z, Huang H, Linke A, Hintze TH (2003) NAD(P)H oxidase-generated superoxide anion accounts for reduced control of myocardial O_2 consumption by NO in old Fisher 344 rats. *Am J physiol Heart Circ Physiol* 285: H1015-1022

[134] Fukaya Y, Yamaguchi M (2004) Regucalcin increases superoxide dismutase activity in rat liver cytosol. *Biol Pharm Bull* 27: 1444-1446

[135] Ichikawa E, Yamaguchi M (2004) Regucalcin increases superoxide dismutase activity in the heart cytosol of normal and regucalcin transgenic rats. *Int J Mol Med* 14: 691-695

[136] Connelly PA, Sisk RB, Schulman H, Garrison JC (1987) Evidence for the activation of the multifunction Ca^{2+}/calmodulin-dependent protein kinase in response to hormones that increase intracellular Ca^{2+}. *J Biol Chem* 262: 10154-10163

[137] Mori S, Yamaguchi M (1990) Hepatic calcium-binding protein regucalcin decreases Ca^{2+}/calmodulin-dependent protein kinase activity in rat liver cytosol. *Chem Pharm Bull* 38: 2216-2218

[138] Kurota H, Yamaguchi M (1997) Inhibitory effect of regucalcin on Ca^{2+}/calmodulin-dependent protein kinase activity in rat renal cortex cytosol. *Mol Cell Biochem* 177: 239-243

[139] Hamano T, Hanahisa Y, Yamaguchi M (1999) Inhibitory effect of regucalcin on Ca^{2+}-dependent protein kinase activity in rat brain cytosol: Involvement of endogenous regucalcin. *Brain Res Brain* 50: 187-192

[140] Hamano T, Yamaguchi M (2001) Inhibitory role of regucalcin in the regulation of Ca^{2+}-dependent protein kinase activity in rat brain tissues. *J Neurol Sci* 183: 33-38

[141] Omura M, Yamaguchi M (1998) Inhibition of Ca^{2+}/ calmodulin- dependent phosphatase activity by regucalcin in rat liver cytosol: Involvement of calmoduling binding. *J Cell Biochem* 71: 140-148

[142] Yamaguchi M, Mori S (1990) Inhibitory effect of calcium-binding protein regucalcin on protein kinase C activity in rat liver cytosol. *Biochem Med Metab Biol* 43: 140-146

[143] Kurota H, Yamaguchi M (1998) Inhibitory effect of calcium-binding protein regucalcin on protein kinase C activity in rat renal cortex cytosol. *Biol Pharm Bull* 21: 315-318

[144] Hunter T (1995) Protein kinases and phosphatases: the Yin and Yang of protein phosphorylation and signaling. *Cell* 80: 225-236

[145] Wang Y, Santini F, Qin K, Huang CY (1995) A Mg^{2+}-dependent, Ca^{2+}-inhibitable seruin/throsine protein phosphatase from bovine brain. *J Biol Chem* 270: 25607-25612

[146] Pallen CJ, Wang JH (1983) Calmodulin-stimulated dephosphorylation of *p*-nitrophenylphosphate and free phosphotyrosine by calcineurin. *J Biol Chem* 258: 850-855

[147] Omura M, Yamaguchi M (1999) Effect of anti-regucalcin antibody on neutral phosphatase activity in rat liver cytosol: Involvement of endogenous regucalcin. *Mol Cell Biochem* 197: 25-29

[148] Omura M, Kurota H, Yamaguchi M (1998) Inhibitory effect of regucalcin on Ca^{2+}/calmodulin-dependent phosphatase activity in rat renal cortex cytosol. *Biol Pharm Bull* 21: 440-443

[149] Omura M, Yamaguchi M (1999) Enhancement of neutral phosphatase activity in the cytosol and nuclei of regenerating rat liver: Role of endogenous regucalcin. *J Cell Biochem* 73: 332-341

[150] Omura M, Yamaguchi M (1999) Regulation of protein phosphatase activity by regucalcin localization in rat liver nuclei. J Cell Biochem 75: 437-445

[151] Morooka Y, Yamaguchi M (2001) Suppressive role of endogenous regucalcin in the regulation of protein phosphatase activity in rat renal cortex cytosol. *J Cell Biochem* 81: 639-646

[152] Morooka Y, Yamaguchi M: (2001) Inhibitory effect of regucalcin on protein phosphatase activity in the nuclei of rat kidney cortex. *J Cell Biochem* 83: 111-120

[153] Morooka Y, Yamaguchi M (2002) Endogenous regucalcin suppresses the ehmancement of protein phosphatase activity in the cytosol and nucleus of kidney cortex in calcium-administered rats. *J Cell Biochem* 85: 553-560

[154] Molkentin JD, Lu JR, Antos CL, Markham B, Richardson J, Robbins J, Grant SR, Olson EN (1998) A calcineurin-dependent transcriptional pathway for cardiac hypertrophy. *Cell* 17: 215-228

[155] Ichikawa E, Tsurusaki Y, Yamaguchi M (2004) Suppressive effect of regucalcin on protein phosphatase activity in the heart cytosol of normal and regucalcin transgenic rats. *Int J Mol Med* 13: 289-293

[156] Hamano T, Yamaguchi M (1999) Inhibitory effect of regucalcin on Ca^{2+}/calmodulin-dependent protein phosphatase activity in rat brain cytosol. *Int J Mol Med* 3: 615-619

[157] Yamaguchi M, Hamano T, Misawa H (2000) Expression of Ca^{2+}- binding protein regucalcin in rat brain neurons: Inhibitory effect on protein phosphatase activity. *Brain Res Bull* 52: 343-348

[158] Tobisawa M, Tsurusaki Y, Yamaguchi M (2003) Decrease in regucalcin level and enhamcement of protein tyrosine phosphatase activity in rat brain microsones with increasing age. *Int J Mol Med* 12: 577-580

[159] Tobisawa M, Yamaguchi M (2003) Suppressive effect of endogenous regucalcin on protein tyrosine phosphatase activity in the nucleus of rat brain: Attenuation with increasing age. *Int J Mol Med* 11: 205-210

[160] Brostrom CO, Bocckino SB, Brostrom MA, Galuska EM (1983) Identification of a Ca^{2+} requirement for protein synthesis in eukaryotic cells. *J Biol Chem* 258:14390-14399

[161] Brostrom CO, Bocckino SB, Brostrom MA, Galuska EM (1986) Regulation of protein synthesis in isolated hepatocytes by calcium-mobilizing hormones. *Mol Pharmacol* 29:104-111

[162] Menaya J, Parvilla R, Ayuso MS (1988) Effect of vasopressin on the regulation of protein synthesis initiation in liver cells. *Biochem* 254:773-779

[163] Thomas AP, Alexander J, Williamson JR (1984) Relationship between inositol polyphosphate production and the increase of cytosolic free Ca^{2+} induced by vasopressin in isolated hepatocytes. *J Biol Chem* 259:5574-5584

[164] Berthon B, Binet A, Mauger J-P, Claret M (1984) Cytosolic free Ca^{2+} in isolated rat hepatocytes as measured by quin 2. Effects of noradrenalin and vasopressin. *FEBS Lett* 167:19-24

[165] Yamaguchi M, Mori S (1990) Effect of calcium-binding protein regucalcin on hepatic protein synthesis: Inhibition of aminoacyl-tRNA synthetase activity. *Mol Cell Biochem* 99:25-32

[166] Tsurusaki Y, Yamaguchi M (2000) Suppressive effect of endogenous regucalcin on the enhancement of protein synthesis and aminoacyl-tRNA synthetase activity in regenerating rat liver. *Int J Mol Med*, 6:295-299

[167] Wang KK, Villalobo A, Ronfogalis BD (1989) Calmodulin-binding protein as calpain substrates. *Biochem J* 262:693-706

[168] Pontremoli S, Melloni E, Salamino F, Sparator B, Michetti M, Horecker BL (1984) Cytosolic Ca^{2+}-dependent neutral proteinases from rabbit liver: Activation of the proenzymes by Ca^{2+} and substrate. *Proc Natl Acad Sci USA* 81:53-56

[169] Yamaguchi M, Tai H (1992) Calcium-binding protein regucalcin increases calcium-independent proteolytic activity in rat liver cytosol. *Mol Cell Biochem*, 12:89-95

[170] Valle B, Williams RJP (1968) Metalloenzymes: the entatic nature of their active sites. *Proc Nsatl Acad Sci USA* 59:498-505

[171] Yamaguchi M, Nishina N (1995) Characterization of regucalcin effect on proteolytic activity in rat liver cytosol: relation to cysteinyl-proteases. *Mol Cell Biochem*, 148:67-72

[172] Mellgren RI (1987) Calcium-dependent proteases: an enzyme system active at cellular membranes? *FASEB J* 1:110-115

[173] Rosser BG, Powers SP, Gores GJ (1993) Calpain activity increases in hepatocytes following addition of ATP. Demonstration by a novel fluorescent approach. *J Biol Chem* 268:23593-23600

[174] Baba T, Yamaguchi M (1999) Stimulatory effect of regucalcin on proteolytic activity in rat renal cortex cytosol: Invovement of thiol proteases. *Mol Cell Biochem* 195: 87-92

[175] Baba T, Yamaguchi M (2000) Stimulatory effect of regucalcin on proteolytic activity is impaired in the kidney cortex cytosol of rats with saline ingestion. *Mol Cell Biochem* 206: 1-6

[176] Bode W, Huber R (1992) Natural protein proteinase inhibitors and their interaction with proteinase. *Eur J Biochem* 204:433-451

[177] Croall DE, Demartino GN (1991) Calcium-activated neutral protease (calpain) system: Structure, function, and regulation. *Physiol Rev* 71:813-847

[178] Melloni E, Pontremoli S, Michetti M, Sacco O, Sparatore B, Horecker BL (1986) The involvement of calpain in the activation of protein kinase C in neutrophils stimulated by phorbol myristic acid. *J Biol Chem* 261:4101-4105

[179] Boynton AL, Whitfield JF, MacManus JP (1980) Calmodulin stimulates DNA synthesis by rat liver cells. *Biochem Biohys Res Commun* 95:745-749

[180] Cruise J, Houck KA, Michalopoulos GK (1985) Induction of DNA synthesis in cultured rat hepatocytes through stimulation of alpha 1 adrenoreceptor by norepinephrine. *Science* 227:749-751

[181] Pujol MJ, Soriano M, Alique R, Carafoli E, Bachs O (1989) Effect of alpha-adrenergic blockers on calmodulin associate with the nuclear matrix of rat liver cells during proliferative activation. *J Biol Chem* 264:18863-18865

[182] Nakagawa T, Yamaguchi M (2008) Nuclear localization of regucalcin is enhanced in culture with protein kinase C activation in cloned normal rat kidney proximal tubular epithelial NRK52E cells. *Int J Mol Med* 21:605-610

[183] Tsurusaki Y, Yamaguchi M (2004) Role of regucalcin in liver nuclear function: Binding of regucalcin to nuclear protein or DNA and modulation of tumor-related gene expression. *Int J Mol Med* 14:277-281

[184] Jones DP, McConkey DJ, Nicotera P, Orrenius S (1989) Calcium- activated DNA fragmentation in rat liver nuclei. *J Biol Chem* 264: 6398-6403

[185] Farber JL (1981) The role of calcium in cell death. *Life Sci* 29: 1289-1295

[186] Yamaguchi M, Sakurai T (1991) Inhibitory effect of calcium-binding protein regucalcin on Ca^{2+}-activated DNA fragmentation in rat liver nuclei. *FEBS Lett,* 279:281-284

[187] Moroianu J, Blobel G (1995) Protein export from the nucleus requires the GTPase Ran and GTP hydrolysis. *Proc Natl Acad Sci USA* 92:4318-4322

[188] Tsurusaki Y, Yamaguchi M (2001) Suppressive effect of endogenous regucalcin guanosine triphosphatase activity in rat liver nucleus. *Biol Pharm Bull,* 24:958-961

[189] Katsumata T, Yamaguchi M (1998) Inhibitory effect of calcium-binding protein regucalcin on protein kinase activity in the nuclei of regenerating rat liver. *J Cell Biochem* 71:569-576

[190] Yamaguchi M, Kanayama Y (1996) Calcium-binding protein regucalcin inhibits deoxyribonucleic acid synthesis in the nuclei of regenerating rat liver. *Mol Cell Biochem* 162:121-126

[191] Tsurusaki Y, Yamaguchi M (2002) Suppressive role of endogenous regucalcin in the enhancement of deoxyribonucleic acid synthesis activity in the nucleus of regenerating rat liver. *J Cell Biochem* 85:516-52

[192] Morooka Y, Yamaguchi M (2002) Suppressive effect of endogenous regucalcin on deoxyribonucleic acid synthesis in the nuclei of rat renal cortex. *Mol Cell Biochem* 229: 157-162

[193] Yamaguchi M, Ueoka S (1997) Inhibitory effect of calcium-binding protein regucalcin on ribonucleic acid synthesis in isolated rat liver nuclei. *Mol Cell Biochem* 173: 169-175

[194] Tsurusaki Y, Yamaguchi M (2002) Role of endogenous regucalcin in nuclear regulation of regenerating rat liver: Suppression of the enhanced ribonucleic acid synthesis activity. *J Cell Biochem* 87: 450-457

[195] Pardo JP, Fernandez F (1982) Effect of calcium and calmodulin on RNA synthesis in isolated nuclei from rat liver cells. *FEBS Lett* 143:157-160

[196] Sturges MR, Peck LJ (1994) Calcium-dependent inactivation of RNA polymerase III transcription. *J Biol Chem* 269:5712-5719

[197] Liu S, Shia D, Liu G, Chen H, Liu S, Hu Y (2000) Roles of Se and NO in apoptosis of hepatoma cells. *Life Sci* 68:603-610

[198] Abou-Elella AM, Siendones E, Padillo J, Montero JL, De la Meta M, Relat JM (2002) Tumour necrosis factor-alpha and nitric oxide mediate apoptosis by D-galactosamine in a primary culture of rat hepatocytes: Exacerbation of cell death by cocultured Kupper cells. *Can J Gastroenterol* 16:791-799

[199] Belloma G, Perotti M, Taddei F, Mirabelli F, Finardi G, Nicotera P, Orrenius S (1992) Tumor necrosis factor α induces apoptosis in mammary adenocarcinoma cells by an increase in intranuclear free Ca^{2+} concentration and DNA fragmentation. *Cancer Res* 52: 1342-1346

[200] Hukkanen M, Hughes FJ, Buttery LD, Gross SS, Evans TJ, Seddon S, Riveros-Moreno V, MacIntyre I, Polak JM (1995) Cytokine stimulated expression of inducible nitric oxide synthase by mouse, rat, and human osteoblast-like cells and its functional role in osteoblast metabolic activity. *Endocrinology* 136:5445-5453

[201] Wink DA, Hanbauer I, Laval F, Cook JA, Krishna MC, Mifchell JB (1994) Nitric oxide protects against the cytotoxic effects of reactive oxygene species. *Ann NY Acad Sci* 738:265-278

[202] Izumi T, Yamaguchi M (2004) Overexpression of regucalcin suppresses cell death in cloned rat hepatoma H4-II-E cells induced by tumor necrosis factor-α or thapsigargin. *J Cell Biochem* 92: 296-306

[203] Alikhani M, Alikhani Z, He H, Liu R, Popek BI, Graves DT (2003) Lipopolysaccharides indirectly stimulate apoptosis and global induction of apoptic genes in fibroblasts. *J Biol Chem* 278:52901-52908

[204] Nalan Y, Vereker E, Lynch AM, Lynch MA (2003) Evidence that lipopolysaccaride-induced cell death is mediated by accumulation of reactive oxygene species and activation of p38 in rat cortex and hippocampus. *Exp Neurol* 184:794-804

[205] Izumi T, Yamaguchi M (2004) Overexpression of regucalcin suppresses cell death and apoptosis in cloned rat hepatoma H4-II-E cells induced by lipopolysaccharide, PD98059, dibucaine, or Bay K 8644. *J Cell Biochem* 93: 598-608

[206] Zelivianshi S, Spellman M, Kellerman M, Kakitelashvilli V, Zhou XW, Lugo E, Lee MS, Taylor R, Daris TL, Hauke R, Lin MF (2003) ERK inhibitor PD 98059 enhances docetaxel-induced apoptosis of androgen-independent human prostate cancer cells. *Int J Cancer* 107: 478-485

[207] Arita K, Utsumi T, Kato A, Kanno T, Kobuchi H, Inoue B, Akiyama J, Utsumi M (2000) Mechanism of dibucaine-induced apoptosis in promyelocytic leukemia cells (HL-60). *Biochem Pharmacol* 60: 905- 915

[208] Giuliano M, Bellacvia G, Lauricella M, D'Anneo A, Vassallo B, Vento R, Tesoriere G (2004) Staurosporine-induced apoptosis in Chang liver cells is associated with down-regulation of Bcl-2 and Bcl-XL. *Int J Mol Med* 13: 565-571

[209] Balakuraman A, Campbell GA, Molsen MT (1996) Calcium channel blockers induces thymic apoptosis *in vivo* in rats. *Toxicol Appl Pharmacol* 139: 122-127

[210] Christensen SB, Andersen A, Kromann H, Treiman M, Tombal B, Denmeads S, Isaacs JT (1999) Thapsigargin analogues for targeting programmed death of androgen-independent prostate cancer cells. *Bioorg Med Chem* 7: 1273-1280

[211] Tombal B, Weeraratna AT, Denmeade SR, Isaacs JT (2000) Thapsigargin induces a calmodulin/calcineurin-dependent apoptotic cascade responsible for the death of prostatic cancer cells. *Prostate* 43: 303-317

[212] Cohen JJ, Duke RC (1984) Glucocorticoid activation of a calcium-dependent endonuclease in thymocytes nuclei leads to cell death. *J Immunol* 132:38-42

[213] Pereira M, Millot J-M, Sebille S, Manfait M (2002) Inhibitory effect of extracellular Mg^{2+} on intracellular Ca^{2+} dynamic changes and thapsigargin-induced apoptosis in human cancer MCF7 cells. *Mol Cell Biochem* 229:163-171

[214] Cano-Abad MF, Villarroya M, Garcia AG, Gabilan NH, Lopez MG (2001) Calcium entry through L-type calcium channels causes mitocondrial disruption and chromaffin cell death. *J Biol Chem* 276: 39695-39704

[215] Fukaya Y, Yamaguchi M (2005) Overexpression of regucalcin suppresses cell death and apoptosis in cloned rat hepatoma H4-II-E cells induced by insulin or insulin-like growth factor-I. *J Cell Biochem* 96:145-154

[216] Spinozzi F, Pagliacci MC, Migliorati G, Moraca R, Grignami F, Riccardi C, Nicoletti I (1994) The natural tyrosine kinase inhibitor genistein produces cell cycle arrest and apoptosis in Jurkat T-leukemia cells. *Leuk Res* 18:431-439

[217] Gamet-Payrastre L, Li P, Lumeau S, Cassar G, Duport NA, Chevolleau S, Gasc N, Tulliez J, Terce F (2000) Sulforaphane, a naturally occurring isothiocyanate, induces cell cycle arrest and apoptosis in HT29 human colon cancer cells. *Cancer Res* 60:1426-1433

[218] Fimognari C, Nusse M, Cesari R, Iori R, Cantelli-Forti G, Hrelia P (2002) Growth inhibition, cell-cycle arrest and apoptosis in human T-cell leukemia by the isothiocyanate sulforaphane. *Carcinogenesis* 23:581-586

[219] Singh AV, Xiao D, Lew KL, Dhir R, Singh SV (2004) Sulforaphane induces caspase-mediated apoptosis in cultured PC-3 human prostate cancer cells and retards growth of PC-3 xenografts in vivo. *Carcinogenesis* 25:83-90

[220] Fukaya Y, Yamaguchi M (2005) Overexpression of regucalcin suppresses apoptotic cell death in the cloned rat hepatoma H4-II-E cells induced by a naturally occurring isothiocyanate sulforaphane. *Int J Mol Med* 15:853-857

[221] Nakagawa T, Yamaguchi M (2005) Hormonal regulation on regucalcin mRNA expression in cloned normal rat kidney proximal tubular epithelial NRK52E cells. *J Cell Biochem* 95:589-597

[222] Nakagawa T, Yamaguchi M (2005) Overexpression of regucalcin suppresses apoptotic cell death in cloned normal rat kidney proximal tubular epithelial NRK52E cells: Change in apoptosis-related gene expression. *J Cell Biochem* 96:1274-1285

[223] Vogelstein B, Lane D, Levine AJ (2000) Surfing the p53 network. *Nature* 408:307 – 310.

[224] Zou H, Hanzel WJ, Liu X, Lutschg A, Wang X (1997) Apaf-l, a human protein homologous to C. elegans CED-4, participates in cytochrome c-dependent activation of caspase-3. *Cell* 90:405 – 413

[225] Widmann C, Gibson S, Johnson GL (1988) Caspase-dependent cleavage of signaling proteins during apoptosis. A turn-off mechanism for anti-apoptotic signals. *J Biol Chem* 273:7141 – 7147

[226] Dieguez-Acuna FJ, Polk WW, Ellis ME, Simmonds PL, Kushleika JV, Woods JS. (2004) Nuclear factor kappaB activity determines the sensitivity of kidney epithelial

cells to apoptosis: Implications for mercury-induced renal failure. *Toxicol Sci* 82:1114-1123

[227] Inagaki S, Yamaguchi M (2001) Suppressive role of endogenous regucalcin in the enhancement of protein kinase activity with proliferation of cloned rat hepatoma cells (H4-II-E). *J Cell Biochem* (Supplement) 36: 12-18

[228] Inagaki S, Misawa H, Yamaguchi M (2000) Role of endogenous regucalcin in protein tyrosine phosphatase regulation in the cloned rat hepatoma cells (H4-II-E). Mol Cell Biochem 213: 43-50

[229] Inagaki S, Yamaguchi M (2000) Enhancement of protein tyrosine phosphatase activity in the proliferation of cloned rat hepatoma H4-II-E cells: Suppressive role of endogenous regucalcin. *Int J Mol Med* 6:323-328

[230] Cohen P, Cohen PTW (1989) Protein phosphatase come of age. *J Biol Chem* 264:21435-21438

[231] Mackintosh C, Malintosh RW (1997) Inhibitors of protein kinases and phosphatases. *Trends Biochem Sci* 19:444-448

[232] Inagaki S, Yamaguchi M (2001) Regulatory role of endogenous regucalcin in the enhancement of nuclear deoxyribonucleic acid synthesis with proliferation of cloned rat hepatoma cells (H4-II-E). *J Cell Biochem* 82: 704-711

[233] Misawa H, Inagaki S, Yamaguchi M (2002) Suppression of cell proliferation and deoxyribonucleic acid synthesis in cloned rat hepatoma H4-II-E cells overexpressing regucalcin. *J Cell Biochem* 84: 143-149

[234] Yamaguchi M, Daimon Y (2005) Overexpression of regucalcin suppresses cell proliferation in cloned rat hepatoma H4-II-E cells: Involvement of intracellular signaling factors and cell cycle-related genes. *J Cell Biochem* 95:1169-1177

[235] Yamaguchi M (2005) Role of regucalcin in maintaining cell homeostasis and function (Review) *Int J Mol Med* 15:371-389

[236] Meijer L, Borgne A, Mulner O, Chhong JP, Blow JJ, Inagaki N, Inagaki M, Delcros JG, Moulinoux JP (1997) Biochemical and cellular effects of roscovitine, a potent and selective inhibitor of the cyclin-dependent kinases cdc2, cdk2 and cdk5. *Eur J Biochem* 243:527-536

[237] Singh SV, Herman-Antosiewice A, Singh AV, Lew KL, Srivastava SK, Kamath R, Brown KD,m Zhang L, Baskaran R (2004) Sulforaphane-induced G2/M phase cell cycle arrest involves checkpoint kinase 2-mediated phosphorylation of cell division cycle 25C. *J Biol Chem* 279:25813-25822

[238] Charollais RH, Buquet C, Mester J (1990) Butyrate blocks the accumulation of CDc2 mRNA in late G1 phase but inhibits both early and late G1 progression in chemically transformed mouse fibroblasts BP-A31. *J Cell Physiol* 145:46-52

[239] Curran T (1991) Fos and Jun: Intermediary transcription factors. In: Cohen P, Foulkes JG, editors. *The hormonal control of gene transcription*. New York: Elservier Science Publisher, pp295-308

[240] Hulla JE, Schneider RP (1993) Structure of the rat *p53* tumor suppressor gene. *Nucleic Acids Res* 21:713-717

[241] Tsurusaki Y, Yamaguchi M (2003) Overexpression of regucalcin modulates tumor-related gene expression in cloned rat hepatoma H4- II-E cells. *J Cell Biochem* 90: 619-626

[242] Tsurusaki Y, Yamaguchi M (2004) Role of regucalcin in liver nuclear function: Binding of regucalcin to nuclear protein or DNA and modulation of tumor-related gene expression. *Int J Mol Med* 14:277-281

[243] Nakagawa T, Sawada N, Yamaguchi M (2005) Overexpression of regucalcin suppresses cell proliferation of cloned normal rat kidney proximal tubular epithelial NRK52E cells. *Int J Mol Med* 16:637-643

[244] Dang ZC, Lowik CW (2004) Differential effects of PD98059 and UO126 on osteogenesis and adipogenesis. *J Cell Biochem* 92:525-533

[245] Tamaoki T, Nomoto H, Takahashi I, Kato Y, Morimoto M, Tomita E (1986) Staurosporine, a potent inhibitor of phospholipids/Ca^{2+}-dependent protein kinase. *Biochem Biophys Res Commun* 135:397-402

[246] Vincenzi FF (1982) Pharmacology of calmodulin antagosism. In: *Calcium Modulators*. Godfranid T, Albertini A, Paoletti R (eds), Elservier Biomedical Press, Amsterdam, pp67-80

[247] Park YC, Lee CH, Kang HS, Chung HT, Kim HD (1997) Wortmannin, a specific inhibitor of phosphatidylinositol-3-kinase, enhances LPS-induced NO production from murine peritoneal macrophages. *Biochem Biophys Res Commun* 240:692-696

[248] Nakagawa T, Yamaguchi M (2006) Overexpression of regucalcin enhances its nuclear localization and suppresses L-type Ca^{2+} channel and calcium-sensing receptor mRNA expressions in cloned normal rat kidney proximal tubular epithelial NRK52E cells. *J Cell Biochem* 99:1064-1077

[249] Tanaka T, Nangaku M, Miyata T, Inagi R, Ohse T, Ingelfinger JR, Fujita T (2004) Blockade of calcium influx through L-type calcium channels attenuates mitochondrial injury and apoptosis in hypoxic renal tubular cells. *J Am Soc Nephrol* 15:2320-2333

[250] Ba J, Friedman PA (2004) Calcium-sensing receptor regulation of renal mineral ion transport. *Cell Calcium* 35:229-237

[251] Nakagawa T, Yamaguchi M (2007) Overexpression of regucalcin suppresses cell response for tumor necrosis factor-α or transforming growth factor-β1 in cloned normal rat kidney proximal tubular epithelial NRK52E cells. *J Cell Biochem* 100:1178-1190

[252] Fan JM, Ng Y-Y, Hill PA, Nikolic-Paterson DJ, Mu W, Atkins RC, Lan HY (1999) Transforming growth factor-β regulates tubular epithelial-myofibroblast trans-differentiation in vitro. *Kidney Int* 56:1455–1467

[253] Zhang YQ, Kanzaki M, Furukawa M, Shibata H, Ozeki M, Kojima I (1999) Involvement of Smad proteins in the differentiation of pancreatic AR42J cells induced by activin A. *Diabetologia* 42:719–727

[254] Hammar EB, Irminger J-G, Richenback K, Parnaud G, Ribaux P, Bosco D, Rouiller DG, Halban PA (2005) Activation of NF-kB by extracellular matrix is involved in apreading and glucose-stimulated insulin secretion of pancreatic beta cells. *J Biol Chem* 280: 30630-30637

[256] Yamaguchi M, Misawa H, Uchiyama S, Morooka Y, Tsurusaki Y (2002) Role of endogenous regucalcin in bone metabolism: Bone loss is induced in regucalcin transgenic rats. *Int J Mol Med* 10: 377-383

[257] Yamaguchi M, Sawada N, Uchiyama S, Misawa H, Ma ZJ (2004) Expression of regucalcin in rat bone marrow cells: Involvement of osteoclastic bone resorption in regucalcintransgenic rats. *Int J Mol Med* 13: 437-443

[258] Uchiyama S, Yamaguchi M (2004) Bone loss in regucalcin transgenic rats: Enhancement of osteoclastic cell formation from bone marrow of rats with increasing age. *Int J Mol Med* 14:451-455

[259] Yamaguchi M, Igarashi A, Uchiyama S, Sawada N (2004) Hyperlipidemia is induced in regucalcin transgenic rats with increasing age. *Int J Mol Med* 14: 647-651

[260] Yamaguchi M, Nakagawa T (2007) Change in lipid components in the adipose and liver tissues of regucalcin transgenic rats with increasing age: Suppression of leptin and adiponectin gene expression. *Int J Mol Med* 20: 323-328

[261] Nakashima C, Yamaguchi M (2006) Overexpression of regucalcin enhances glucose-utilization and lipid production in cloned rat hepatoma H4-II-E cells: Involvement ofinsulin resistance. *J Cell Biochem* 99:1582-1592

[262] Nakashima C, Yamaguchi M (2007) Overexpression of regucalcin suppresses gene expression of insulin signaling-related proteins in cloned rat hepatoma H4-II-E cells: Involvement of insulin resistance. *Int J Mol Med* 20:709-716

[263] Yamaguchi M (2010) Regucalcin and metabolic disorder: osteoporosis and hyper-lipidemia are induced in regucalcin transgenic rats. Mol Cell Biochem 341:119-133

[264] Ishigami A, Fujita T, Handa S, Shirasawa T, Koseki H, Kitamura T, Enomoto N, Sato N, Shimosawa T, Maruyama N (2002) Senescence marker protein-30 knockout mouse liver is highly susceptible to tumor necrosis factor-alpha-and Fas-mediated apoptosis. *Am J Pathol* 161:1273-1281

[265] Ishigami A, Kondo Y, Nanba R, Ohsawa T, Handa S, Kubo S, Akita M, Maruyama N (2004) SMP30 deficiency in mice causes an accumulation of neutral lipids and phospholipids in the liver and shortens the life span. *Biochem Biophys Res Commun* 315:575-580

[266] Kondo Y, Inai Y, Sato Y, Handa S, Kubo S, Shimokado K, Goto S, Nishikimi M, Maruyama N, Ishigami A (2006) Senescence marker protein 30 functions as gluconolactonase in L-ascorbic acid biosynthesis, and its knockout mice are prone to scurvy.
Proc Natl Acad Sci USA 103:5723–5728

In: Calcium Signaling
Editor: Masayoshi Yamaguchi

ISBN: 978-1-61324-313-8
©2012 Nova Science Publishers, Inc.

Chapter 9

CALCIUM ($[CA^{2+}]_I$) SIGNALING AS A KEY-TARGET FOR CANCER TREATMENT

Ana-Maria Florea[1] and Dietrich Büsselberg[2]

[1] Universitätsklinikum Düsseldorf, Institut für Neuropathologie, Düsseldorf, Germany
[2] Weill Cornell Medical College in Qatar, Qatar Foundation – Education City, Doha, Qatar

ABSTRACT

Cancer represents a major field of investigation for the modern medicine. Anti-cancer drugs induce tumor cell death. While some anti-cancer drugs are genotoxic, directly targeting nucleic acids, others interfere with the signaling pathways to trigger programmed cell death. For the induction and the execution of apoptosis, intracellular calcium signals are important. Much interest is now oriented in discovering new combinatorial strategies that are now tested in clinical trials in order to improve standard procedures. Our results indicate that cisplatin (cis-di-amino-dichloride-platin) and arsenic trioxide modify intracellular Ca^{2+} by different mechanisms, supporting the hypothesis that calcium signaling plays an important role in chemical-induced neuroblastoma cell toxicity. In addition, understanding how specific drug combinations work to result in enhanced effects on neuroblastoma cell death might represent a new therapeutic strategy for neuroblastoma as well as ways to overcome drug resistance. There may be a strong link between drug resistance, cell death and changes in intracellular Ca^{2+}. Future experiments have to identify the specific mechanisms to illuminate possible targets and to override drug resistance.

Keywords: Calcium signaling, cancer, cancer therapy, apoptosis, cisplatin, arsenic trioxide

INTRODUCTION

Cancer represents a major field of investigation for the modern medicine. In the last years major interests have been oriented to discover novel drugs and treatment strategies that are able to fight cancer and drug resistance. Cancer cells have a dysfunctional apoptotic program.

Modulations of the intracellular calcium concentration play a central role in the induction of apoptosis. This represents a major therapeutic target, and a better understanding of calcium dependent signaling pathways and molecules that govern apoptosis in cancer cells and will open new directions for anti-cancer therapies. To reduce the risk of drug resistance, combinations of anti-cancer drugs are clinically employed. Much interest is now oriented in discovering new combinatorial strategies. The recent topics indicate that the two anti-cancer drugs cisplatin (cis-di- amino-dichloride-platin) and arsenic trioxide modulate intracellular calcium, which induces neuroblastoma cell toxicity are reviewed in this chapter.

CALCIUM SIGNALING AND CANCER DISEASE

Cancer represents a major field of investigation for the modern medicine. In the last years major interests have been oriented to discover novel drugs and treatment strategies that are able to fight cancer and drug resistance.

Several human diseases, including cancer, are attributed to the malfunction of apoptosis. The malfunction could be related to (i) an impairment of cell death that could result in imbalance between cell proliferation and apoptosis that in turn will result in cell accumulation followed by uncontrolled cell growth and eventually tumor formation, or, (ii) cell loss when the apoptotic program is abnormally activated (Zhivotovsky and Orrenius, 2006; Fadeel and Orrenius, 2005; Fulda, 2009). Therefore, evading apoptosis (cell survival) is a pivotal mechanism in both carcinogenesis and resistance to anticancer therapy.

In general, cancer cells have a dysfunctional apoptotic program. Modulations of the intracellular calcium concentration ($[Ca^{2+}]_i$) play a central role in the induction of apoptosis. Thus, modulation of $[Ca^{2+}]_i$ could represent a major therapeutic target, and a better understanding of the $[Ca^{2+}]_i$ dependent signaling pathways and molecules that govern apoptosis in cancer cells and will open new directions for anti-cancer therapies.

Generally, intracellular Ca^{2+} is a universal second messenger; therefore, its concentration ($[Ca^{2+}]_i$) is strictly regulated. Calcium signals are involved in the regulation of vital cell functions including: cell proliferation, gene expression, activation of oncogenesis and cell death, etc. while the trigger of programmed cell death (apoptosis), plays an important role in maintenance of tissue homeostasis (Berridge, 1995, 1997; Orrenius et al., 2003, García et al., 2006).

CALCIUM SIGNALING AND ANTICANCER DRUGS

Anti-cancer drugs induce tumor cell death. While some anti-cancer drugs are genotoxic, directly targeting nucleic acids, others interfere with the signaling pathways to trigger programmed cell death. For the induction and the execution of apoptosis, intracellular calcium signals are important.

The relevance of the $[Ca^{2+}]_i$ – signal in cancer cells and for cancer treatment has been recently highlighted by two nature review articles (Monleith et al, 2007, Roderick and Cook, 2008). Several anti-cancer drugs modulate $[Ca^{2+}]_i$ homeostasis *in vitro*: e.g. cisplatin (cis-di-amino-di-chloride-platin = CDDP; Splettstoesser et al., 2007; for review see Florea and

Büsselberg, 2011), carboplatin (Splettstoesser et al., 2007); topotecan (Li et al., 2006) and arsenic trioxide (As_2O_3) (Pettersson et al., 2007; Florea et al., 2007; Florea and Büsselberg, 2008, Günes at al., 2008). Expanding the knowledge of the molecular "cross-talks" that regulate tumor cell demise is crucial in guiding the successful design of future anti-cancer therapeutics. The understanding of $[Ca^{2+}]i$ dependent processes of cancer cells will give new perspectives for improving the current anti-cancer treatment.

While the pharmacology of most chemotherapeutics is still not fully understood, the therapeutic goal in cancer treatment is to trigger tumor- specific cell death. Elevated intracellular Ca^{2+}-levels could have a major impact on the induction of cell death. Generally, anticancer drugs increase $[Ca^{2+}]_i$ by at least three different mechanisms: (1) Ca^{2+} entry from the extracellular space (through channels or transporters) or (2) the release of calcium from the intracellular stores. (3) inhibition of active calcium transporters (which move the Ca^{2+} against its concentration gradient back to the extracellular space or the calcium stores). To induce an increase in $[Ca^{2+}]_i$ the anticancer drug could: (a) activate a receptor at the cell membrane (which in turn opens a calcium selective channel protein or triggers intracellular pathways that cause the opening of calcium channels or, (b) pass through the cell membrane (directly or via a transport protein) and interfere with the Ca^{2+}-regulating proteins (channels and/or transporters) at the store sites (endoplasmic reticulum, mitochondria, nucleus) (for review see Florea and Büsselberg, 2005, 2006, 2009a, b, 2011).

CALCIUM SIGNALING AND DRUG RESISTANCE

Overall, drug resistance is the most important cause of cancer treatment failure. Drug resistance mechanisms are either **innate** (intrinsic) or **acquired** as a result of adaptive changes upon therapy, e.g. the selection of cancer cells which survive the impact of the drug and multiply. At present, there is only limited knowledge in regard to the underlying mechanisms of simple drug therapies, and even less is known about the interactions, which occur in multiple drug therapies (for review see Maris et al., 2007; Ishola and Chung, 2007; Johnsen et al., 2009).

Nevertheless, drug resistance could be related to changes of $[Ca^{2+}]_i$ homeostasis in tumor cells, such changes have been found in human lung adenocarcinoma A549 cells (Liang and Huang, 2000; Padar et al., 2004) and human breast cancer cells (DeBernadi and Brooker, 2006). In breast cancer cells the decay kinetics of $[Ca^{2+}]_i$ were delayed resulting in an increase of $[Ca^{2+}]_i$ in the drug resistant cells (DeBernadi and Broker, 2006), while a decrease of $[Ca^{2+}]_i$ to one third compared to the concentration of non-resistant cells was reported for human lung adenocarinoma cells (Liang and Huang, 2000). This decrease was linked to altered calcium influx pathways and changes of the calcium management of the calcium stores in the drug resistant cells (Padar et al., 2004).

CALCIUM SIGNALING IN CONBINATORIAL THERAY

To reduce the risk of drug resistance, combinations of anti-cancer drugs are clinically employed. The clinically relevant agents for chemotherapy are **platinum compounds** (cisplatin, carboplatin), alkylating agents (cyclophosphamide, ifosfamide, melphalan), topoisomerase II inhibitor (etoposide), anthracycline antibiotics (doxorubicin), vinca alkaloids (vincristine), topoisomerase I inhibitors (topotecan and irinotecan) etc. (for review see Maris et al., 2007; Ishola and Chung, 2007; Johnsen et al., 2009).

Much interest is now oriented in discovering new combinatorial strategies that are now tested in clinical trials in order to improve standard procedures.

Recently we demonstrated that the two anti-cancer drugs cisplatin (cis-di-amino-dichloride-platin = CDDP) and arsenic trioxide (As_2O_3) modulate $[Ca^{2+}]_i$ *in vitro* (Splettstoesser et al., 2007, Florea and Büsselberg, 2005, 2006, Günes et al., 2008). To monitor the $[Ca^{2+}]_i$ dynamics we employed live cell imaging combining confocal fluorescent microscopy and Ca^{2+}-sensitive dyes (fluo-4, rhod2). For the experiments we used low, clinically relevant concentrations of As_2O_3 (0.1 nM to 1 μM) and/or cisplatin (1 nM to 10 μM) which were extremely effective in interacting with $[Ca^{2+}]_i$ homeostasis of tumor cells (SH SY-5Y, HeLa) as well as non-tumor cells (HEK). Transient and sustained $[Ca^{2+}]_i$ signals were observed in response to the application of As_2O_3 or CDDP. While we could not link the short time transient $[Ca^{2+}]_i$ signals to apoptosis (Günes et al., 2008), the sustained increase was clearly related to this event. It must be pointed out that the sustained $[Ca^{2+}]_i$ rise derived from different sources: while As_2O_3 triggers the release of Ca^{2+} from intracellular stores (Florea et al., 2007, Florea and Büsselberg, 2008; Florea and Büsselberg, 2009 a,b) the application of cisplatin induces a Ca^{2+}-influx from the extracellular space (Splettstoesser et al., 2007; Florea and Büsselberg, 2009 a, b) while calcium currents through voltage activated channels are reduced (Tomaszewski and Büsselberg, 2007; Florea and Büsselberg, 2009 a, b). As mentioned, the increase of $[Ca^{2+}]_i$ was directly related to the induction of apoptosis.

Since As_2O_3 and CDDP elevate $[Ca^{2+}]_i$ by different mechanisms, we investigated whether the combination of As_2O_3 and CDDP would result in an additional increase of $[Ca^{2+}]_i$ and, therefore, in a higher rate of apoptosis on neuroblastoma cells than each drug applied alone. Our experiments showed that co-application of As_2O_3 and CDDP have additive or even synergistic effects on $[Ca^{2+}]_i$ homeostasis, depending on the order of application (Günes at al., 2008). The highest increase of $[Ca^{2+}]_i$ occurred when the cells were first incubated with CDDP followed by a co-application with As_2O_3. The magnitude of the $[Ca^{2+}]_i$ increase was accompanied directly by an increased cytotoxicity and apoptosis (Splettstoesser et al., 2007, Florea et al., 2007, Florea and Büsselberg, 2008; Günes at al., 2008; Florea and Büsselberg, 2009a,b). Therefore, this increase in cytotoxicity by opening a calcium conductance simultaneously at the stores and the cell membrane (as established by the *in vitro* co-application of CDDP and As_2O_3), suggests that co-administration of these drugs clinically may be a more effective anti-cancer therapy than either one alone (Florea and Büsselberg, 2009b).

OUTLOOK

In conclusion, our results indicate that CDDP and As_2O_3 modify $[Ca^{2+}]_i$ by different mechanisms. Furthermore it strongly supports the hypothesis that $[Ca^{2+}]_i$ signaling plays an important role in CDDP and As_2O_3 induced neuroblastoma cell toxicity. A deep understanding of the mechanism of drug mediated calcium signaling and toxicity of cancer cells could dramatically improve the knowledge regarding the biology of tumor cells. In addition, understanding how specific drug combinations work to result in enhanced effects on neuroblastoma cell death might represent a new therapeutic strategy for neuroblastoma as well as ways to overcome drug resistance. Overall, there is a strong link between drug resistance, cell death and changes in $[Ca^{2+}]_i$. Future experiments have to identify the specific mechanisms to illuminate possible targets and override drug resistance.

REFERENCES

Berridge MJ. Calcium signalling and cell proliferation. 1995; *BioEssays* 17: 491 - 500.

Berridge MJ. Elementary and global aspects of calcium signalling. 1997; *J Physiol* 499: 291 - 306.

DeBernardi MA, Brooker G. High-content kinetic calcium imaging in drug- sensitive and drug-resistant human breast cancer cells. *Methods Enzymol.* 2006; 414: 317-35.

Fadeel B, Orrenius S. Apoptosis: a basic biological phenomenon with wide- ranging implications in human disease. *J Intern Med* 2005; 258(6): 479-517.

Florea AM, Büsselberg D. Toxic effects of metals: modulation of intracellular calcium homeostasis. *Materialwissenaschft und Werkstofftechnik 2005;* 36: 757 – 760.

Florea AM, Büsselberg D. Occurrence, use and potential toxic effects of metals and metal compounds. 2006; *Biometals* 19: 419-427.

Florea AM, Splettstoesser F, Büsselberg D. Arsenic trioxide (As_2O_3) induced calcium signals and cytotoxicity in two human cell lines: SY-5Y neuroblastoma and 293 embryonic kidney (HEK). *Toxicol Appl Pharmacol.* 2007 May 1; 220(3): 292-301.

Florea AM, Büsselberg D. Arsenic trioxide in environmentally and clinically relevant concentrations interacts with calcium homeostasis and induces cell type specific cell death in tumor and non-tumor cells. *Toxicol Lett.* 2008 Jun 10; 179(1): 34-42.

Florea AM, Büsselberg D. Anti-cancer drugs interfere with intracellular calcium signaling. *Neurotoxicology.* 2009a; 30(5): 803-10.

Florea AM, Büsselberg D. Metal compounds elevate intracellular calcium by different mechanism inducing cell death in "in vitro" systems. *Materialwissenschaft und Werkstofftechnik* 2009b; 4: 1-2.

Florea AM, Büsselberg D. Cisplatin as an anti-cancer drug: cellular mechanism of activity, drug resistance and induced side effects. *Cancers.* 2011; 3(1), 1351-1371.

Fulda S. Apoptosis pathways and neuroblastoma therapy. *Curr Pharm Des.* 2009; 15(4):430-5.

Garcia AG, Garcia-De-Diego AM, Gandia L, et al. Calcium signaling and exocytosis in adrenal chromaffin cells. 2006; *Physiological reviews* 86: 1093- 1131.

Günes DA, Florea AM, Splettstoesser F, Büsselberg D. Co-application of arsenic trioxide (As2O3) and cisplatin (CDDP) on human SY-5Y neuroblastoma cells has differential effects on the intracellular calcium concentration ($[Ca^{2+}]_i$) and cytotoxicity. *Neurotoxicology.* 2009 Mar; 30(2): 194-202.

Ishola TA, Chung DH. Neuroblastoma. *Surg Oncol.* 2007 Nov; 16(3): 149-56.

Johnsen JI, Kogner P, Albihn A, Henriksson MA. Embryonal neural tumours and cell death. *Apoptosis.* 2009 Apr; 14(4): 424-38.

Li X, Ruan GR, Lu WL, Hong HY, Liang GW, Zhang YT, Liu Y, Long C, Ma X, Yuan L, Wang JC, Zhang X, Zhang Q. A novel stealth liposomal topotecan with amlodipine: apoptotic effect is associated with deletion of intracellular Ca^{2+} by amlodipine thus leading to an enhanced antitumor activity in leukemia. 2006; *J Control Release*. May 15; 112(2): 186-98.

Liang X, Huang Y. Intracellular free calcium concentration and cisplatin resistance in human lung adenocarcinoma A549 cells. *Biosci Rep*. 2000 Jun; 20(3): 129-38.

Maris JM, Hogarty MD, Bagatell R, Cohn SL.Neuroblastoma. *Lancet*. 2007 Jun 23; 369(9579): 2106-20.

Monteith GR, McAndrew D, Faddy HM, Roberts-Thomson SJ. Calcium and cancer: targeting Ca^{2+} transport. *Nat Rev Cancer*. 2007 Jul; 7(7): 519-30.

Orrenius S, Zhivotovsky B, Nicotera P. Regulation of cell death: the calcium- apoptosis link. *Nat Rev Mol Cell Biol*. 2003 Jul; 4(7): 552-65.

Padar S, van Breemen C, Thomas DW, Uchizono JA, Livesey JC, Rahimian R. Differential regulation of calcium homeostasis in adenocarcinoma cell line A549 and its Taxol-resistant subclone. *Br J Pharmacol*. 2004 May; 142(2): 305-16.

Pettersson HM, Karlsson J, Pietras A, Øra I, Påhlman S. Arsenic trioxide and neuroblastoma cytotoxicity. *J Bioenerg Biomembr*. 2007 Feb; 39(1): 35-41.

Roderick HL, Cook SJ. Ca^{2+} signalling checkpoints in cancer: remodelling Ca^{2+} for cancer cell proliferation and survival. *Nat Rev Cancer*. 2008 May; 8(5): 361-75.

Splettstoesser F, Florea AM, Büsselberg D. IP_3 receptor antagonist, 2-APB, attenuates cisplatin induced Ca^{2+}-influx in HeLa-S3 cells and prevents activation of calpain and induction of apoptosis. *Br J Pharmacol*. 2007 Aug; 151(8): 1176-86.

Tomaszewski A, Büsselberg D. $SnCl_2$ reduces voltage-activated calcium channel currents of dorsal root ganglion neurons of rats. *Neurotoxicology*. 2008 Nov; 29(6): 958-63.

Zhivotovsky B, Orrenius S. Carcinogenesis and apoptosis: paradigms and paradoxes. *Carcinogenesis* 2006; 27(10): 1939-45.

INDEX

D

E

F

G

H

U

V

W